Building and Land Management Law for Students

Consulting Editor
C.R. Bassett, BSc., FCIOB, FFB

Building and Land Management Law for Students

Third Edition

Anne Galbraith, LLB
and
Michael Stockdale, PhD, LLB

Newnes
An imprint of Butterworth–Heinemann Ltd
Linacre House, Jordan Hill, Oxford OX2 8DP

℞ A member of the Reed Elsevier group

OXFORD LONDON BOSTON
MUNICH NEW DELHI SINGAPORE SYDNEY
TOKYO TORONTO WELLINGTON

First published as *Building Law for Students* 1989
Second edition 1991
Reprinted 1992
Third edition 1993

© Anne Galbraith 1989, 1991
© Anne Galbraith and Michael Stockdale 1993

All rights reserved. No part of this publication
may be reproduced in any material form (including
photocopying or storing in any medium by electronic
means and whether or not transiently or incidentally
to some other use of this publication) without the
written permission of the copyright holder except in
accordance with the provisions of the Copyright,
Designs and Patents Act 1988 or under the terms of a
licence issued by the Copyright Licensing Agency Ltd,
90 Tottenham Court Road, London, England W1P 9HE.
Applications for the copyright holder's written permission
to reproduce any part of this publication should be addressed
to the publishers

British Library Cataloguing in Publication Data
Galbraith, Anne
    Building and Land Management Law for
    Students. — 3 Rev. ed
    I. Title II. Stockdale, Michael
    344.20378624

ISBN 0 7506 1623 7

Printed in England by Clays Ltd, St Ives plc

Contents

Preface	ix
Table of statutes	xi
Table of cases	xiv

1 The nature of law

Introduction	1
The UK and the European Community	4
New legislation	5
Pressure groups	9
Delegated legislation	10
Statutory interpretation	11
Judicial precedent	13
Common law and equity	16

2 Settlement of disputes

Self-help	17
Court procedures	18
Civil cases	20
Criminal cases	23
Tribunals	29
Arbitration	29
Administrative control	31
Affording the law	31
The legal profession	32

3 Central and local government

The structure of central government	34
Procedure of the House of Commons	37
The committee system of the House of Commons	39
Financial procedures of the House of Commons	41
The Civil Service	42
The structure of local government	43
Control of local authorities	44
Bye-laws	46
Public corporations	47
Challenging decisions made by public bodies in the courts	48

4 Business organizations
Partnership	61
Company incorporation	62
Creating a company	64
Public and private companies	66
Management of a company	67
Company finance	69
Winding up companies	71
Single member private limited companies	74

5 The law of contract
The nature of a contract	75
Formation of the contract – agreement	76
Consideration	81
Form of the contract	90
The contents of the contract	92
Exclusion and limitation clauses	95
Misrepresentation and mistake	101
Other factors affecting validity and enforceability	109
Illegal contracts	110
Discharge of the contract	114
Enforcing the contract – remedies for breach	124

6 Special contracts
Agency	131
Sale of goods	139
Supply of goods and services	147
Hire purchase, credit sales and conditional sales	149
Other methods of consumer protection	151
Insurance contracts	152

7 Special contracts – building contracts
Control of work under a building contract	161
Tendering for building contracts	163
The JCT standard form contracts	164
Sub-contracts	171

8 Employment law
Employment and self-employment	173
Creation of a contract of employment	176
Duties of the parties	178
Statutory rights for employees	184

Contents vii

Protection against discrimination	186
Termination of contract of employment	188
Unfair dismissal	189
Remedies for unfair dismissal	193
Redundancy payments	195
Employees and trade unions	197
Trade disputes	204

9 The law of tort
Introduction	208
Trespass	211
Nuisance	213
The rule in *Rylands v. Fletcher*	220
Liability for the spread of fire	222
Negligence	223
Occupier's liability	244
Breach of statutory duty	251
Defective products	252
Vicarious liability	255

10 Health and safety
The background to legislation	259
The Robens Report	261
The Health and Safety Commission	261
The Health and Safety at Work etc. Act 1974	263
The Control of Substances Hazardous to Health Regulations 1988	274
The Factories Act 1961	274
Construction regulations	276
Compensation for injuries at work	279

11 Land law
Classification of property	281
What is land?	282
Freehold and leasehold estates	285
Interests in land	286
The conveyancing process	287
Trusts	293
Joint ownership	293
Rights and duties of owners of land	295
Access to Neighbouring Land Act 1992	296
Easements	296

	Profits *a prendre*	299
	Licences	299
	Restrictive covenants	300
	Mortgages of land	302
	Adverse possession	303
12	**Landlord and tenant**	
	Leases and licences	306
	Obligation to repair	312
	Security of tenure	319
	Business tenancies	323
	Residential tenancies	335
	Residential tenancies of flats	351
	Council and other public sector tenants	352
13	**Planning law**	
	Central and local planning responsibilities	356
	Development	358
	Change of use	359
	Permission under the General Development Order 1988	360
	Making a planning application	361
	Appeals	364
	Planning enforcement	365
	Planning and the preservation of amenity	366
	Financial problems – compensation and betterment	368
14	**Highways**	
	Introduction	370
	Acquisition and ownership of highways	371
	Upkeep of the highway	373
	Streets and street works	374
	Standard of the highway	376
	Heavy traffic	378
	Closing and obstruction of highways	378
	Building and improvement lines	381
15	**Building regulations**	
	Purpose and scope of the Building Regulations	382
	Enforcement procedures	383
	Defective and dangerous buildings	384
Index		386

Preface

When I approached the task of preparing a third edition of this book, it seemed sensible to look back to remind myself of the original aim, which was to provide a clear and readable account of those areas of law that are of principal relevance to the 'built environment'. It was always obvious that such a wide-ranging book could only cover basic principles but, thus armed, the reader will hopefully be sufficiently well informed to know when to seek advice. Legal advice given in good time may prove to be very cost effective in the long term.

In some areas of law, the rules are derived from the cases, which are real disputes between real parties. These cases not only determine what the rules should be, but also flesh out our understanding of how those rules work in practice. In other areas, Parliament has chosen to provide a framework of laws in statutory form. Acts of Parliament and Regulations provide the bare bones of the law, which must then be interpreted and applied by the courts.

This edition breaks new ground in two important respects. First, in the assistance I have received from Dr Michael Stockdale, a colleague at the University of Northumbria at Newcastle. Secondly, we have taken a decision to incorporate substantial treatment of some new topic areas. These new sections on landlord and tenant, agency and administrative law have been largely prepared by Dr Stockdale. We hope that the wider content will encourage students on courses in surveying and estate management to use the book as a basic text. We have sought to cover the material on landlord and tenant in some depth as the law is complex and a superficial treatment could be misleading.

The opportunity to publish a new edition has also permitted us to overhaul chapters on land law and planning, and to undertake some updating in employment, health and safety and highways. The impact of European law creeps inexorably on, reminding us of Gladstone's comment that 'we are part of the community of Europe, and we must do our duty as such'.

Inevitably, laws that govern so many people and so much diverse activity tend to be complex. The temptation will often be felt to leave the complexities to the professional lawyer, but we should all remember that ignorance is not bliss. Almost four centuries ago, John Selsdon reminded us that 'ignorance of the law excuses no man'. No doubt in these days of non-discrimination, he would need to reword his comments.

The experience of working with a new partner has proved to be rewarding. Michael and I would both like to pay tribute to the many colleagues who have been supportive and encouraging, and from whom we have learnt so much.

Newcastle upon Tyne
1993

Anne Galbraith
Michael Stockdale

Table of statutes

Access to Neighbouring Land Act 1992 296
Agricultural Holdings Act 1986 .... 336, 343
Animals Act 1971 221
Arbitration Acts 1950–1979 30

Building Act 1984 .... 252, 382
Business Names Act 1985 64

Civil Aviation Act 1949 282
Civil Evidence Act 1968 240
Civil Liability (Contribution) Act 1978 243
Common Law Procedure Act 1852 320
Companies Act 1985 .... 5, 63
Companies Act 1989 .... 64, 68
Consumer Credit Act 1974 149
Consumer Protection Act 1987 .... 140, 253
Control of Pollution Act 1984 219
Copyright Designs and Patents Act 1988 183
Courts Act 1971 .... 22, 27
Courts and Legal Services Act 1990 .... 20, 33
Criminal Justice Act 1982 .... 26, 383
Criminal Law Act 1977 322

Defective Premises Act 1972 .... 235, 250
Development Land Tax Act 1976 369
Drought Act 1976 9

Employer's Liability (Compulsory Insurance) Act 1969 156
Employment Act 1980 .... 189, 196, 200
Employment Act 1982 .... 189, 202
Employment Act 1988 202
Employment Act 1990 .... 189, 200
Employment Protection Act 1975 197

xii Table of statutes

Employment Protection (Consolidation) Act 1978 .... 177, 185
Enduring Powers of Attorney Act 1985 137
Environmental Protection Act 1990 219
Estate Agents Act 1979 .... 137, 134
Equal Pay Act 1970 .... 179, 186
European Communities Act 1972 4

Factories Act 1961 .... 251, 260, 264, 274
Fires Prevention (Metropolis) Act 1774 222

Health and Safety at Work etc. Act 1974 ... 3, 6, 10, 23, 196, 185, 251, 260, 263
Highways Act 1980 370
Housing Act 1985 .... 336, 340, 343, 352
Housing Act 1988 .... 306, 319, 322, 335
Housing and Planning Act 1986 361
Housing and Urban Development Bill 1993 .... 350, 452

Insolvency Act 1986 71
Interpretation Act 1978 12

Land Registration Act 1925 286
Landlord and Tenant Act 1927 .... 308, 313, 332
Landlord and Tenant Act 1954 .... 306, 319, 323, 335, 351, 353
Landlord and Tenant Act 1985 .... 315, 316
Landlord and Tenant Act 1987 .... 315, 351
Landlord and Tenant Act 1988 308
Latent Damage Act 1986 237
Law of Property Act 1925 .... 90, 160, 285
Law of Property (Miscellaneous Provisions) Act 1989 .... 90, 288
Law Reform (Contributory Negligence) Act 1945 .... 208, 243
Law Reform (Frustrated Contracts) Act 1943 .... 117, 121
Leasehold Property (Repairs) Act 1938 .... 314, 320
Leasehold Reform Act 1967 .... 324, 350
Legal Aid Act 1988 32
Limitation Act 1980 .... 130, 236, 303
Local Government Act 1972 .... 44, 46
Local Government and Housing Act 1989 351

Misrepresentation Act 1967 .... 102, 229, 231

New Road and Street Works Act 1991 373

Occupiers' Liability Acts 1957 and 1984 .... 8, 208, 224, 244, 247, 257, 266
Offices, Shops and Railway Premises Act 1963 275
Official Secrets Act 1911 27
Offshore Safety Act 1992 263

Parliament Acts 1911 and 1949 35
Partnership Act 1890 61
Patents Act 1977 182
Policies of Assurance Act 1867 160
Prescription Act 1832 218

Private Street Works Act 1961 370
Property Misdescription Act 1991 138
Protection from Eviction Act 1977 322
Public Health Act 1936 219
Public Order Act 1986 207

Race Relations Act 1976 188
Refuse Disposal (Amenity) Act 1978 368
Rent Act 1977 .... 306, 319, 336, 339, 343
Rent (Agriculture) Act 1976 339
Road Traffic Act 1972 210

Sale of Goods Act 1979 .... 8, 76, 81, 118, 125, 129, 139, 209, 252
Sex Discrimination Acts 1975 and 1986 186
Social Security Act 1975 29
Supplementary Benefits Act 1976 13
Supply of Goods and Services Act 1982 147
Supreme Court Act 1981 315

Tort (Interference with Goods) Act 1977 .... 208, 211
Town and Country Planning Act 1990 .... 46, 295, 356
Trade Descriptions Act 1968 151
Trade Union Act 1984 .... 197, 206
Trade Union and Labour Relations Act 1974 198
Trade Union & Labour Relations (Consolidation) Act 1992 . 198, 199, 204, 206
Trade Union Reform & Employment Rights Bill 1993 .. 177, 179, 185, 196, 197
Truck Acts 1831–1896 179

Unfair Contract Terms Act 1977 .... 11, 75, 97, 108, 143,
                                                              149, 155, 231, 247, 249, 252

Wages Act 1986 179

Regulations
Building Regulations 1991 382
Construction (General Provisions) Regulations 1961 276
Construction (Head Protection) Regulations 1990 260
Construction (Health and Welfare) Regulations 1966 276
Construction (Lifting Operations) Regulations 1961 276
Construction (Working Places) Regulations 1966 .... 261, 276
Control of Advertisements Regulations 1989 368
Control of Substances Hazardous to Health Regulations 1988 274
Management of Health & Safety at Work Regulations 1992 267
Town and Country Planning (General Development) Order 1988 360
Town and Country Planning (Use Classes) Order 1987 359
Transfer of undertakings (Protection of Employment) Regulations 1981 .. 185

Table of cases

Allen v. Gulf Oil Refining Ltd 1981 218
Aluminium Industrie Vaassen BV v. Romalpa Aluminium 1976 145
Amalgamated Investment Co. Ltd v. John Walker & Sons Ltd 1976 .. 120, 367
AMF International Ltd v. Magnet Bowling Ltd 1968 244
Anchor Brewhouse Developments Ltd v. Berkley House Ltd 1987 .... 212, 283
Andreae v. Selfridge and Co. Ltd 1938 214
Andrews v. Schooling 1991 251
Annamunthodo v. Oilfields' Workers Union 1961 55
Anns v. Merton London Borough Council 1977 .... 224, 234
Applegate v. Moss 1971 130
Associated Dairies v. Hartley 1979 264
Atlantic Baron, 1979 84
Atlas Express Ltd v. Kafco Ltd 1989 110
Attorney General v. Overton 1980 285

Barnard v. National Dock Labour Board 1953 57
Barnett v. Chelsea and Kensington Hospital Management Committee
  1968 241
BBC v. Hearn 1978 205
Bernstein v. Skyviews and General Ltd 1978 282
Beswick v. Beswick 1968 126
Bickerton v. North West Regional Hospital Board 1969 172
Bissett v. Wilkinson 1927 103
Bolton v. Mahadeva 1975 114
Borden UK Ltd v. Scottish Timber Products Ltd 1979 145
Boston Deep Sea Fishing Co. v. Ansell 1888 181
BP Exploration Co. Ltd v. Hunt 1982 122
Bradbury v. Enfield London Borough Council 1967 44
Bradford v. Robinson Rentals Ltd 1967 242
Branca v. Cobarro 1947 92
British Railways Board v. Herrington 1972 247
British Steel Corporation v. Cleveland Bridge and Engineering Co. Ltd
  1984 130

Table of cases   xv

Bull v. Pitney Bowes 1966 113
Bunge Corporation v. Tradex Export SA 1981 95
Butler v. Ex Cello Machine Tool Co. 1979 78

Caparo Industries plc v. Dickman 1990 230
Central London Property Trust v. High Trees House 1947 86
Chappell & Co. v. Nestlé & Co. 1960 83
Clay v. Crump 1964 229
Collins v. Godefory 1831 83
Commercial Plastics Ltd v. Vincent 1965 113
Curran v. Northern Ireland Co-Ownership Housing Association Ltd 1982 235

D & C Builders v. Rees 1966 87
D & F Estates Ltd v. Church Commissioners for England 1988 .... 228, 250
Davis Contractors Ltd v. Fareham UDC 1956 118
Dimmock v. Hallett 1866 103
Donoghue v. Stevenson 1932 .... 15, 223, 224, 234, 253
Doughty v. Turner Manufacturing Co. Ltd 1964 242
DPP v. Broome 1974 207
Dutton v. Bognor Regis UDC 1972 .... 226, 233

Eastham v. Newcastle United 1964 113
Errington v. Minister of Health 1935 56

Ferguson v. Dawson and Partners Ltd 1976 175
Ferguson v. Walsh and others 1987 246
Froom v. Butcher 1975 243

Glassbrook Bros v. Glamorgan CC 1925 83
Grant v. Australian Knitting Mills 1936 16
Greater Nottingham Co-op Society Ltd v. Cementation Piling and
  Foundations Ltd 1988 228

Hadley v. Baxendale 1854 126
Hall v. Shoreham-by-Sea UDC 1964 58
Hedley Byrne v. Heller 1964 .... 105, 229
Herne Bay Steamship Co. v. Hutton 1903 118
Holwell Securities v. Hughes 1974 79
Home Counties Dairies v. Skilton 1977 112
Hong Kong Fir Shipping Co. Ltd v. Kawasaki Kisen Kaisha Ltd 1962 94
Howard Marine and Dredging Co. Ltd v. Ogden & Sons Ltd 1978 105
Howard v. Secretary of State for the Environment 1975 53

Jarvis v. Swans Tours Ltd 1973 128
Jones v. Padavatton 1969 89
Junior Books Ltd v. Veitchi Co. Ltd 1983 .... 227, 253

Kennaway v. Thompson 1981 215
Krell v. Henry 1903 118

Latimer v. AEC 1952 239

xvi   Table of cases

Lauritzen A. S. v. Wijsmuller B. V. 1990 120
Lavender v. Minister of Housing and Local Government 1970 59
Leigh and Sillivan v. Aliakmon Shipping 1986 227
Leigh v. Taylor 1902 284
Loweth v. Minister of Housing and Local Government 1971 57
Lyons Co. v. London City Bank 1903 283

McArdle Re 1951 82
McLoughlin v. O'Brien 1983 233
Massey v. Crown Life Insurance Co. 1978 175
Merritt v. Merritt 1970 89
Metropolitan Water Board v. Dick Kerr & Co. 1918 119
Midland Bank v. Green 1980 290
Midland Bank Ltd v. Shepherd 1988 110
Mihalis Angelos, The 1971 95
Miller v. Jackson 1977 215
Mitchell, George Ltd v. Finney Lock Seeds Ltd 1983 100
Murphy v. Brentwood D. C. 1990 .... 224, 228, 236, 250

Nagle v. Fielden 1966 113
National Carriers v. Panalpina Ltd 1981 .... 120, 322
National Westminster Bank Ltd v. Morgan 1983 110
Nettleship v. Weston 1971 240
New Zealand Shipping Co. v. A. M. Satterthwaite & Co. 1975 85

Pannett v. McGuiness & Co. 1972 249
Pao On v. Lau Yiu Long 1970 109
Parker v. Clark 1960 89
Parsons Ltd v. Uttley Ingham & Co. Ltd 1978 126
Pharmaceutical Society of Great Britain v. Dickson 1970 112
Pilkington v. Wood 1953 128
Pirelli General Cable Works Ltd v. Oscar Faber and Partners 1983 236
Polkey v. A. E. Dayton Services 1987 191
Price v. Civil Service Commission 1978 186
Proudfoot v. Hart 1890 313

R v. Barnsley Council ex p Hook 1976 55
R v. Hendon RDC ex p Chorley 1933 54
R v. Medical Appeal Tribunal, ex parte Gilmore 1957 49
R v. Secretary of State for the Environment ex p Ostler 1977 52
R. v. Sec. of State for Transport, ex p. Factortame Ltd 1990 5
R v. Sevenoaks District Council ex p Terry 1985 55
R v. Shoreditch Assessment Committee ex parte Morgan 1910 50
R. v. Swan Hunter Shipbuilders Ltd 1981 265
R v. Thames Magistrates' Court ex p Polemis 1974 56
R v. Willesdon JJ ex p Utley 1947 56
Raffles v. Wichelhaus 1864 108
Ready Mixed Concrete Ltd v. MPNI 1968 175
Rickards v. Lothian 1913 221
Roberts v. Hopwood 1925 58
Rose and Frank Co. v. Compton Bros 1925 89

Royscot Trust Ltd *v.* Rogerson 1991 107
Rylands *v.* Fletcher 1866 .... 220, 257

Sigsworth Re 1935 12
Simaan General Construction Co. *v.* Pilkington Glass Ltd 1988 228
Simpkins *v.* Pays 1955 89
Smith *v.* Eric S. Bush 1989 .... 230, 288
Smith *v.* Land and House Property Corporation 1884 103
Somma *v.* Hazelhurst 1978 306
Spartan Steel and Alloys Ltd *v.* Martin & Co. (Contractors) Ltd 1972 226
Spiro *v.* Glencrown Properties Ltd 1990 92
Startup *v.* McDonald 1843 116
Stilk *v.* Myrick 1809 84
Street *v.* Mountford 1985 .... 305, 323, 335
Sutcliffe *v.* Thackrah 1974 .... 31, 169, 232

Taylor *v.* Caldwell 1863 118
Trollope and Colls Ltd *v.* Atomic Power Construction Ltd 1973 164
Tweddle *v.* Atkinson 1861 88
Tynan *v.* Balmer 1967 207

Wagon Mound, The 1961 241
Watt *v.* Hertfordshire County Council 1954 239
Webb *v.* Minister of Housing and Local Government 1965 60
Weller *v.* Foot and Mouth Disease Research Institute 1965 226
Wentworth *v.* Wiltshire County Council 1992 377
Westminster Corporation *v.* LNWR 1905 60
Wheeldon *v.* Burrows 1879 298
White and Collins *v.* Minister of Health 1939 56
Williams *v.* Roffey Bros 1990 84
Wilson *v.* IDR Construction Ltd 1975 184
Woolcott *v.* Sun Alliance of London Insurance 1978 153
Woollerton and Wilson Ltd *v.* Richard Costain Ltd 1970 282
Wright *v.* McAdam 1949 297

1
The nature of law

Introduction

Any society quickly finds that it needs rules or laws to enable it to function smoothly. Consider the results if you had decided to drive to work today on the right instead of the left. People need to be able to live and go about their business in certainty, knowing that they can expect others to abide by the same rules. It is probably true that most people obey most of the rules most of the time. They may not have any precise knowledge of the actual rules which they are obeying, but that precise knowledge only becomes important when disputes arise or problems have to be solved.

As society evolves and becomes more complex, the number of laws increases and the types of laws become more sophisticated. Initially people need laws to protect their person or their property. Some rules accord with basic ideas of morality, e.g. not killing or not stealing but it is not essential for all the rules to have a moral base. For instance, when we create a rule for motorists that they must drive on the left the rule commands obedience simply because it is necessary and certain.

Although certainty is an important principle in any legal system, the rules of law should also aim to achieve justice. This is sometimes accomplished by giving a discretion. Some areas of English law are subject to discretionary principles known as equity. It should be remembered that laws are equally binding even if they are thought to be unjust. The remedy then is to seek to reform the law, not to violate it.

Sanctions

If a law is disobeyed, there must be some effective means by which it can be enforced. This may be described as a sanction and could take the form of a fine, imprisonment, an award of damages or an injunction, which is an order of the court forbidding certain actions or behaviour. Although such sanctions may deter people from breaking the law, in fact the reasons for obeying the law may be far more complex. A person may obey the law because he fears the disapproval or hostility of others, or because he believes in the 'rightness' of the rule, or because he sees obedience of the law as a duty he owes as a citizen.

Divisions of the law

It should be noted that all the contents of this book relate to English law. Although it is appropriate to talk about the British Constitution it should be remembered that Scotland has a different system of law to England. Whereas some of the rules are common to both countries, there are still significant differences. These are most apparent in areas like the system of land holding and in the structures and procedures of the courts.

English law may be sub-divided in a number of ways. A simple division would be into criminal and civil law. It is quite common for the layman to believe that the bulk of our law consists of criminal law. This inevitably results from the publicity given in the media to major criminal cases. In this sense the law is like an iceberg – the criminal law is the part seen above the water and the civil law is the mighty bulk that lies below. The state has an interest in preserving our society and upholding law and order, and it makes criminal laws to secure those objectives. A person who infringes those laws commits a criminal offence for which he can be prosecuted by the state. The principal objectives of the criminal law may, therefore, be seen to be to deter and to punish. The civil law protects rights and creates obligations between individuals, although sometimes those individuals will be large public corporations, government departments, local authorities etc.

There is no simple way to classify civil law, but it may be subdivided into the law of contract, the law of tort (i.e. civil wrongs), the law relating to property and the law relating to persons. The rules of civil law must deal with matters as diverse as contracts for the sale of goods, actions for damages for negligence, planning applications and compulsory purchase, divorce, tenancy, creation of companies, contracts of insurance, making a will, money lending, recovery of debts, defective workmanship and unfair dismissal.

Language of the law

Every profession has its language or jargon, which is often a convenient shorthand method of communication between people engaged in that profession. With its long history, the language of the law is especially rich as it frequently uses Latin expressions such as *res ipsa loquitur* (the facts speak for themselves,) or *caveat emptor* (let the buyer beware). Different terminology applies to criminal and civil law. In criminal proceedings a person is arrested, charged with an offence and prosecuted in summary or indictable proceedings by a prosecutor. If the accused person is found guilty he will be sentenced and punished. If he is found not guilty he will be acquitted. In a civil action a plaintiff sues a defendant. If the defendant is found to be liable he may be ordered to pay damages or an order of the court such as an injunction may be made against him. Where the terminology is correctly used it is possible to tell whether a case concerns criminal or civil law.

Although it is convenient to break down the mass of English law into sub-divisions for the purposes of study, it should be realized that these various divisions of the law are not mutually exclusive. It is possible for one set of facts to give rise to both civil and criminal liability. The simplest example is the motorist who drives carelessly and injures a pedestrian. A motorist may both be prosecuted in criminal proceedings for an offence under the Road Traffic Acts and be sued in civil proceedings by the injured pedestrian hoping to recover compensation. Similarly, an employee injured at work may wish to bring civil proceedings against his employer to recover damages where the employer has been negligent in caring for the employee's safety. Arising out of the same set of facts, the Health and Safety Executive may prosecute the employer for breaches of the Health and Safety at Work Act 1974.

Making the law

Unless some catastrophic event like a war or a revolution occurs, which may cause a country to adopt a completely new system of law, the law-making processes will have developed over a period of centuries. Many of the rules of English law which are still in force are of considerable antiquity. Laws do not become ineffective merely because they are very old. Indeed some old rules have been used very imaginatively by the judges in the courts to create principles relevant to modern life.

In the early stages of law making, custom will usually play a large part. As the needs of a society become more sophisticated, custom as a source of law tends to be superseded by more formal sources. Local custom may still be upheld as a valid part of the law where it can be shown that the custom is reasonable and certain and has been continued without interruption since time immemorial. For practical purposes this normally means that the custom must be shown to have existed during living memory. Occasional examples still come before the courts.

The two main sources of law in our system today are legislation (laws made or approved by Parliament) and judicial precedents (binding decisions of the judges). Both of these sources of law must now take account of the fact that the United Kingdom is a member of the European Community.

The UK and the European Community

The UK joined the European Community in 1972. The legal power for its membership is contained in the European Communities Act 1972. There had been previous abortive attempts to join the Community. Even when our membership became a reality, doubts were expressed which were sufficient to prompt a unique referendum in 1975 which confirmed the view that the UK should stay in the Community. Since our entry other countries have joined, most recently Spain and Portugal.

Many of the doubts about joining the Community related to the loss of sovereignty which would result. To what extent would it be true to say after 1972 that Parliament is supreme? At the time many people in the UK believed that our first loyalties lay with the Commonwealth and that inevitably membership of the European Community would result in a diminution of those ties.

The European Community operates through a number of institutions, principally the Council of Ministers, the Commission and the European Assembly or Parliament. There is also a Council of Europe which consists of heads of government and foreign ministers who meet several times a year. The Council is now recognized by the Single European Act and because of the stature of the persons who attend it is obviously very significant. Vital issues like the size of our financial contribution to the European budget are thrashed out in the council.

Additionally there is a Council of Ministers composed of one representative of each member state, who will usually be the Minister concerned with the subject matter under discussion. This is the principal decision-making body and member states hold its presidency in turn. Usually the Council of Ministers takes decisions based on the proposals put forward by the Commission. The Commission is the executive arm of the Community. Its members become Commissioners, each with responsibility for a specific aspect of work of the Community, for example agriculture. The Commission initiates legislation, usually after consulting the Civil Service of the various member states. In deciding on its response to the Commission's proposals, the Council will have consulted the European Parliament. The Parliament has been a directly elected body since 1979. The UK has eighty-one Euro MPs who are elected on a party political basis. Once elected the Euro MPs sit in political, not national, groupings. The role of the Parliament is gradually being extended.

Both the Council of Ministers and the Commission can make regulations which have direct force in the UK. In other cases where they issue 'directives' or 'decisions', our own Parliament must implement the rules for this country, although then it may have some discretion about the method of implementation. The European Communities Act 1972 provides that the various Community Treaties shall have effect in UK law and that decisions of the European Court are applicable in UK courts. Where relevant, therefore, Community law prevails over UK law.

This thorny question of 'supremacy' has caused much argument. One view is that European law is supreme and takes precedence over domestic legislation. In joining the European Community, the argument runs, the United Kingdom conceded that the British Parliament can no longer pass inconsistent legislation. A counter argument has always insisted that what Parliament has accepted – in the European Communities Act 1972 – it could also reject. Recent cases, however, confirm that the former view is correct. In *R. v. Secretary of State for Transport, ex parte Factortame Ltd*, the European Court of Justice indicated that where there is a national rule in conflict with a Community rule, the national rule becomes inapplicable. Cases involving equal pay and sex discrimination have been a rich field for development of these principles (see page 186).

New legislation

Law-making authority is vested in the Crown and Parliament, although the role of the Crown is now almost entirely formal. Our complex modern society requires rule-making techniques which are capable of coping with economic, social and welfare problems. Parliament, consisting of the House of Lords and the House of Commons, is said to be supreme, but in practice its ability to make rules must be viewed subject to a number of limitations. By virtue of the European Communities Act 1972 a new element of European Community law was introduced into our system. European Community law takes precedence over the national law of any member state. This, therefore, imposes a limit on the ability of Parliament to make whatever laws it wishes. If English law is found to conflict with European Community law the Community law will prevail. Other factors which would limit the power of Parliament are the existence of an opposition party, or parties, whose duty is to seek to curb or limit government proposals; Parliamentary question time which is held daily, when ministers, including the Prime Minister, must justify their activities; the two chamber system whereby the House of Lords can at least delay the passing of legislation; and public opinion, pressure groups, freedom of speech and publicity through the media. Once laws are enacted by Parliament the interpretation of those laws is carried out by the judges in the courts. If Parliament has passed a law which is seen to be too harsh or too extensive in its application it may be possible for the judges, by restrictive interpretation of the words of the statute, to limit the scope of the new rule.

Acts of Parliament, which can also be called statutes, contain the main laws made by Parliament, acting in its legislative role. Until the statute or Act has passed through all its stages in both Houses and received the royal assent, it is referred to as a Bill.

The inspiration for new legislation may come from a number of sources. A new government will have made manifesto commitments and will have outlined its own policies. During its first year or so in office it will be keen to push through those changes which were outlined in its election manifesto. Inevitably this source of new legislation becomes comparatively less significant as the government's term of office, a maximum of five years, progresses. Each of the major government departments will have a programme of legislation which it would like to introduce. Parliamentary time is at a premium and this may well produce competition between departments. Where a small measure is needed a department may seek to persuade a private member who has won a high place in the private members' ballot to introduce a bill on its behalf.

New laws may also be needed to implement recommendations of the Law Commission, a body established in 1965 to review the law with a view to its systematic development and reform. Occasionally the government of the day sets up Royal Commissions or other special committees of enquiry. These are usually established to investigate one specific topic area and they will be disbanded once their reports are published. Many of these reports gather dust and do not result in their recommendations being implemented in legislation. In other cases the recommendations may be implemented in their entirety. One such example was the Robens Report on Health and Safety which very quickly became the Health and Safety at Work etc. Act 1974. There is a limited scope for individual politicians to introduce private members' bills, by which means small changes may be introduced into our law. Occasionally new laws will be required to meet a sudden emergency and, where necessary, Parliament can act with considerable speed. It is possible for an Act of Parliament to pass through all its stages in both houses and receive the Royal Assent in a day if necessary.

Once an idea for a piece of legislation has been accepted by the Legislation Committee of the Cabinet there will usually follow an intensive period of deliberation and consultation. Parliamentary Counsel (the draftsmen of the Bills) will then be required to draw up the Bill clause by clause. Many Bills will be significantly redrafted before they are enacted in their final version. The problems which beset the draftsmen include the lack of precision of our language; trying to reconcile many conflicting demands; attempting to cover situations which can be envisaged but which have not yet arisen in practice; and pressure of time. Not surprisingly many Acts of Parliament are subsequently found to create interpretive difficulties which must be resolved in the courts.

Acts of Parliament begin life as bills, which may be either public, private or private member's bills. Public bills comprise the vast majority passed in each parliamentary session. An example of a private bill may be one promoted by a local authority to authorize some activity specifically in its own local area. Private member's bills provide the limited opportunity available in each parliamentary session for an MP to introduce some proposals for change in the law on a topic of his choice. Public bills are usually introduced by government departments, and may be introduced in either House. The likelihood is that most of these public bills will be passed. A government, particularly one with a large majority, has an effective stranglehold on procedures in Parliament. Non-controversial bills are often introduced in the House of Lords but all money bills must be introduced in the Commons.

The normal procedure in both Houses is that when the bill is introduced it will have its first reading. Normally the sponsors of the bill present it in dummy form at the table of the House and one of the clerks reads out the title. The bill is then deemed to have been read for the first time. It is next ordered to be printed and published, and a date fixed for the second reading. Its second reading is the occasion for a parliamentary debate on the principles of the bill. The bill is not considered clause by clause at this stage. There will be a vote at the end of the debate and the bill could be rejected at this stage. Assuming that the bill is not lost it is then referred to a committee, to undergo its committee stage. It will usually go to a standing committee of between sixteen and fifty members, chosen to reflect the relative strength of the political parties in the House, and having regard to the special qualifications, concerns and interests of the MPs in question. These standing committees usually sit in the morning, and they will consider the bill in detail, examining it clause by clause, trying to produce a result which is unambiguous. The committee will also deal with proposed amendments to the bill. Where a bill is of major constitutional importance, the committee stage may be taken in a Committee of the Whole House. However, the drawback of this system is that, while it sits as a committee, no other business can be conducted by the House.

Many of the amendments proposed to the bill at the committee stage may be put down by the minister in overall charge of the bill himself. These proposed amendments may reflect afterthoughts by him or his officials, or they may be the result of concessions to outside pressure groups. Once the committee stage is completed, the bill is then reported back to the House (the report stage) when the changes introduced in committee will be outlined. There is then a third reading of the bill, often done immediately after the report stage is concluded, and the debate on the third reading may culminate in a vote.

Where the measures proposed are controversial, both the government and the opposition will be anxious to have as many of their supporters voting as possible, and each may have issued a three line whip, which is an instruction to their supporters to attend and vote in accordance with instructions. The whip system is the means by which political parties control and organize their members in Parliament. There is a system of whipping in the Lords, but its principal importance is in relation to the activities of the House of Commons. The name comes originally from the world of hunting. The whipper-in is a hunt official charged with the control of the hounds. The word has now transferred itself to the Parliamentary context. A whip in Parliament is a person whose job it is to give information to his party members and to maintain discipline among them.

Once the bill successfully passes all its stages in both Houses, it receives the Royal Assent and becomes an Act of Parliament. It takes effect from the date on which it receives Royal Assent, or from the date of commencement set out in the Act itself, or from a date to be fixed in the future. Power to fix that date will usually by given to an appropriate minister. The Act then remains in force until it is repealed. Repeals are effected by exactly the same process. It is usual to refer to the Act by its short title, for example, the Sale of Goods Act 1979, or the Occupiers' Liability Act 1984. Unless the Act states to the contrary, it will apply throughout the UK.

Nowadays Acts of Parliament follow a fairly standard pattern. Before setting out any of the main text there is a preamble which establishes the purpose of the Act. For example, the Drought Act 1976 begins: 'an Act to confer fresh powers to meet deficiencies in the supply of water due to exceptional shortage of rain and for connected purposes.' The main text is then divided into sections and sub-sections and, if appropriate, the whole Act will be set out in parts. Where the Act contains detailed lists, these will often be contained in a schedule to the Act. Most modern Acts have sections which deal with definitions, repeals, date of commencement and area of operation. The definition section in an Act of Parliament is important because within it, the draftsman of the Act can set out the precise meaning of a word for the purpose of that Act only. This can be of considerable help to a judge when he subsequently needs to interpret the exact meaning of the legislation. The repeals section will indicate which earlier Acts or regulations or parts of Acts and regulations have been repealed by the present Act.

Pressure groups

It is undoubtedly true today that in drawing up and seeking to implement its legislative programme, no government can afford to ignore the views of pressure groups. Joining a pressure group is one of the ways in which individuals have an opportunity to influence government decisions and future legislation. There are two main types of pressure groups – interest groups, where the group exists primarily to promote the interests of its own members (for example the Automobile Association or the National Farmers Union or the various trade unions) and cause groups, which are usually formed for the purpose of lobbying for a specific cause (for example Help the Aged, The National Society for the Prevention of Cruelty to Children and the Committee for Nuclear Disarmament).

The word 'lobbying' means influencing members of the legislature. The word itself derives from the practice of people meeting their MPs in the lobby of the House of Commons. The process of lobbying can be extended over a very long period and requires considerable energy, organization and resources to carry it out effectively. A large organization may employ professional lobbyists to pursue its cause.

The government of the day is likely to be influenced by pressure groups for a number of reasons – the pressure group may well be expert within its particular field or the government may need active cooperation from the members of the pressure group in order to implement its policies. Some pressure groups are immensely wealthy, and can promote their cause by advertising and other forms of publicity, while other pressure groups are powerful in terms of the significant proportion of the community which they represent.

Pressure groups can seek to exercise influence in a number of ways and at a number of points in the legislative process. Before any legislation is introduced, members of pressure groups may be represented on advisory bodies connected with appropriate government departments. If legislation is proposed, consultation with appropriate groups will frequently take place before a bill is introduced. A pressure group will always be working to advertise and promote its cause, possibly by liaising with sympathetic MPs, even when no immediate legislation is proposed. Once a bill has been introduced, MPs sympathetic to particular causes may participate at the detailed committee stage, when numbers of amendments to a bill may be proposed. At that point, the pressure group will wish to stay in touch with its 'tame' MPs. Even after an Act has been passed, a watching and monitoring role can be played by pressure groups who may seek changes by regulations if an Act is seen to have loopholes. Much of the success of a pressure group may depend on making and maintaining good personal contact with MPs, peers who sit regularly in the Lords, civil servants and ministers.

Delegated legislation

It can easily be seen that pressure on parliamentary time is so great that all the rules necessary in a complex modern society cannot be made in the laborious and time consuming way previously outlined. It is therefore common for Parliament to lay down a framework in an Act (an 'enabling Act') and then grant power to some other person or organization to make the detailed regulations. For example, by s.15 of the Health and Safety at Work etc. Act 1974, the Secretary of State has power to make regulations for any of the general purposes of that Act.

Delegated legislation can take one of three forms:

1 *Orders in Council* – where the rule-making power is vested in the Privy Council. This is normally confined to emergency situations and was much used during the Second World War.
2 *Statutory instruments* – this is the form commonly used when ministers make regulations by virtue of delegated powers.
3 *Bye-laws* – made by local authorities or other public bodies (e.g. British Rail) under powers granted to them by Parliament.

Although delegated legislation can be seen to be necessary and to have some advantages as a rule-making technique, there must be adequate control of the process. The advantages include:

- *Speed* – where there is an emergency or a temporary or rapidly changing situation.
- *Flexibility* – out-dated rules can be more easily changed.
- *Expertise* – consultation with outside bodies may harness the experience of technical and other experts.

Despite these advantages, it will be realized that rule making by these methods gives great power to persons who have not been elected. Critics of delegated legislation would argue that it vests too much power in the hands of the executive, and that scrutiny and control of the process are vital. That control may come through the procedures themselves, for example by delegating powers on strict terms to prevent abuse, or by subjecting any regulations proposed to an affirmative resolution procedure of one or both Houses of Parliament, or by requiring prior consultation with an advisory body.

If the power which had been granted is exceeded, control can be exercised by the application of the *ultra vires* principle. This would require a person aggrieved by the abuse of power to claim in court that the rule maker has acted outside the scope of his authority i.e. that his action has been *ultra vires*. If so found by the court, the *ultra vires* action will be declared to be void i.e. of no legal effect. Parliament itself is aware of the danger of abuse of power and has established a select committee of statutory instruments, usually known as the Scrutiny Committee, to act as a watchdog. Parliament might pass only ninety Acts of Parliament in any one year, yet in the same period more than 3000 Statutory Instruments might be made, so the potential size of the problem can be seen.

Statutory interpretation

When an Act of Parliament comes into force it may cause disputes as to its true meaning. The written word is not an exact form of communication and it will ultimately be the duty of the judges, when cases arise on disputed points in an Act, to determine the 'intention of Parliament'.

Faced with such a problem, the judge starts by looking at the Act itself. Most Acts contain a definition section, in which words will be given a specific meaning for the purposes of that Act only. For example in s.14 of the Unfair Contract Terms Act 1977, definitions are given of the words 'business', 'goods' and 'notice', thereby giving those words a specific and restricted meaning confined to that Act. 'Business' is defined there as including a profession and the activities of any government department or local or public authority.

Where no definition is given within the Act itself, a judge may take the meaning of a word from a dictionary, but if he does so, he must bear in mind the principle expressed in Latin, *'noscitur a sociis'* which means 'a word is known by the company it keeps'. The meaning of a word may well be governed by the context in which it is used, e.g. in the phrase used in another Act, 'interest, annuities or other annual payments', the word 'interest' means annual interest. The judge may also turn to the Interpretation Act 1978 which gives meanings for standard phrases and expressions like day, week and month. And again, the word or phrase in question may have been interpreted and applied in an earlier case heard by the court, so that the present judge would be obliged by the rule of judicial precedent to accept that meaning in the case before him.

Assuming that these aids do not solve the problem, the judge may then adopt one of the following approaches to the question:

1 *The literal approach* – this approach dictates that words should be given their ordinary, plain and natural meaning, even if this produces an absurd result. Judges favouring this approach take the view that it is then Parliament's responsibility to put the situation right by amending the law. In using this approach, two general rules of construction should be remembered:

   (a) The *euisdem generis* rule – when specific words are followed by general words, the general words must be read in the light of the specific words. For example, 'on a pond, lake, river, canal or other water'. By applying this rule, these words would be read to include a reservoir but not the open sea. It is clearly intended by the specific words to confine the stretches of water to inland water.

   (b) The rule *'expressio unius est exclusio alterius'* – if one thing is specifically mentioned in the Act, the Act is presumed not to apply to another e.g. some sections of an Act may refer to the seller and the buyer. A subsequent section mentioning only the seller is presumed not to apply to the buyer.

The literal approach to interpretation is not favoured by some judges. One judge has said: 'When a defect appears a judge cannot simply fold his hands and blame the draftsman.'

2 *The golden rule* – this allows the judge to interpret so as to avoid the absurdity, or to avoid a result which is inconsistent with the Act itself or common law principles. This approach was used in *Re Sigsworth*, in a case where a man claimed his mother's property, despite having murdered her. Under the Act which governed the distribution of her property, this ought to have been shared amongst her children. The son was the only child. The judge held

that the common law rule that a murderer cannot benefit from the person he has murdered, prevailed over the clear words of the Act.
3 *The mischief rule* – another approach adopted where the judge looks to see which defect of the law is intended to be corrected by the Act and then interprets to correct the defect.

A simple example from a case may illustrate some of the problems of interpretation faced by judges. In a case concerning single payments under the old Supplementary Benefit scheme, a claimant requested a single payment to help towards the purchase of wallpaper to redecorate her lounge and kitchen. The wording of the appropriate regulation required that payments could only be made in respect of expenses of *essential* internal redecoration. The court had to consider what was the meaning of the word 'essential'. It was not defined anywhere in the Supplementary Benefits Act 1976 or the relevant Regulations and it fell, therefore, to be given its ordinary meaning in everyday use. When a dictionary was consulted, the *Shorter Oxford English Dictionary* gave two rather differing meanings of the word 'essential'. It can mean 'material' or 'important' in the sense of 'your work is essential to the success of this project', or it can mean 'indispensably requisite', as for example, 'water is an essential ingredient of a cup of tea'. Relying on the stricter meaning of the word, the benefit officer had refused a single payment. On appeal to the tribunal, however, the judge favoured the less strict interpretation of the word. He was guided in doing so by the fact that the same word 'essential' was used elsewhere in the Regulations, where its meaning tended to be 'material' and 'important' rather than 'indispensably requisite'.

No one instructs a judge as to which of these approaches he should use. It is clear that the process of interpreting and applying Acts of Parliament gives considerable power to the judges. But their role in the law-making process is not limited to the interpretation of statutes because the other principal source of law is judicial precedent (binding decisions of the judges).

Judicial precedent

When a judge gives his decision in a case before him, this has two elements:

1 The actual decision affecting the parties, e.g. he finds X to be liable and that X must pay damages to Y of £1000,
2 The principles of law which have caused the judge to arrive at that decision.

Where the case concerns facts or situations which are comparable with earlier cases heard by the courts, the judge will normally apply the principles from the earlier cases. In one well-known case, a soft-drinks manufacturer was held to be liable to the ultimate consumer who had suffered physical harm by drinking a bottle of the manufacturer's ginger beer which was contaminated by the remains of a decomposed snail. If, on a future occasion, a court hears a case brought by a person made ill by eating a pie containing a dead mouse, the same principles can be applied. By analogy these principles may be extended further. The principle in the 'ginger beer' case was actually applied to a later case where a person contracted dermatitis because of wearing underpants where chemicals had been left in the material during their manufacture.

Although many systems of law in practice apply this principle of deciding cases in the same way as previous cases of a similar nature, the English system goes further. The judge is bound by the earlier decisions, hence the system is often referred to as the system of binding precedent. One difficulty presented by this technique lies in finding the actual part of the case which is binding. When a judge gives a decision, his legal reasoning, when set out in the law reports, can be several pages long. Much of what he says may relate to why he rejected other arguments put to him by counsel, or what would have influenced him if the facts had been slightly different. These parts of his judgement are not binding in later cases. They are 'said by the way' or in legal terms, *'obiter dicta'*. The judge in a later case must search to find the rule of law upon which the decision is based, properly called the *'ratio decidendi'*.

The process of identifying the ratio is a vital part of the training of a lawyer. In each case the first step is to establish the material facts. For example, Mrs Smith, a mother of four aged forty-five, knocked down and injured Mr Brown last Friday. The accident was caused because Mrs Smith had gone to sleep at the wheel. The only material facts here are that a motorist injured a pedestrian by driving negligently. Once the material facts have been established, if they correspond with the material facts of an earlier decision, the judge will usually be bound by the earlier case. However, a judge is normally only bound by decisions given in a court superior to his own. This allows the system to be flexible (as it is open to a higher court, on appeal, to overrule or reverse his decision) but certain (as it obliges lower courts to follow the decisions of higher courts). The structure of the courts is considered in Chapter 2.

Advantages and disadvantages of the system of precedent

One of the greatest advantages of the system of binding precedent is that the rules have evolved from real-life cases and are, therefore, essentially practical. Again, the 'binding' feature of the system makes it reasonably certain. However, two major criticisms are often levelled against the system. The first is that it creates a bulky system of very detailed rules which requires the production, at considerable expense, of large numbers of law reports to enable the previous cases to be checked and quoted. A more significant problem is that the system can be very uncertain, due to the powerful role played by the judges. Consider the case where a judge is faced with a precedent which in theory is binding on him, but which he prefers not to apply. There is a technique called 'distinguishing' one case from another which may allow him to avoid applying the earlier decision. The judge is allowed to distinguish (and therefore not apply) any case where the material facts are not the same as in the case before him. When this technique is properly used it allows the law to be flexible. When improperly used it results in hair-splitting distinctions, which can leave two apparently conflicting decisions reported in the casebooks.

So that the system of binding precedent can function effectively there must be a well-organized procedure for reporting cases and there must be an established hierarchy of the courts. Generally, lower courts are bound by their own previous decisions and by those of higher courts.

If a person is dissatisfied with the outcome of his case he may have grounds for appeal. The final court of appeal in the English system is the House of Lords. Until 1966, even the House of Lords was bound by its own previous decisions, but it was then announced that: 'Their Lordships propose to modify their present practice, and while treating former decisions of this House as normally binding, to depart from a previous decision when it appears right to do so.' The most recent example of such a departure can be seen in Murphy v. Brentwood D. C. in 1990 (see page 225).

Applying a precedent

Examples taken from actual cases demonstrate how the case law system works. To see how the normal processes of applying a precedent operate, look first at a case from 1932, *Donoghue v. Stevenson*. A manufacturer of ginger beer which was marketed in an opaque bottle was held to be liable to the ultimate consumer of the drink in the tort of negligence. (A tort is a civil wrong.) The consumer had become ill on discovering the remains of a decomposed snail in the bottle. If this precedent had subsequently been confined by judges to sets of facts involving drinks sold in opaque bottles its value as a precedent would have been very limited. However, in 1936, another court seized the opportunity to extend the scope of the precedent. In the case of *Grant v. Australian Knitting Mills*, the plaintiff, Dr Grant, had purchased a pair of long woollen underpants, manufactured by the defendants. He contracted a severe form of dermatitis after wearing them, shown to have been caused by excessive amounts of sulphur which had not been successfully washed out during the manufacturing process. Were the defendants liable? The judges examined the Donoghue case, and extracted from it the following statement of principle:

> A manufacturer of products which he sells in such form that he intends them to reach the ultimate consumer with no reasonable possibility of intermediate examination, and where the absence of reasonable care in the preparation of the products will result in injury to the consumer, owes a duty to the consumer to take reasonable care.

Undoubtedly a manufacturer of underpants must realize that potential customers cannot carry out scientific tests to measure sulphur levels before purchasing, and the courts had no difficulty in applying the Donoghue principle. If the court in the Grant case had thought the principle was unsound, it could have sought to distinguish the two cases on the ground that the material facts were not similar. For example, the snail case concerned goods to be consumed internally, whereas the underpants case concerned goods to be worn externally. Such fine distinctions could have been justified if the earlier precedent had been unpopular for any reason, but the law develops best when the scope of a sound principle is gradually extended.

Common law and equity

The expression 'common law' is frequently used and can have a number of meanings. It may mean the law which is common throughout England. Indeed the phrase was originally used in that sense to distinguish it from local rules and customs. The expression may also mean rules developed through precedents, rather than created by Acts of Parliament. It may further be used to mean rules which are not derived from equity. An unusual system emerged in England between the twelfth century and 1875 whereby a completely separate system of courts developed, administering quite separate rules – the principles of equity. These rules were usually developed to meet situations where the common law had no remedy to offer or where the common law was in some way deficient or unjust. The equitable principles are discretionary in nature and, since 1875, have been administered side by side with the rules of common law. If the two sets of rules should ever conflict, the principles of equity will prevail.

2
Settlement of disputes

In any situation where things go wrong, causing a dispute or conflict between people or organizations, the parties involved may think of turning to the 'law' for a remedy or a solution. So if you have bought faulty goods, or want to claim against a motorist who has damaged your car, or are owed money and want to recover the debt, you may wish to take proceedings in the courts against the relevant party. It would be wrong to think only of the courts when considering the settlement of disputes. Some measures of self-help can be very effective, and in other cases the parties may prefer the privacy of arbitration proceedings. Sometimes legislation has provided that a forum other than the courts is more appropriate for the settlement of disputes, for example those cases where any rights must be pursued through specialized tribunals. There are also occasions where a dispute reveals no cause of action capable of being pursued through the courts, and in those cases it may be that conciliation procedures or intervention by the Ombudsman may be more appropriate. The scope of these various procedures is considered separately.

Self-help

Where a dispute arises between parties, it may be possible to settle matters by negotiation without the need to go to court. For example, in the law of trespass the person in possession of land can ask the trespasser to leave, and if he does not do so, may then use a reasonable amount of force to eject him. What is reasonable will vary with the circumstances of each case. More specialized examples of self-help can be found in the rules relating to set-off and lien. Set-off can occur where one party owes money to another, and is in turn owed money by that other. So if A owes B £10, and B owes A £5, then A may set-off the £5 he is owed by B and thereby reduce the amount he must pay to B.

A lien is the right of a person in possession of goods which belong to someone else to retain them until some demand, usually a demand for payment, has been met. This remedy can be particularly useful to people such as garage owners who have carried out expensive repairs on a vehicle. They can hold on to the vehicle, i.e. exercise their right of lien, until their bill is paid.

These are legally recognized forms of self-help, but less formal methods may be equally effective. Large numbers of pressure groups now exist which may support individuals who have grievances or who are involved in disputes. The power of the press and the media can sometimes achieve more than the use of formal legal procedures. Moreover, the intervention of an advice agency may be sufficient to prompt the other party to a dispute to resolve matters without resorting to court action.

Court procedures

If proceedings in court become necessary, it will quickly be realized that there is an elaborate system of courts, both civil and criminal. However, the vast majority of cases are disposed of by the inferior, or lower, courts at first instance. In criminal law, for example, more than 90 per cent of all cases are dealt with by the Magistrates Courts.

Courts of first instance are where cases are first heard. The complicated system of courts in England results from their haphazard development over hundreds of years. In an attempt to rationalize the system and concentrate courts in the places where they were most needed, major reforms of the system took place in 1970 and 1971. The present structure can be seen in Figure 1 where possible rights of appeals are indicated by the arrows.

Figure 1 shows that the House of Lords is the highest court of appeal for both civil and criminal cases in our English legal system. It is also the final court of appeal for civil cases from Scotland. The court can sit with a minimum of three judges, but cases are usually heard by five Lords of Appeal in Ordinary. One judge will be appointed from Scotland and another from Northern Ireland. Other judges who may sit in the House of Lords are the Lord Chancellor, any ex Lord Chancellors, the Master of the Rolls and other peers who have held

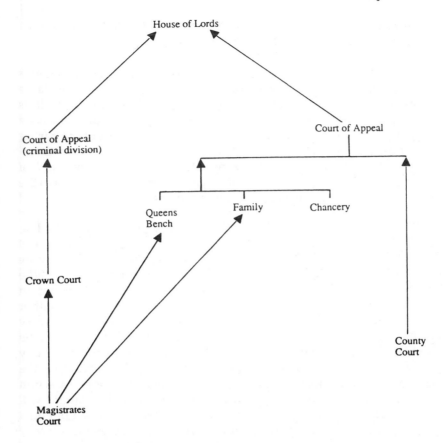

Figure 1 *Court procedures structure*

high judicial office. Hearings of the court take place in the House of Lords at Westminster and all cases coming before the court will involve points of law of general public importance. Leave of the court below is necessary before an appeal can be brought. Not surprisingly, only a very small proportion of cases can reach this level of the court structure, not least because of the expense involved. In Figure 1, courts which principally hear criminal cases are shown on the left side. Whereas some courts have exclusively criminal jurisdiction, for example the Court of Appeal Criminal Division, other courts have a mixed jurisdiction to hear both civil and criminal cases, for example, the Magistrates Courts. Others hear only civil cases, for example the County Courts.

Civil cases

The County Court

The Country Court is a local court involved with settling civil disputes between individuals. However, one of those individuals may be a large company or public corporation or a nationalized industry. It is sometimes said that when it becomes necessary to use a court to settle a dispute, this is a sign of failure. The failure is on the part of the parties to compromise or reach some satisfactory settlement between themselves. It is often much more satisfactory to accept some kind of compromise rather than embark on litigation (bringing an action to the court). Litigation involves stress, expenditure of time and money and uncertainty about the final outcome.

The jurisdiction of the County Court was previously limited to claims worth less than £5,000. Changes introduced by the Courts and Legal Services Act 1990 mean that, in future, there will be a more flexible system for allocating work between the County Court and the High Court. The County Court is a local court, and to commence an action there, one of the following conditions must be satisfied: either the defendant must reside in the district; or the defendant must carry on business in the district; or the cause of action must have arisen in the district. Cases are started by giving the County Court the necessary details so that a summons can be served. The person bringing the action is called the plaintiff, while the person against whom the action is brought is the defendant. To serve the summons the plaintiff will have to pay the appropriate fee to the Court. This is worked out on a sliding scale, depending on the amount claimed. If a plaintiff is totally successful in his claim he will ultimately recover his fee.

Once the defendant receives the summons he may accept that he owes the money involved and pay up in full. It is more likely that he will wish to dispute the claim in whole or in part, and then he must send in his defence and any counter-claim he may have. The forms sent to him provide him with places to fill in these details. If the defendant does nothing at all in response to a summons, then the plaintiff will obtain judgement, and may then want to use the various procedures available through the County Court to enforce the judgement.

Once the defendant sends in his defence and/or his counter-claim, a date will be fixed for a pre-trial review in front of the District Judge of the County Court. Very often a settlement between the parties can be reached at this stage. Even if no settlement is reached, the value of the pre-trial review is that it is possible for the District Judge to see on which points the parties agree, and on which they remain in dispute.

He can decide what evidence it will be necessary for the parties to produce, and he can assess how long must be set aside for the trial of the case. Where necessary a date will be fixed for the trial. That may take place before the District Judge, where the amount claimed is small, or before a Circuit Judge.

Where the amount claimed is less than £1,000 the parties are encouraged to appear without professional representation and the arbitration (by the District Judge) will be conducted extremely informally. The rules of evidence will not be strictly adhered to. Arbitration in this context is quite different from commercial arbitrations which will be considered separately (see page 29).

Where there is to be a formal trial, this will be a proper hearing in open court, with 'wigs and gowns', and subject to the operation of the rules of evidence, such as giving evidence on oath. This is in contrast with the less formal atmosphere of the pre-trial review, and those cases which are set down for arbitration. If the plaintiff succeeds in his case, and a judgement is given in his favour, then he must hope that the defendant will comply with the terms of that judgement. Very often the judgement directs the payment of money, sometimes in a lump sum, or sometimes by stated instalments. If the defendant defaults, the plaintiff will need to use the procedures provided through the County Court to enforce his judgement. He may consider using execution. This consists of sending the court bailiff to seize sufficient goods of the defendant to satisfy the judgment against him. When the bailiff arrives, the defendant is given a further chance to pay, and many defendants do so at this stage. If it does become necessary for the bailiff to seize goods, they will be removed and sold by auction. All the costs involved in doing this will be added to what the defendant owes.

Where the defendant is in employment, a plaintiff may consider an attachment of earnings order as a method of enforcing his judgement. In essence this consists of the court informing the employer that he must deduct a specified weekly or monthly sum from the earnings of the defendant and pay that sum directly to the court. There is a protected earnings amount fixed by the court so that, in any week or month, the defendant must always be permitted to retain that amount of his wages or salary.

Where a defendant is beset by multiple debts and is being harassed by a number of creditors, he may consider asking the court to make an administration order. Effectively the result of this order is to consolidate all the debts, and the court will the decide the amount which is to be distributed pro rata amongst the creditors. The value of an administration order is that it relieves the debtor of the burden of dealing with all his separate creditors and ensures some fairness and uniformity of treatment between those creditors. It is a step short of bankruptcy, so of course it is essential that the debtor must be able to pay something regularly, as the order can only run for three years.

Where the defendant is hopelessly in debt, with no means to pay off the debts, the law provides the bankruptcy procedure to allow him to rid himself of those debts. Inevitably there is a price to be paid by the debtor, as there is still considerable social stigma attached to bankruptcy and an undischarged bankrupt is under a number of penalties. For example, it is a criminal offence for him to obtain credit of any kind, such as a bank loan or hire purchase, without revealing that he is an undischarged bankrupt.

Where the defendant owns property, for example his own house, the plaintiff may consider enforcing a judgment by means of a charging order on the property. If the defendant then sells the property, the plaintiff will have a claim against the proceeds of the sale. The drawback of this method of enforcement is that the plaintiff may have a long wait for his money.

In certain types of case coming before the County Court, for example bad neighbour disputes, the court may make an order called an injunction ordering the defendant to do, or stop doing, a particular thing. If the defendant ignores the injunction, his behaviour is then said to be in contempt of court, and he could be sent to prison. This is the ultimate sanction which ensures the injunctions are usually obeyed.

Much of the work of the County Court relates in one way or another to problems of debt. Of the 2.6 million cases commenced in the County Court in one year, just over 90 per cent related to claims for money. Possession of land and personal injuries were the next largest categories. The Court also has extensive family and domestic jurisdiction covering divorce, maintenance, distribution of property of the parties to a divorce, access to children, wardship and adoption. The County Court also has the power to grant injunctions in domestic violence situations, the effect of which is, where necessary, to exclude one party from the matrimonial home.

The High Court

Where a civil case is outside the jurisdiction of the County Court, proceedings must be commenced in the appropriate division of the High Court. The High Court sits mainly in London but, since the passing of the Courts Act 1971, a number of centres have been designated throughout the country for High Court hearings. There are three divisions of the court. The Family Division hears cases involving divorce and children; Chancery hears cases involving trusts and wills; Queen's Bench hears cases involving breach of contract or negligence.

One particular aspect of the work of the Queen's Bench Division of the High Court is its supervisory jurisdiction over inferior courts and tribunals. The Queen's Bench Division has the power to grant orders of mandamus, for example, to compel the performance of some public duty imposed by the law on a particular person or body. Where an inferior court or tribunal has acted unlawfully, or in some way exceeded its power, the Queen's Bench Division may quash the decision of the inferior court or tribunal by making an order of certiorari. This procedure could be used when the inferior court or tribunal has acted in ignorance of the principles of natural justice.

The Court of Appeal and the House of Lords

Appeals from the County Court and the High Court in civil cases are heard by the Court of Appeal, which is presided over by the Master of the Rolls, and which normally sits with a quorum of three Lords Justices of Appeal. Hearings take place in London. If cases reach this level of appeal, it is likely that they involve some important point of law or a new application of existing principles, and the case may well be important as a precedent for the future. As a last resort, there is a final appeal, with leave, to the House of Lords. Appeals are usually heard by five Lords of Appeal in Ordinary. Exceptionally, some appeals go direct to the House of Lords from the High Court, using a 'leap-frog' procedure.

Criminal cases

The Magistrates Court

The scope of the criminal law is much wider than merely dealing with cases like murder and theft. Today no business, industrial or commercial organization can ignore the criminal law as there is a wide range of statutory offences e.g. trade descriptions or offences under the Health and Safety at Work Act 1974.

All criminal cases begin in the Magistrates Court. The process used there will depend upon whether the offence is less serious (a summary offence) or more serious (an indictable offence) or an offence triable either way (either summarily or on indictment depending on a number of factors). With summary offences the trial takes place in the Magistrates Court, usually before a bench of three magistrates. With indictable offences, the role of the Magistrates Court is to hold a preliminary hearing to establish if there is a sufficient case made out for it to be committed to the Crown Court for trial. These are committal proceedings. If the offence is an 'either way' offence, then it could be tried in the Magistrates Court in appropriate cases, or otherwise committal proceedings will be held to send that case to the Crown Court.

The work in the Magistrates Court is undertaken by a bench of magistrates or justices of the peace. Historically their office is a very ancient one, and can be traced back to the thirteenth century. In those early days, much of their work was administrative rather than judicial. They were selected for their local knowledge and that still continues to be a feature of their appointment today. Magistrates are appointed by the Lord Chancellor in the name of the Queen. The only qualification for office is that a magistrate must live within fifteen miles of the commission area for which he is appointed. There is an explanatory booklet published by the Lord Chancellor: *The Appointment and Duties of Justices of the Peace in England and Wales*. This booklet makes clear that certain persons will not be appointed, for example, those over the age of sixty, persons convicted of certain offences, undischarged bankrupts, persons whose sight or hearing is impaired, persons who are serving members of the armed forces or members of the police forces. Magistrates must be British subjects, both men and women are eligible, and there are no property qualifications. Names of potential magistrates are put forward to the Lord Chancellor by advisory committees whose constitution is kept secret. This is to prevent the committees being pressured or lobbied. Candidates can have their names put forward by various groups or organizations, but it is also possible for a candidate to put himself forward. In making appointments to a particular bench, the Lord Chancellor must aim to achieve a balance in terms of sex and politics, as well as seeking to reflect different occupations and backgrounds. This balance is important for maintaining general public confidence in Magistrates Courts.

Before appointment, every person selected to be a magistrate must undertake to complete the required training, as well as agreeing to undertake a fair share of the duties of a magistrate. Preliminary training has been obligatory since 1966. This is designed to ensure that the magistrate learns to act judicially; obtains a sufficient knowledge of the law to follow normal cases; obtains a working knowledge of the law of evidence; understands the nature and purpose of the sentences a magistrate imposes; and understands the relationship which should exist between magistrates and court officials. Once the preliminary training has been completed a magistrate may then sit in court, initially as an observer. Training will also include instruction from the clerk of the court, and there will be visits to institutions like prisons. Further refresher training is available throughout the period of service as a magistrate which may include training for work in the juvenile court or for sitting in the Crown Court.

A new magistrate will sit with at least one and usually two other more experienced magistrates, and although he has an equal voice in the decision making, it is likely that he will learn initially from his more experienced colleagues. None of the training is designed particularly to turn magistrates into legal experts. It has always been regarded as a particular strength of the magistracy that people are judged by their fellow men who bring sound judgment and common sense to the task. On questions of law the magistrates will be guided by the clerk to the justices.

The clerk is in charge of the administration of the Magistrates Court, as well as being its chief legal adviser. In large urban courts the clerk will usually have a sizeable staff including deputy clerks, clerical staff, finance and accounting staff and court ushers. The range of the clerk's functions will be extensive, covering the organization of courts, the sitting of magistrates and their training. When advising the magistrates, the clerk is restricted to advising on law and procedure and should not participate in their decision on the facts.

Where an accused person pleads guilty before the magistrates the hearing will consist largely of any plea in mitigation, together with enough information about the offence and the circumstances in which it was committed to enable the magistrates to impose an appropriate sentence. Where an accused pleads not guilty, a full trial takes place. The accused may be represented by a solicitor. It is for the prosecution to prove its case beyond all reasonable doubt. A person is presumed to be innocent until proved guilty. Each side may call witnesses to give evidence who can then be cross-examined by the other side. Other evidence may also be available, for example clothing, weapons, photographs. The press may be present to report the case and the court has a public gallery where any member of the public may come to hear the proceedings. If the magistrate finds the accused not guilty, he is acquitted. If he is found guilty he must then be sentenced to an appropriate punishment. The magistrates may order an absolute discharge at one end of the scale or may commit the guilty person to prison at the other. Between these extremes they may fine the offender or put him on probation or sentence him to a community service order. In deciding the appropriate sentence magistrates will have regard to the needs of society as well as the needs of the offender himself. It should be borne in mind that as well as ordering a suitable punishment, courts will be anxious to seek to rehabilitate offenders.

Where the penalty imposed is a fine, the court maintains systems whereby such moneys can be collected. An offender may ask for time to pay. If a fine remains unpaid, one of the ways to remedy that situation is for the court to employ a bailiff. He will then be empowered to seize goods from the debtor ('levying distress') sufficient to cover the amount of the unpaid fine, together with the costs involved. This can be a very effective procedure where the debtor is a company or business with substantial assets, but it tends to be less effective when used against private individuals who may be unemployed and without any worthwhile assets. There are limits on the penalties which magistrates can impose. Usually they may not sentence a person to longer than six months in prison, or impose a fine greater than £5,000. In some cases, the offence in question may carry a maximum penalty fixed by Parliament. Inevitably, that maximum will become out of date, because of the effects of inflation. To counter this problem, the Criminal Justice Act 1982 fixes standard scales of fines, from scale 1 to scale 5, and the maximum represented by these scales can be adjusted from time to time by order of the Secretary of State.

Some of the main criticisms levelled against Magistrates Courts relate to the magistrates themselves. They are sometimes perceived as middle aged, middle class, professional people. There has been a recent trend to try to appoint younger magistrates, but inevitably as they can serve until they are seventy, some will be elderly. It is generally desirable that a bench should be well balanced, and no doubt the Lord Chancellor would like to appoint more wage earners, but many feel that taking the necessary time off work to fulfil their magisterial obligations (probably a minimum of one half day per fortnight) could interfere with their position at work.

The strength of the system of lay magistrates undoubtedly lies in the appointment of local people with local knowledge who give their services free. An alternative would be to appoint the necessary number of stipendiary magistrates (legally qualified, full-time paid magistrates) but this would be costly, and would deprive the system of a feature, the lay element, in which the public appears to have faith. However, in large conurbations it is common to appoint a number of stipendiaries.

The Crown Court

The Crown Court was established in 1971, replacing the former Assizes and Quarter Sessions. The Crown Court was created as a direct result of the recommendations made in a Royal Commission Report under the chairmanship of Lord Beeching, to attempt to deal more conveniently, efficiently and economically with the workload of cases. The new court was designed to simplify the previous structure of courts; to use judges as flexibly as possible; and to position the courts in places where they were most needed and where the public had best access to them. In order to create maximum flexibility within the Crown Court it was decided to create centres operating at three different levels. First-tier centres can undertake all types of criminal work, and additionally can be used by High Court judges to hear civil cases. Second-tier centres can hear the full range of criminal cases, but undertake no civil work. Third-tier centres are visited only by circuit judges and District Judges, which inevitably curtails the range of criminal work that they can undertake.

Cases in the Crown Court can be heard by a number of different types of judge. These are either the High Court judges who are not permanently present in the Crown Court, but visit at intervals and then hear the most serious cases; the circuit judges, a new rank created by the Courts Act 1971 whose task is to staff the County Courts and to undertake a significant part of the relatively serious crime workload of the Crown Courts; or the recorders, who are part-time judges appointed from the ranks of practising barristers or solicitors, who generally undertake to sit for a minimum of four weeks each year.

The Crown Court has considerable flexibility in the range of judicial manpower available to it, so cases can be listed according to their degree of seriousness for hearing by the most appropriate type of judge. Cases are classified in four categories. Class 1 offences are of the most serious type, for example murder or offences under the Official Secrets Act 1911. These cases are tried by a High Court judge. Class 2 offences, for example manslaughter or rape, are usually tried by a High Court judge, but may be released to a circuit judge. Class 3 offences, which are all crimes not falling into any of the other categories, can be tried by a High Court judge, circuit judge or recorder, but would usually be heard by a circuit judge. Class 4 offences include all offences triable either way where the defendant has elected to be tried at the Crown Court. Class 4 offences will usually be heard by a recorder.

The role of the court listing officer is vital in ensuring a smooth flow of cases through the court. He will be aiming to have courts sitting in all of his available court rooms and will need to gauge the approximate length of the cases. He will be guided by his knowledge of the number of witnesses to be called and his experience of past cases. The administrators in the courts are well aware that judicial time is an expensive commodity, and that sittings need to be organized to maximize the use of the judges available. An effective system of justice should aim to bring cases to trial as early as possible, bearing in mind the constraints of preparing a case and getting together for the hearing all the advocates and witnesses involved.

In cases in the Crown Court where the accused pleads not guilty, the court will sit with a jury to try the case. The jurors will have been summoned at random. Those eligible must be registered as electors, aged between eighteen and sixty-five, and ordinarily resident in the UK. Some people are ineligible for jury service, for example police officers. Some are disqualified from jury service, for example certain people with criminal records. Some people may seek to be excused from jury service, for example MPs or doctors. There are no property qualifications nowadays. Jurors are paid travelling and subsistence allowances and compensation for loss of earnings.

The use of the jury is another example of the use of laymen in the English legal system. The rules used to demand that a jury of twelve recorded a unanimous verdict but now a judge can accept a ten to two majority verdict. If the jury cannot reach a majority verdict it must be discharged, a fresh jury sworn in and the case reheard.

The practice of using juries is often criticized, on the grounds that it is an expensive system to operate; that jurors may not have the necessary intellectual skills, particularly in long and complex fraud cases; that juries may be swayed by professional advocates and make their decisions for the wrong reasons; that jurors are pressed into service and some are very reluctant; and that jurors could more easily be approached and bribed than judges. Apart from small changes in the rules about eligibility and some changes regarding the number of jurors who may be challenged without cause and required to step down, there are no plans to radically alter or dispense with the system of jury trial.

If an accused is found guilty in the Crown Court he will be sentenced by the judge on the basis of guidelines suggested by the Court of Appeal, which include taking into account any previous convictions. The judge will be aware of the maximum penalties which can be imposed for various offences. A sentence will be imposed in respect of each offence for which the accused is convicted but where there are several sentences these are frequently made to run concurrently. Where the sentence is one of imprisonment the court has the power to make it suspended, so that the person convicted will only serve the sentence if he re-offends during the period of suspension.

Appeals from the Crown Court lie to the Court of Appeal and may be against conviction and/or sentence. Where the appeal is against conviction, it may be allowed if the Court of Appeal considers that the verdict of the jury was unsafe; or that there has been a wrong decision on a question of law; or that there has been some material irregularity in the trial. Further appeal lies to the House of Lords.

Tribunals

A significant feature of dispute-solving since 1945 has been the rise in the number of tribunals created by Act of Parliament to deal with specific questions. Common examples of tribunals are the Industrial Tribunals, Rent Tribunals, Social Security Appeal Tribunals and Medical Appeal Tribunals. The reasons for this trend are:

- The volume of work could not be given to the ordinary courts as the system would have become overloaded.
- The questions to be resolved by tribunals are frequently specialized, and the expertise of specialists can be used in the decision making.
- Tribunals can dispose of cases quickly, cheaply and informally.

Each type of tribunal has its constitution fixed by the statute creating it. For example, under the Social Security Act 1975 a Medical Appeal Tribunal (which determines questions of disablement arising from industrial injuries) consists of a legally qualified chairman and two medical specialists. The tribunal may have its own appeal structure created by the statute. For example, appeals from the Medical Appeal Tribunal lie to a Social Security Commissioner.

The outcome of the type of case being heard by tribunals can have important financial consequences for a claimant, e.g. if he is refused a pension. Many of the rules of law applied are complex, yet there is no legal aid available to pay for a claimant to be legally represented at the hearing. Despite the intention that the proceedings should be informal, lack of proper representation may put a claimant at a considerable disadvantage. If he considers that the tribunal has given the wrong decision he may be able to appeal. However, if he considers that the proceedings have been improperly conducted, he may apply to the High Court, asking it to exercise its supervisory powers over the tribunal. The High Court can interfere if it can be shown that the tribunal has failed to abide by the principles of natural justice, which require that each party is given the opportunity to put his case forward, and that the tribunal is not biased in favour of one side.

Arbitration

Going to court can be an expensive, public and time-consuming activity for the parties to a dispute. It may be better to agree to arbitrate, when the parties themselves can exercise some control over the choice of arbitrator. This process is particularly useful where a dispute involves points of a technical nature. In some types of contract (e.g. building contracts) it is common for the parties to agree not to refer disputes arising from those contracts to the court until they have submitted to arbitration. Then, if a dispute does arise, it will be a breach of contract if one of the parties tries to take the case to court instead of using the arbitration agreement. Although the parties must finance the arbitration themselves, the process has obvious appeal because of its speed and privacy. Once the arbitrator gives his decision (award), it can be enforced like a court judgment.

The law governing arbitrations is to be found in the Arbitration Acts 1950–1979 together with the general principles of the law of contract. For the Act to apply there must be a 'written agreement to submit present or future differences to arbitration'. The written agreement need not name the arbitrator, and if necessary the court can appoint a suitable person. The most usual course is for the parties to agree that the arbitrator will be nominated by some outside body or organization e.g. Association of British Travel Agents, or the local Chamber of Trade or the President of the RIBA. The arbitration can be undertaken by one person but sometimes, as a variation, each side may be required by the terms of the agreement to appoint an arbitrator. The two arbitrators may then appoint an umpire who would decide the issue if the two arbitrators should be unable to agree. The JCT standard form building contract provides for the appointment of a single arbitrator.

On normal contractual principles, the two parties to an arbitration agreement are free to fix the terms of their own agreement. They are precluded from ousting the jurisdiction of the court, as this is regarded as contrary to public policy. In effect, this means that they cannot insert a clause stating that the arbitration is final and that neither party may have any recourse to the courts. The courts have power to supervise arbitrations, and may remove an arbitrator who has misconducted himself.

A number of terms are implied into an arbitration agreement by the Arbitration Act 1950 unless the parties have agreed otherwise. These include that the parties to the agreement must be prepared to be examined on oath by the arbitrator; that they must produce all documents required by the arbitrator; that the arbitration will be final and binding on the parties; and that the arbitrator can award costs.

Once the arbitrator makes his decision, 'the award', it can be enforced just like a court order or judgment. The parties are more likely to find the decision of an arbitrator acceptable but they pay a high price for this alternative to court proceedings as they are responsible for financing the arbitration proceedings.

The distinction between arbitration and valuation has always been regarded as problematic. In arbitration the aim is to settle an existing dispute. A valuation seeks to prevent a dispute arising, by allowing a third party to fix the value or price. The distinction is not always clear and the third party may be anxious to be sure of his status, as a valuer can be sued for negligence but an arbitrator cannot. This was an important point in issue in the case of *Sutcliffe v. Thackrah* in 1974, where an architect issued an interim certificate on the basis of which his employer paid the contractor. The employer subsequently sacked the contractor and wanted to recover damages from him for proven shoddy work. The contractor went into liquidation, so the employer then sued his architect for his negligence in certifying the poor work. In issuing the certificate, does the architect act as a valuer or an aribitrator? Once the architect issues a certificate, the employer is then obliged to pay. But the architect is *not* at that point determining a dispute between the employer and the contractor, although his duty when certifying is to act impartially between them. In consequence the architect in this case could be held liable for his negligent certification.

Administrative control

Where a dispute involves allegations of maladministration by a local or central government department or agency it may be possible to refer the dispute to the appropriate Commissioner for Administration (often referred to as the Ombudsman), who can investigate the complaint and make recommendations for putting the matter right. Unfortunately, the Commissioners have no power to order anyone to do anything, but the publicity attached to their reports may be sufficient to make a defaulting person or body act differently.

Affording the law

Where the ordinary man in the street is a party to a dispute, many of the procedures outlined above may seem inhibiting to him, either because he fears what it may cost to be involved with 'the law', or because he doubts his ability to cope with complex rules and procedures. If he wants to engage a lawyer to advise him, what will that cost him? To what extent can he be helped financially to afford the services of a lawyer? The answer lies in the legal aid and advice scheme. The scheme provides:

1 *For the provision of advice and assistance* – under the rules of the Legal Advice scheme, a solicitor may do most things that are normally regarded as part of his work, short of actually starting legal proceedings. He can write letters, negotiate, draw up documents, take statements, or prepare a written case for a client to use himself

at a tribunal. Recent restrictions on the Green Form Scheme now restrict work and advice in connection with conveyancing and wills. There is, however, a twofold financial restriction:

(a) The client requiring advice must be financially eligible. The solicitor will make his calculation as to this on a special form (the Green Form). He can determine whether the client is within the scope of the legal advice scheme, and if so, what contribution if any he must make towards the cost of the solicitor's service. The client's 'disposable income' and 'disposable capital' are calculated by reference to fixed criteria.

(b) The solicitor who wishes to aid a client by using the scheme may not undertake more than two hours of work unless he obtains prior permission from the Legal Aid Area Office to increase that figure.

2 *For the provision of actual representation in court proceedings* – the court case may be a criminal prosecution or a civil action. Legal aid in these two areas has always been administered separately.

Civil legal aid is based on the system of financial eligibility rather similar to that for the legal advice scheme. As before, the client may be assessed as liable to pay a contribution. A person seeking civil legal aid must also show to the satisfaction of a local committee that he has reasonable grounds for taking, defending or otherwise being a party to the proceedings. An important point to remember is that costs incurred in granting legal aid may be recovered out of any money received following a successful action. The administration of the legal aid scheme has now been transferred to a newly constituted Legal Aid Board, under the Legal Aid Act 1988. Major changes in the provision of legal services are now possible as s.4 confers a power to enter into contracts with persons and bodies other than solicitors (e.g. CABs) for the provision of advice and assistance.

Criminal legal aid is somewhat different in that applications are made direct to the court and there are no strict financial limits.

The legal profession

Even if a person is financially eligible to be legally aided, he may be uncertain how to find a lawyer to look after his case. The term 'lawyer' is a general word embracing all those who are learned in the law. Practising lawyers are divided into solicitors and barristers. The two professions are entirely separate, with their own entrance requirements and examinations. Solicitors are usually compared to general medical practitioners with barristers being seen as the equivalent of consultants. This can be rather misleading, however, as a solicitor, more usually in a large firm, may have an extremely specialized practice, while a young barrister seeking to make his name as an advocate may have to be prepared to be a 'jack of all trades'. No one may consult a barrister except through a solicitor. If no solicitor is known, public libraries or Citizen's Advice Bureaux keep referral lists of solicitors practising in their localities.

The legal profession was subjected to an intensive scrutiny in a Royal Commission chaired by Lord Benson in 1979, but little has happened as a result of its recommendations. It was basically in favour of the present division of the legal profession. However, in January 1989, the Lord Chancellor published three Green Papers (consultative documents) containing far-reaching proposals for changes in the legal profession, including:

- Ending the monopoly of the bar in the higher courts.
- Alternative ways of funding litigation, with contingency fees as a possibility.
- Mixed practices of different professionals.
- Allowing banks and building societies to undertake conveyancing work.
- The appointment of solicitors to the High Court bench.

The Green Papers were unfavourably received by many practitioners, both solicitors and barristers, and even senior judges spoke out against the proposals. Many of their proposals have now been implemented by the Courts and Legal Services Act 1990. Additionally, the Act has appointed a new Legal Services Ombudsman, who took up office in 1991, and created a new Advisory Committee on Legal Education and Conduct. The new Ombudsman is able to order payment of compensation to complainants.

3
Central and local government

The structure of central government

When society develops to the point where the activities of its members need to be directed and controlled, it will require a government. The tasks of government are: to formulate and carry out policies, which is its executive function; to frame laws, which is its legislative function; and to enforce those laws, which is its judicial function. At the same time as it grants power to a government to fulfil these functions, society wants to see control exercised over the government so that it does not become too powerful or dictatorial. Such control is exercised through the rules of the constitution, rather as a club or association is controlled by means of its rules. A striking feature of the British Constitution is that it is unwritten. Unlike countries such as France or the USA, we cannot point to one document embodying our guaranteed rights. Nevertheless, important constitutional laws are contained in a number of historic Acts of Parliament. These Acts are not invested with any special protection and could be revoked or altered by the same processes as any other Act of Parliament. By contrast, in certain foreign constitutions there are entrenched provisions which can only be changed by special procedures.

The leader of the political party which commands a majority in the House of Commons will be invited by the monarch to form a government. The House of Commons is one part of Parliament which consists of the monarch, the House of Lords and the House of Commons. The Queen is a constitutional monarch who exercises her powers only on the advice of her ministers. The House of Lords, which is the upper chamber of Parliament, is a non-elective assembly. Its members include hereditary peers, life peers, Lords of Appeal in

Ordinary (a maximum of eleven judges who are specifically appointed to the House of Lords to hear appeal cases from the civil and criminal courts but who are also entitled to take part in all business of the House of Lords) and the Lords Spiritual. The Lords Spiritual are not peers. They comprise the Archbishops of Canterbury and York, the Bishops of London, Durham and Winchester and the next twenty-one most senior bishops of the Church of England. Inevitably the major criticism of this element of representation in the Lords is that it is exclusive to the Church of England. Indeed, over the years, the make-up of the House of Lords has been subjected to regular criticism. This mainly centres on the fact that the House is largely aristocratic and non-elective, with an inevitable inbuilt permanent Conservative majority. It is often thought that the House of Lords has little power today since the Parliament Acts of 1911 and 1949 restricted its role to the delaying of legislation rather than its out and out rejection. In answer to the criticisms of the House of Lords, however, it is worth remembering that the quality of membership of the Lords overall is high, reflecting a wide range of backgrounds and experience, with life peers being selected from trade unions, commerce and industry, public life and the armed forces. Standards of debate are generally accepted to be high. The part which the House of Lords can play in improving and refining legislation sent from the Commons is regarded as an important and significant part of its functions.

The House of Commons is the elected chamber of Parliament with 650 Members of Parliament. Each MP represents a constituency and elections must be held at least once every five years, at a time chosen by the Prime Minister when he or she believes it will be politically opportune.

Once the result of the general election is known, the leader of the political party which commands a majority of the House of Commons will be invited to form a government. Many of the MPs will be hoping to hold office in the government. The new Prime Minister will then select the Cabinet team and ministers for the various government departments. The Cabinet represents the most important departments but is of no fixed size and the selection of ministers to be in the Cabinet may reflect the government's policies and priorities. Traditionally the Cabinet will always include the Chancellor of the Exchequer, the Home Secretary, the Foreign Secretary and the Lord Chancellor. To be of a reasonable working size, the Cabinet is likely to consist of about twenty-four ministers. Outside the Cabinet, however, there are other ministers and junior ministers and in total the government is likely to have over 100 members. Some ministers will be members of the House of Lords but it would be an unpopular move to appoint too many from the Lords because they cannot be questioned in the House of Commons, a corrective and control feature much prized by MPs.

The government, once formed, is Her Majesty's Government, and the ministers are Her Majesty's Ministers. The new session of Parliament will be formally opened by the Queen, reading the Queen's Speech from the Throne in the House of Lords. Originally all executive power was vested in the monarch, but a series of historical events reduced the power and produced the modern figurehead monarchy of today. All acts of government are done in the name of the monarch but not necessarily with the monarch's personal participation. The Queen does, however, preside at meetings of the Privy Council. All bills going through the two Houses of Parliament require the Royal Assent to become Acts of Parliament. In theory the Queen could refuse her Assent but in this, as in most of her activities, she is bound by constitutional conventions. The principal convention is that the Queen exercises her formal legal powers only on the advice of her ministers. It is said that the Queen has 'the right to be consulted, the right to encourage and the right to warn.'

There is very little formal law governing our constitution and, therefore, many of the 'rules' under which central government and the Crown operate are simply conventions. A constitutional convention is a rule of political conduct, not a strict law in the sense that there is no sanction to enforce such a convention. Conventions can change imperceptibly over a period of time and it can therefore be argued that they keep the constitution flexible. Conventions have been described as 'the flesh which clothes the dry bones of the law, they make the legal constitution work, they keep in touch with the growth of ideas.'

It has been seen that the government of the day will emerge largely from members of the House of Commons. The House of Commons itself has three main functions; to make laws; to control national expenditure and taxation; and to criticize policy. Its law-making function has already been considered. In order to carry on its business the first task of the newly elected members of the Commons is to elect their Speaker who is an impartial chairman of the proceedings in the House. Usually this will mean the re-election of the previous Speaker if he is prepared to stand. Proceedings in the House of Commons take place in the Chamber with the government side sitting to the right of the Speaker's chair and the opposition to the left. The Chamber is not large enough to accommodate all the members at once, but usually they only want to be present in force for a limited number of great debates, or when there is going to be a division (i.e. a vote). MPs are paid salaries and have an allowance for secretarial expenses. They enjoy free travel within the UK on parliamentary business, and telephone calls and postage from the Commons are also free.

There is no formally recognized career path towards becoming an MP. The 650 MPs in the House of Commons represent a wide range of backgrounds and many have worked previously in the fields of journalism, the law, teaching, social work, finance and accountancy. There are comparatively few MPs now with shop floor experience from factories, mines or steel works. Many will have prepared themselves for their careers at Westminster by service with local authorities or trade unions.

Procedure of the House of Commons

There is a 'bible' of procedure in the House of Commons, a book called *Parliamentary Practice* by Erskine May. It started off as a small handbook, but as the procedure of the House of Commons has become more complex, the book has grown and grown. The procedure is complex because of the range of activites and interests with which the House must deal. Inevitably rules are important where the Speaker needs to control the activities of 650 volatile and articulate MPs of strong conviction, many of whom hold violently opposing views. If there are disputed points of procedure, the final word lies with the Speaker.

After numerous assorted items of business including question time, private notice questions, ministerial statements, introductions of new members and requests for emergency debates, the House eventually gets down to the main business of the day which might be a major debate or the second reading of an important bill. When an issue is put to the vote in the House of Commons, the Speaker puts the question and he must then weigh up whether the 'ayes' or the 'nos' have succeeded. If his assessment is challenged, he orders the lobbies to be cleared and the division bells will be rung, tellers are appointed and members must file through the appropriate lobby to record their votes. The tellers give the results to the Speaker who announces the outcome to the House.

Parliamentary questions

Question time takes place in the House of Commons on Mondays to Thursdays and lasts about forty-five minutes. On Tuesdays and

Thursdays fifteen minutes are devoted to Prime Minister's questions. Other ministers take it in turn to be questioned on a rota basis. Questions in Parliament have been a feature of the House of Commons for about 200 years but the present format and style of question time can be traced to the turn of this century, when it was decided to fix a definite time each day at which questions would end and the public business of the House would begin.

There are a number of reasons why an MP may wish to ask a question. He may wish to elicit factual information. He may well be asked by a minister to table a question to give the minister an opportunity to make a public statement in the House. The question may seek redress of a grievance for a constituent. Very often in such a case the question will only be asked if initial approaches to the minister or department concerned have drawn a blank. Questions in Parliament are not the only option open to an MP acting for a constituent. He can also refer the matter to the Parliamentary Commissioner for Administration. The question may be designed to embarrass a minister. This can be achieved particularly successfully in the follow-up to the oral question in what are called supplementaries. Although the minister will have notice of the question tabled, a cunning supplementary might find him unprepared. It is likely, however, that he will be well briefed by his civil servants with information to anticipate any supplementaries. Again, the question may be designed to enhance the reputation of the MP concerned, as opportunities to speak in the House are extremely limited.

The value of the parliamentary question can be considerable, especially if it ensures that a minister gives his personal attention and consideration to issues or problems which would otherwise be handled by officials in his department. Often decisions will have been taken at quite a low level within the department. On a closer look at the file, the minister may have occasion to reconsider his decision or redefine the policy within his department.

Question time can be regarded as an important time for the ordinary member of Parliament, the backbencher, particularly when it is remembered that this is part of the day when the parliamentary whips are not in control. Moreover, starred questions for oral reply are not usually asked by leading frontbenchers (members of the Government or Shadow Cabinet) although supplementaries may be put by them. The publicity potential is very great as question time takes place at a fixed time each day at the start of the day's business when the chamber is likely to be at its fullest and the press are in attendance, as there is always the prospect that a minister may be caught out or shown up in some way. A leading writer on constitutional law, De Smith, has said of question time:

Parliamentary reputations have been made and ruined in the rapid cut and thrust of question time. For a few minutes the House comes to life audibly and visibly; wit, feigned or genuine outrage, cheers and jeers intrude upon the solemnity of the proceedings; government and opposition are briefly locked in verbal combat; the Prime Minister and the Leader of the Opposition may gain or lose a point or two in the public opinion polls; a backbencher shows his ministerial potential, and the House wonders how much longer the Minister of Cosmology can last.

Private member's bills

If the opportunities to ask questions are restricted, then the opportunities to introduce a private member's bill are even more limited. The number of private member's bills passed in each session is always small, often not running into double figures. As a general rule private members will seek to introduce bills which do not require public expenditure, because if this were necessary the bill would face further hurdles. The government would then need to be persuaded to put forward a financial resolution. If a member wishes to introduce a private member's bill there are three possible ways of doing it: by winning a high place in the annual ballot; under the ten minute rule; or under Standing Order No. 37. The safest of these three courses is winning a high place in the ballot for promoting a private member's bill. This takes place annually. Successful MPs may have some personal preference for the measure which they wish to introduce, or an MP may be solicited by various pressure groups anxious to promote some particular measure. Occasionally the government itself has small measures that it wishes to promote. Where it has no time in the government programme, it may approach an MP who has been successful in the ballot asking him to take on a particular bill. Once a member knows what the subject matter of his bill is to be, he must then see to its drafting. This can be a costly process, especially if the bill is lengthy. The first ten placeholders in the ballot do, however, receive a small allowance towards drafting costs.

The committee system of the House of Commons

Whenever there is a very large organization with significant amounts of business to transact, it is common to find delegation of some of that business to specialized committees. In the House of Commons, delegation can be to a Committee of the Whole House, when the committee consists of all of the MPs. But, usually, two main types of committee are used – the standing committee and the select committee. It has been suggested that the names of these two types of committees are misleading. Standing committees are those specifically set up to consider the committee stage of a particular bill, after which the committee will be suspended, i.e. it does not 'stand'. Select committees on the other hand will usually operate for the duration of a Parliament.

Standing committees

These consist of sixteen to fifty members who are chosen to reflect the political strength of each party in the House. A government with a large majority will, therefore, enjoy a large majority in standing committees. The actual members for each committee are chosen by the Committee on Selection, which is itself a select committee. The chairman of the standing committee will be chosen by the Speaker from a panel of members which he selects at the start of the parliamentary session. The main task of the standing committee is to consider amendments to individual clauses of the bill before them. This is the first time that the specific terms of the bill will be given detailed clause by clause consideration. Standing committees sit in the mornings. Because of the number of votes on amendments, many of them of critical importance to the government if the substance of its policy is to be actively carried through, the whip system operates fully over the activities of these committees.

Select committees

Some of the select committees are permanent features of the life of the House of Commons but with extremely varying functions. At one end of the scale is the Services Committee (previously the Kitchen Committee) which controls the staff and services of the House of Commons and is responsible for catering there. At the other end is the prestigious Committee on Privileges which considers any breaches of privilege referred to it. The most significant recent development in the field of select committees occurred in the early 1980s, following recommendations made by the Select Committee on Procedure. That committee reviewed the procedures of the House of Commons and made proposals for changes. It was thought necessary to establish a new system of departmental select committees, fourteen in all, to scrutinize the work of the major government departments. These have been in operation since 1980 and are established with a permanent membership for the duration of a Parliament. The committees are small, seven to eleven members and the chairmanships are divided between the political parties, although the government would tend to retain the chairmanship of the more important committees like Home Affairs. They are free to determine for themselves which areas they will investigate.

Once a committee decides upon an area of investigation it will usually appoint its own expert advisers. It can call before it witnesses from industry and commerce, from the government and the civil service. It publishes its findings in a report to the House, to which the government must respond. Because of lack of parliamentary time, only a very small proportion of these reports is ever debated in the House, but there is, nonetheless, considerable value in the work of these committees. Their work is a means by which scrutiny and control of the executive is provided. They produce a considerable amount of information with which MPs may arm themselves for debate and questions. That information can be useful to add to the armoury of material used by pressure groups. Because issues are considered in some detail, MPs can develop considerable expertise. Membership of these committees is highly prized amongst MPs and they foster a co-operative sense of cross-party unity.

Financial procedures of the House of Commons

The Treasury, under the Chancellor of the Exchequer, is the government department with responsibility for the management of the economy. Inevitably, as Treasury approval is required for all large-scale expenditure, it is a significant department, but its influence may wax and wane depending on the forcefulness of individual Chancellors.

In order to run the essential services of the country, money must be raised and collected, and some priorities have to be established about how it should be paid out. There must also be mechanisms for controlling how that money is used. The government has substantial revenues from trading and commercial activities, but significant amounts have to be raised by taxes, which may be annual taxes like income tax, or longer term taxes like stamp duty.

The major financial statement of the year in the House of Commons is the Budget Speech delivered by the Chancellor of the Exchequer. The proposals contained in the Chancellor's speech will ultimately be implemented in the Finance Act, but it may be several months before the Act receives Royal Assent. In the meantime, authority to collect annual taxes like income tax needs to be renewed, and the gap between the Budget and the Finance Act will be bridged using the authority contained in an annual Provisional Collection of Taxes

Act. The appropriate resolutions will be passed as soon as the Chancellor has finished delivering his Budget speech.

In former times, it was common to raise taxes specifically for particular items, but now all moneys raised are lumped together into one fund, the Consolidated Fund, which is effectively just an account at the Bank of England.

There is a Treasury and Civil Service Committee, one of the new breed of select committees established in 1980, with eleven members. Its terms of reference are to examine the expenditure, administration and policy of the Treasury, Board of Inland Revenue, and Boards of Custom and Excise. There is also a Public Accounts Committee which has the task of ensuring that parliamentary grants have been applied to the objects which Parliament has laid down. That committee considers financial irregularities, wasteful and extravagant expenditure and imprudent contractual transactions. Its reports are debated annually and command great respect in Whitehall.

There are many ancient conventions with regard to financial matters in Parliament. Most significant amongst these are that the granting of public money and the imposing of taxes is the function of Parliament, not the government; that the granting of public money and imposing of taxes must begin in the House of Commons and be finally determined by the Commons; and that the redress of grievances must precede any grant of public money. There must be no taxation without representation.

The Civil Service

Initially, when a minister takes over a new department or ministry, it might seem a daunting task to have to quickly learn all that is necessary for him to be able to make appropriate decisions. However, the role played by the Civil Service should not be forgotten. Civil servants are the permanent officials employed in the ministries. Unlike the American system where all officials change when the President changes, our Civil Service is permanent. The civil servants work for the government of the day whatever its political complexion. Civil servants cannot stand for election as MPs, and may only play a limited role in local politics. The civil servants in most senior positions are in close touch with their ministers, and play a part in the formation of policy, watching over bills on their way through Parliament and keeping their ministers supplied with information to enable them to answer questions in Parliament. They will frequently make decisions by applying rules and policy to individual cases, and when this happens, such a decision is regarded as having been made on behalf of the minister, who must accept responsibility for what has been done.

The structure of local government

There has been a long tradition of local administration in England, dating back to 1066. With poor networks of roads and the dangers of travel, the advantages of control and decision making resting with local organizations were obvious. But as those justifications have gradually disappeared, it may be questioned whether it is necessary in the twentieth century to divide responsibilities between central and local government. Two of the main advantages are:

1 The variations necessary for different local conditions can be introduced.
2 Local people will be involved, and those living in the area will see some benefit from the effort which they contribute.

Local authorities were reorganized in 1972. Basically, the structure is a division of the country into counties which in turn are divided into districts. In large conurbations, the counties are called metropolitan counties; the districts within them are metropolitan districts. Below district level are the parishes. Counties, districts and parishes all have councils. The type of council will determine the range of functions, although some functions may be shared. County councils have responsibility for matters such as highways, fire, police and social services. District councils have responsibility for housing, public health, rates and refuse collection. Parish councils have responsibility for footpaths, allotments, burial grounds and bus shelters. Planning is an example of a responsibility shared between county and district councils.

Local authorities are examples of bodies enjoying corporate personality, i.e. the local authority is a 'person' recognized by the law. It is run by a council of elected members within the powers granted to the local authority by the Act of Parliament which created it. The members of the council may change at each election, but the council has permanent officials, called its officers, who fulfil a role rather like that of the civil servants. Commonly a local authority has a Clerk or Chief Executive, a Medical Officer, an Education Officer and a Treasurer. The business of a local authority is transacted at council meetings, sometimes meetings of the whole council, but more often meetings of smaller committees set up for specific purposes, and with powers delegated to them, e.g. the planning committee. These committees must now reflect the political balance of the council as a whole.

Meetings are open to the public, unless the council resolves to meet in private, usually because publicity about a particular matter would be prejudicial to the public interest.

Council members may represent a particular political viewpoint and it is quite often the case at local government elections that the votes of the electorate will swing against the political party in power in central government. This can produce conflicts between local and central government, e.g. on issues such as the sale of council houses or the size of rate increases. This naturally raises the question, to what extent is a local authority truly independent? The answer to this lies in the amount of control which can be exercised over the local authority.

Control of local authorities

Every local authority must act within the scope of the powers granted to it. A local authority is a statutory corporation, which can only do such things as are authorized expressly or impliedly by the Act of Parliament which created it. If the local authority acts in excess of its powers, its actions are said to be *'ultra vires'* (beyond its powers) and void (of no legal effect). Obviously, it would require immensely detailed legislation to spell out everything a local authority could do. Consequently, by s.111 of the Local Government Act 1972, authorities are empowered to do anything which is reasonably incidental to their authorized activities and designed to facilitate the discharge of any of their functions.

Where a local authority has acted in a way which causes someone to be aggrieved, it will be necessary to check whether the authority acted under some express power, some implied power, or in some way reasonably incidental to an existing power. Not only must the local authority show that it had power to carry out a particular act, it must also be able to show that, procedurally, it carried out the act in accordance with the prescribed rules. A typical example of how things can go wrong occurred in 1967, in *Bradbury v. Enfield London Borough Council*. This case involved the reorganization of schools and the introduction of the comprehensive school system. Where the local authority intended to 'cease to maintain' a school, it was obliged to submit its proposals to the minister and give public notice of its intentions, to allow affected persons to make objections. The local authority took the view that what it was proposing to do in its reorganization plans did not amount to ceasing to maintain any of the schools involved. An aggrieved person applied to the court to stop the authority going ahead with its plans. An injunction was granted, because the behaviour of the local authority was *ultra vires* as it had failed to observe a mandatory procedural requirement.

Central and local government 45

When the help of the courts has to be sought, it is on the basis of judicial review of the local authority's decision or actions. The courts have a number of remedies available: including an order of certiorari, which is used to quash an *ultra vires* decision; prohibition, which is used to prevent an *ultra vires* action which is about to take place; mandamus, which is used to compel the performance of a public duty; and declaration, which simply makes clear what the court determines the legal position to be. For a more detailed treatment, see page 48.

One possible way to give greater scope for action to a local authority, without rendering it constantly liable to claims of *ultra vires*, is to give the authority discretion as to whether, or how, it will act. Although this may appear at first sight to be a welcome prospect, the exercise of discretion is fraught with problems. Complaints may arise from the way in which the discretion has been exercised and in making its decisions the local authority will usually be obliged to exercise its discretion reasonably and to observe the rules of natural justice. These rules are usually stated to be *audi alterem partem* (hear both sides) and *nemo judex in causa sua* (no one should be a judge in his own case). In practice, these rules often require a local authority to give prior notice of their intentions to persons who may be affected by their decisions, so that such persons have an opportunity to put forward their own case; and a local authority should exclude from the decision-making process anyone who has some pecuniary interest in the decision, or who has some other bias.

Where it is suggested that an application to the High Court may be needed to control the activities of local authorities, many individuals would be deterred by the expense involved or by the formality of the procedures. There are, however, Commissions for Local Administration, which have created ombudsmen for local government. A local commissioner can investigate complaints of maladministration resulting in injustice. If a member of the public wants to take a case to the local commissioner, the procedure is to refer the case to a member of the authority concerned. The word 'maladministration' is not defined, but had been said to cover instances of bias, neglect, inattention, delay, incompetence, ineptitude, perversity and arbitrariness. If the commissioner finds that there has been maladministration, he will make a report, and the local authority concerned is then under a duty to consider the report, and notify the commissioner of the action it proposes to take. Its actions could include compensating persons who have suffered injustice, but the commissioner has no direct power to order or award compensation. Nor does the commissioner usually propose a remedy, but in those cases where he is dissatisfied with the response to his report by the authority, he may issue a second report with more positive proposals about remedies.

Control by central government

The main type of control which central government can exercise is financial. The control can consist of either withdrawing grants or refusing loans or capping expenditure. Without financial support from central government, and without the ability to raise loans, no local authority could function for long. Admittedly, there are other forms of income available to it, especially the rates levied on business property within the area. There are also less significant amounts, such as rents from council houses, profits from undertakings run by the local authority (e.g. swimming pools, markets, buses) and fees from the granting of various licences. However, the inadequacy of their income is apparent when it is realized that their funds have to be supplemented annually by central government rate support. The domestic rating system has been withdrawn, and was superseded by a community charge, commonly called a poll tax. This tax proved to be immensely unpopular, and it is replaced by a new council tax.

When considering the role of central government in controlling local authorities, it is important to emphasize that much of the control is informal, exercised in the process of consultation which takes place with various government departments. Circulars, codes of practice and memoranda issued by central government departments form the basis of much decision making at local authority level. What is more, a local authority is often unable to proceed without some formal approval being given by a Secretary of State, e.g. approval of its structure plan under the Town and Country Planning Act 1990.

The range of services and amenities provided by local authorities is vast. Not surprisingly, there will be some people living in the area covered by a council who will be dissatisfied with the council's efforts. It should be remembered that ultimate control lies with the electorate, whether at the time of the elections, or by the strength of opinion they express to their councillors, or through the media. It is clear from the low level of votes cast at local elections that some of these controls are not fully used.

Bye-laws

Within its own area, a local authority will have power to regulate activities by means of bye-laws, either under general legislation or by virtue of private Acts of Parliament. When making bye-laws, a local authority is usually governed by the procedures of s.236 of the Local Government Act 1972. A bye-law must be made under the seal of the council and must be confirmed by the appropriate authority, which will usually be the relevant Secretary of State. Before confirmation, the local authority must publish notice of its intention to submit the bye-law. Once submitted, the confirming authority will usually check to ensure that the bye-law is *intra vires*, but this is not the only test which it must satisfy. Unlike laws made by statute, bye-laws are subject to the requirements that they must be reasonable; they must be certain; and they must be consistent with the general law. If valid and effective, a bye-law may impose penalties for breach. It is usually possible for anyone to institute proceedings for breach.

Public corporations

We have just considered the situation in which central government devolves some of its power to local authorities. It has become a trend of the twentieth century for central government to devolve other administrative tasks onto specially created bodies, which are known as public corporations. Among well-known examples are British Coal, the BBC, the Atomic Energy Authority, Health Authorities, the English Tourist Board, the Horse Race Betting Levy Board and the Advisory, Conciliation and Arbitration Service.

Many of these public corporations are established to manage commercial undertakings. As far as possible, they are expected to be self-sufficient and make a profit, although the statute creating the public corporation often expressly requires the undertaking to be run in the public interest, and that may not be compatible with making a profit!

All public corporations are created by Act of Parliament or Royal Charter, and the constitution of each is governed by its particular creating instrument. In general, however, such an Act will provide for an appropriate minister to appoint a chairman and members, usually after consulting appropriate bodies. Once established, the public corporation is a separate legal 'person', and may carry on those activities for which its creating instrument provides.

Control over public corporations is exercised either by use of the *ultra vires* rule, or, in some cases, by ministerial control. Although it is important that the activities of public corporations should be subject to control, they can only operate effectively if there is as little interference from central government as possible. Central government has created the public corporations in question and should trust them to carry out their specific tasks. Nonetheless, ultimate control does lie with Parliament, as it is always possible for a public corporation to be wound up by statute. The public also has some opportunity to play a part in controlling a public corporation, either by seeking publicity for adverse opinions about the corporation or by using one of the various consultative councils established to act as sounding boards for public opinion.

Over recent years, there has been a significant trend towards privatization of the public utilities. This move has resulted in water, gas, electricity, generation of power and telephone services becoming public companies, with their shares quoted on the Stock Exchange (see page 66).

Challenging decisions made by public bodies in the courts

A person who is unhappy with a decision made by a public body, such as a local authority or a government department, may be able to challenge the decision in court. There are three ways in which it may be possible to raise an issue of public law before the courts. These are by appeal, by application for judicial review, or in the course of an ordinary 'private law' action.

Appeal

There is no automatic right to appeal to the courts against a decision made by a public body. Such an appeal is only possible where an Act of Parliament has created the right to appeal against decisions made by a particular type of public body. For example, the Town and Country Planning Act 1990 permits persons aggrieved by planning decisions made by local planning authorities to appeal to the Secretary of State. If the aggrieved person is not then satisfied with the Secretary of State's decision, the Act allows him to appeal to the High Court.

Where an Act of Parliament creates a right of appeal, it will also lay down the grounds of appeal, i.e. it states the reasons which, if established, justify the courts in substituting their decision for that of the original decision-maker. An appeal against a decision will not succeed simply because the appellant can show that he has good reason to be unhappy with the decision. In order to bring a successful appeal, it is necessary to establish one or more of the specified grounds of appeal. An Act of Parliament which creates a right of appeal also specifies the nature of those persons who are entitled to appeal.

Application for judicial review

Where there is no right to appeal against a decision, it may still be possible to challenge the decision by application for judicial review. Basically, the High Court may hear an application for judicial review of decisions made by public bodies.

1 Remedies of judicial review

An applicant for judicial review may apply for one or more of the following remedies.

(a) *Certiorari*. An order of certiorari can be used to quash an unlawful act done by a public body. It may be awarded if the act can be shown to be illegal or unauthorized. Following the making of the order the act ceases to have legal effect though, if appropriate, a new decision may be made by the public body. An order of certiorari will only lie in respect of a decision which affects the rights or interests of individuals.

An example of the award of an order of certiorari is provided by *R. v. Medical Appeal Tribunal, ex parte Gilmore*. The tribunal misinterpreted regulations made under an Act of Parliament with the result that the award of compensation which they made to an injured workman was less than that to which he was legally entitled (an error of law). An order of certiorari was awarded to quash the tribunal's decision. The court recognized that the workman's case would then go back to the tribunal where a valid decision would be made. (If the court had had reason to believe that the tribunal would not have made a fresh decision, they could have obliged them to do so by awarding an order of mandamus (see below).)

Certiorari is a discretionary remedy. The court may refuse to award it where it feels that its award would be inappropriate. Examples are where an appropriate right of appeal is available, where the results of making an order would be undesirable, or where the applicant's behaviour does not justify the award of a remedy.

(b) *Prohibition*. An order of prohibition can be used to prevent a public body from doing an unlawful act. It may be awarded if the act can be shown to be illegal or unauthorized. Following the making of an order, the act cannot lawfully be carried out though, if appropriate, a new decision may be made by the public body. An order of prohibition will only lie in respect of a decision which affects the rights or interests of individuals.

Like certiorari, prohibition is a discretionary remedy.
(c) *Mandamus*. An order of mandamus can be used to force a public body to do something which it is legally required to do. It may be awarded if it can be established that a public body has failed to perform its public duties.

In *R. v. Shoreditch Assessment Committee, ex parte Morgan*, the committee refused to initiate a process whereby the rateable value of a house could be reduced to take account of a fall in its value. Since the committee had a duty to initiate this process, mandamus was awarded in order to oblige them to do so.
(d) *Declaration*. This is a court order which declares what are the legal rights of the parties to an application for judicial review.
(e) *Injunction*. An injunction is a court order which may prohibit an act from taking place or, more unusually, may require an act to take place. The grant of an injunction is discretionary.
(f) *Damages*. Where one or more of the five remedies referred to above is applied for by way of judicial review, the applicant may also claim damages in respect of any matter to which the application relates. The court may award damages if an ordinary civil action for damages (e.g. a contract or tort action) would have succeeded.

2 *Applications for judicial review and 'private law' matters*

The application for judicial review process is only available in relation to public law matters, not in relation to private law matters. Issues of private law should normally be raised in the form of an ordinary private law action, such as an action for damages in contract or tort, and not by application for judicial review. The distinction between public and private law matters may be difficult to draw in certain cases. The following guidelines may be helpful.

1 Where the legal issue involved is one of contract, such as employment law, the issue will normally be one of private law even if the employer is a public body (such as the BBC).
2 Where power is given to a body by an Act of Parliament, decisions made in relation to the exercise of the power will normally be public law matters (e.g. the exercise of planning powers possessed by local authorities or by the Secretary of State).
3 Where power is given to a body by a contract, decisions made in relation to the exercise of the power will normally be private law matters (e.g. the powers of the National Greyhound Racing Club).
4 Where a power stems from the Royal Prerogative (ancient common

law powers of the Sovereign) decisions made in relation to the exercise of the power will normally be public law matters (e.g. refusal or withdrawal of a passport).
5 When determining whether the exercise of a power by a body is a public law matter, the courts take into account not just the source of the power (i.e. Act of Parliament, contract or Royal Prerogative) but also consider whether the function performed by the body is of a public or private law nature. For example, decisions of the Panel on Take-overs and Mergers, a body appointed by City of London Institutions such as the Stock Exchange, which devises and polices the code on take-overs and mergers, have been held to be subject to the application for judicial review process.

3 The process of application for judicial review

(a) *Leave to apply*. Before an application for judicial review can be made, leave to apply must be obtained. Applications for leave are made *ex parte* (i.e. by only one side to the action) to a single judge. There need not be a hearing in open court unless the applicant requests one. If leave is refused, the application may not be made, though it is possible to appeal against refusal. Basically, leave will be granted if the applicant has an arguable case. If leave is granted, the application for judicial review will be heard in open court, usually by one judge.

(b) *Delay*. An application for review must be made within three months from the date when the grounds for the application arose, unless the court considers that there is good reason for extending the period. Even if made within three months, the court may refuse an application if it has not been made promptly.

(c) *Standing*. An applicant for judicial review must have standing (or *locus standi*). This means that he must have a sufficient interest in the issues which he wishes to raise before the court. The test of 'sufficient interest' varies to some extent depending upon the nature of the particular remedy or remedies applied for. Basically the test is whether the applicant has been or may be affected by the act or decision which he wishes to question, as opposed to a busybody who is meddling in matters which do not concern him.

Ordinary 'private law' actions

Basically, a public law matter should be raised by appeal or application for judicial review and not by bringing an ordinary private law action. It may, however, be permissible to raise a public law matter in the course of an ordinary private law action as a defence to that action. Further, there are exceptional cases in which it may be appropriate to bring an ordinary private law action in order to raise a public law matter.

Privative clauses

Privative clauses (or ouster clauses) are clauses in Acts of Parliament which remove or restrict the courts' power to hear matters. The normal approach of the courts is that such clauses will be interpreted very narrowly, so that if there is any doubt as to their effect, the courts' powers will not be reduced. Where there is no such doubt, however, the effect of the clause will be to restrict or remove the court's power to hear the matter in question.

An example is provided by *R. v. Secretary of State for the Environment, ex parte Ostler.* Ostler discovered that a trader had withdrawn his objection to a compulsory purchase order after the Department of Transport had secretly promised him that a road would be widened. The road widening affected Ostler but he had not been told about it so he had not objected to the compulsory purchase order. Ostler claimed that the order was invalid, as it was made in breach of the fair hearings rule of natural justice (see page 55). The compulsory purchase order was protected by a statutory clause which provided that it could only be challenged in the courts within six weeks of notice of its making being published. Ostler had not discovered the secret promise, and consequently had not challenged the order, until more than a year after the order was made. Thus the court could not interfere with the order.

Application for judicial review: the grounds of challenge

Upon application for judicial review of a decision, the High Court will not award a remedy simply because it disagrees with the decision. It is not for the court to substitute its decision for that of the decision maker. Rather, a remedy will only be awarded if the decision is unlawful or unauthorized. The High Court may be prepared to award a remedy upon proof of procedural impropriety, excess of power, or abuse of power. These grounds will be considered in more detail.

1 Procedural impropriety

(a) *Failure to observe procedural requirements laid down by Parliament.* Procedural requirements specify the process which a public body must adopt in acting or making decisions. If, in the course of making a decision, a public body fails to observe procedural requirements laid down either by Parliament or under statutory authority, this may render its decision invalid but will not necessarily do so. Basically, the approach of the courts is that breach of a mandatory procedural requirement normally makes a decision invalid. Breach of a merely directory procedural requirement will not usually have this effect.

Mandatory procedural requirements are those which serve an important purpose. Directory procedural requirements are those which are not so important.

An example is provided by *Howard v. Secretary of State for the Environment*, which concerned a requirement of the Town and Country Planning Act 1968. Notice of appeal against an enforcement notice had to be served on the Secretary of State within 42 days of service of the enforcement notice. (The nature and purpose of enforcement notices is considered on page 365.) The Court of Appeal classified this requirement as mandatory. They did so because without a time limit the enforcement provisions of the Act would not have worked effectively, so the time limit was of great practical importance. The Act also required that the notice of appeal contain certain specified information relating to the grounds of appeal and the facts of the appeal. The Court of Appeal classified this requirement as directory. They did so because once the notice of appeal had been served, the appellate mechanism began to operate. Failure to provide the specified information could easily be remedied at a later date and thus it was not essential that the information be contained in the notice. On the facts of the case, notice of appeal had been served within 42 days but a letter containing the required information had not arrived until more than 42 days had passed. The Court held that the procedural defect did not render the appeal invalid.

At times the courts may regard the breach of a minor procedural requirement, which appears to be merely directory, as rendering the decision to which it relates invalid. This may be the case, for example, where the court believes that the requirement has been deliberately ignored. Equally, the courts may at times regard the breach of an important procedural requirement, which appears to be mandatory, as not rendering the decision to which it relates

invalid. This may be the case where, for example, the court believes that the breach has not caused any harm.
(b) *Breach of the rules of natural justice (or the requirements of fairness).* The rules of natural justice are common law rules of procedure which the courts impose both on themselves and on other public bodies. The rules do not apply to all public bodies and sometimes apply differently to different bodies. It is possible, however, to draw some general conclusions as to what the rules are and how they work. Basically there are two rules. First, the rule that a decision-maker must not be biased and secondly, the rule that a person whose rights, interests or legitimate expectations are affected by a decision has a right to a fair hearing before the decision is made.

(i) *The bias rule.* The bias rule itself has two elements. The first requires that the decisions of public bodies must not be tainted by pecuniary bias. The second requires that the decisions of public bodies must not be tainted by non-pecuniary bias.

Pecuniary bias. The decision of a public body is vitiated by pecuniary bias if a member of the body possesses a personal financial interest in the subject-matter of the decision. This rule is applied strictly in the sense that even if the court is satisfied that the person who had the interest was not influenced by it, the decision will still be quashed.

An example is provided by *R. v. Hendon RDC, ex parte Chorley.* The case concerned a councillor who was acting as estate agent in relation to a land transaction. The transaction would only go ahead if planning permission was granted in relation to a development on the land. The councillor took part in the meeting which decided to grant planning permission for the development and voted in favour of granting permission. The court quashed the grant of planning permission.

Non-pecuniary bias. The decision of a public body is vitiated by non-pecuniary bias if a member of the body has a non-financial interest in it. This could occur, for example, where the decision-maker is a close relative or professional colleague of a person who is affected by the decision. This rule is not always applied as strictly as the pecuniary bias rule, in the sense that the mere possession of a non-pecuniary interest by a decision-maker does not automatically invalidate a decision.

Normally, in order to show that a decision is invalidated by non-pecuniary bias, it is not necessary to show that the decision was actually influenced by the possession of a non-pecuniary interest. Rather, what is necessary is to persuade the court that a reasonable and fair minded person knowing all of the relevant facts would have a reasonable suspicion that the decision was biased.

An example of a case in which non-pecuniary bias arose is provided by *R. v. Barnsley Council, ex parte Hook*. A market trader urinated in a side street. The market manager reported this to the council who revoked the trader's licence. The market manager was present whilst the relevant council committee made its decision and thus acted both as prosecutor and judge. The decision was held to be vitiated by non-pecuniary bias and was quashed.

Sometimes, the courts apply a different test, and require proof of actual bias, in the sense that they will only quash a decision if satisfied that the decision was so influenced by possession of the non-pecuniary interest that a proper decision could not be reached. They do this where the operation of the statutory powers which the case concerns is inconsistent with the strict application of the bias rule. In *R. v. Sevenoaks District Council, ex parte Terry*, a local authority granted permission to develop land which it owned. The court refused to regard the decision as vitiated by non-pecuniary bias. Since Parliament had given the council the power to own land, to develop land, and to grant permission in relation to land which it owned, the strict application of the bias rule was not appropriate.

(ii) *The requirement of a fair hearing.* Basically, fairness requires that a person whose rights, interests or legitimate expectations are affected by a decision is entitled to a fair hearing before that decision is made. (A legitimate expectation of a fair hearing may arise where a promise or a regular practice on the part of a decision-maker leads a person to believe that he will be given a fair hearing before a decision is made.) The rule has three basic requirements.

First, the person so affected by the decision must be made aware of the charges, complaints, grounds, arguments or evidence against him. In *Annamunthudo v. Oilfields Workers Union*, A was accused of breaking four union rules, none of which could result in expulsion from the union. He did not attend the meeting at which the charges were to be considered. At the meeting, a new and more serious charge was raised and he was expelled. The expulsion was invalid because, since A had not been made aware of the charge, he had not been given a fair hearing.

Secondly, the person so affected by the decision must be provided with an opportunity to prepare his own defence or arguments. In *R. v. Thames Magistrate's Court, ex parte Polemis*, a ship's captain was charged with discharging oil into navigable waters. His summons was served at 10.30 and he was to appear in court at 14.00. This did not provide time for chemical analysis which might have shown that the oil did not come from his ship. The magistrates refused to allow an adjournment to allow the analysis to be carried out as the ship was sailing that day. They convicted him and fined him £5,000. Thus the magistrates, in failing to allow the captain time to prepare his defence, did not give him a fair hearing.

Thirdly, the person so affected by the decision must be provided with an opportunity to present the defence or arguments which he has prepared. This does not necessarily mean that he must be allowed to appear in person before the decision-maker. It may be sufficient to allow him to hand in written arguments, provided that both sides to a dispute are treated in the same way.

In *Errington v. Minister of Health*, following a public inquiry, the Secretary of State received further information from a local authority without allowing objectors to the local authority's scheme to comment upon it. Since the objectors had not been given an opportunity to put their arguments forward, they had not been given a fair hearing.

2 Excess of power

A public body may only do those things which it possesses the power to do. (Normally such power will be given by or under an Act of Parliament or, more unusually, by the common law.) If a public body does not possess power to make a decision then, in making such a decision, it exceeds its powers: its decision is thus *ultra vires* and invalid.

In *White & Collins v. Minister of Health*, a local authority made a compulsory purchase order in respect of land which formed part of a park. The authority possessed power to make compulsory purchase orders but did not possess power to make such orders in respect of land which formed part of a park. Consequently, the order was quashed.

R. v. Willesden JJ, ex parte Utley, concerned an offence of failing to stop a motor vehicle when requested to do so by a police officer. The maximum penalty for the offence was a fine of £5. The magistrates imposed a fine of £15, which was quashed.

Incidental matters. Where a public body is expressly empowered to do something, it may be implied that the public body can also do things incidental to that which it is expressly empowered to do. In *Loweth v. Minister of Housing and Local Government* a council was empowered to make a compulsory purchase order to acquire land on which to build a civic centre. The empowering Act did not specify whether the order could also include land on which to construct an access road to the civic centre. The court held that it could, as this was a matter incidental to development of the civic centre.

Unlawful delegation. Power should only be exercised by the body to which it is given. Only where a decision-maker is given the power to delegate can it lawfully delegate its powers to another person or body.

In *Barnard v. National Dock Labour Board*, a local dock labour board, without the power so to do, attempted to delegate disciplinary powers to a port manager. The port manager suspended some dock workers. The suspensions were *ultra vires*.

A common example of lawful delegation is encountered in relation to local authorities. Section 101 of the Local Government Act 1972 empowers a local authority to delegate its powers to a committee, a sub-committee or an officer or to another local authority. Delegation to an individual councillor is not authorized, however.

3 Abuse of power

Where a public body exercises a power which it does possess, it does not exceed its powers but it may abuse them. Power is abused when its exercise is not in accordance with law. Thus, where in the course of making a decision a decision-maker makes an error of law, he abuses his power.

Basically, an error of law is an error of legal reasoning. Such an error may take a variety of forms. In particular it may consist of:

(a) The making of a decision which is so unreasonable that no reasonable person properly understanding the law would make it;
(b) a failure to take legally relevant matters into account or the taking into account of legally irrelevant matters;
(c) the use of a power for a purpose which is legally improper.

(a) *Unreasonableness.* The courts may be prepared to quash a decision on the basis of unreasonableness if it is so unreasonable that no reasonable decision-maker would ever have made it. The fact that the court does not agree with a decision does not necessarily mean that it is unreasonable because, sometimes, two reasonable people can come to different conclusions on the same set of facts. Consequently, not every mistake is necessarily an unreasonable mistake. Thus, for a decision to be classified as unreasonable, it should be overwhelmingly clear that no reasonable decision-maker would have regarded it as one which he was legally entitled to make in the circumstances.

In *Hall v. Shoreham-by-Sea UDC*, a council granted conditional planning permission to an applicant. The condition imposed was that the applicant construct a road on his land and dedicate it to the public. No compensation was to be paid by the council to the applicant for the loss of his land. The court held that the condition was invalid. It was clearly unreasonable of the council to seek to force the applicant to give up his property without compensation by making the grant of planning permission dependent upon this.

(b) *Failure to take legally relevant matters into account/taking legally irrelevant matters into account.* The courts may be prepared to quash a decision if satisfied that, in reaching it, the decision-maker failed to take a legally relevant matter into account. Equally, they may be prepared to quash a decision if satisfied that, in reaching it, the decision-maker took a legally irrelevant matter into account.

Legally relevant matters are matters which, according to law, the decision-maker should take into account in reaching his decision. Basically, these include all matters which a reasonable decision-maker would take into account. They also include all matters which an Act of Parliament, expressly or by necessary implication, indicates should be taken into account.

Legally irrelevant matters are matters which, according to law, the decision-maker should not take into account in reaching his decision. Basically, these include all matters which a reasonable decision-maker would not take into account. They also include all matters which an Act of Parliament, expressly or by necessary implication, indicates should not be taken into account.

Roberts v. Hopwood concerned the validity of the wage rates paid by a council to its employees. The council was empowered to pay its employees 'such wages as it thought fit'. The council fixed wage levels far higher than those paid either in the public sector or by other councils. The council fixed these wage levels because it sought to become a model employer. In so doing, it failed to take into account ordinary economic considerations, such as the market price for labour. Thus, the council based its decision on a legally irrelevant matter, its desire to become a model employer, and failed to take a legally relevant matter into account, namely the market price of labour. The importance of the latter matter becomes obvious when it is realized that the effect of paying extremely high wages to council workers was that workers doing the same or similar jobs in the private sector, and being paid the lower market rate, would additionally be required to face higher rate bills in order to meet the increased public sector wage bill.

Sometimes, a public body adopts a policy by which to exercise its powers. If such a policy is rigid, in the sense that exceptions to it will not be considered, the public body is said to 'fetter its discretion'. This means that by rigidly applying the policy it fails to take into account legally relevant considerations, namely, the individual facts of each separate case that comes before it. An example is provided by *Lavender v. Minister of Housing and Local Government*, in which the Minister of Housing adopted the policy of refusing to allow gravel working on agricultural land if the Minister of Agriculture objected.

Where a policy is adopted but it is made clear that exceptions will be considered and, consequently, the individual facts of each separate case are examined in order to determine whether they are exceptional, discretion is not fettered.

(c) *Improper purposes.* The courts may be prepared to quash an exercise of power if they are satisfied that the power has been exercised for a purpose other than that for which it can lawfully be exercised. In order to decide whether power has been exercised for an improper purpose the courts must first determine the purpose for which Parliament intends the power to be exercised. Having determined this they must then decide whether power has been exercised for this or for a different purpose.

In *Webb v. Minister of Housing and Local Government*, a local council possessed a power to acquire land by compulsory purchase for the purpose of building sea walls for coastal protection. They used this power to acquire land to build a promenade along the sea front. Their decision was invalid because the purpose for which the power had been used was improper.

Sometimes a power may be exercised for more than one purpose. As long as the main purpose for which it is exercised is proper, it does not matter that a subsidiary purpose is also achieved.

In *Westminster Corporation v. LNWR*, a council was empowered to construct public conveniences on or under any road. It constructed conveniences under a road with access from both sides of the road. It was asserted that the decision to build the conveniences achieved an unlawful purpose, namely the building of a subway. The House of Lords held, however, that the council's primary object was to build conveniences and thus the fact that the development also achieved the secondary objective of constructing a subway did not make the development unauthorized.

4
Business organizations

Partnership

When a person wants to set himself up in business there is nothing to prevent him simply trading on his own account, with only his own money at risk. As such, the law would describe him as a sole trader. Trading in this way, however, may involve a number of drawbacks:

- He may lack adequate funds to develop the business.
- There is no one with whom he can share responsibility for running the business.
- If the business gets into financial difficulties, he will have unlimited liability for its debts. Creditors will not be restricted to seizing the assets connected with the business. Even his home may have to be sold.

The sole trader may be able to overcome the first two of these disadvantages by forming a partnership. This relationship is described by the Partnership Act 1890 as 'existing between persons who carry on business in common with a view to profit'. The arrangements for a partnership can be very informal, but it is advisable to have a properly drawn up agreement, in case disputes arise between the parties.

Partnership is a very suitable form of business association for professional people, e.g. accountants, architects, quantity surveyors and consulting engineers. Under the 1890 Act all partners are legally entitled to take part in the running of the business, but in practice the parties can make whatever arrangements they like. All partners are agents of each other for the purposes of partnership business (e.g. making contracts, paying money and engaging staff on behalf of the firm). For this reason, it may be sensible to limit the number of partners. By law, the maximum number is usually twenty. There is no limit on the number in partnerships of certain professional people like accountants and solicitors. A partnership is often referred to as a firm e.g. a firm of solicitors, but legally that term has no significance. The law is concerned to know, is this a partnership or a company? The word 'firm' might be equally appropriate to a firm of builder's merchants trading as a limited company, but the legal consequences of dealing with a partnership and a company can be entirely different.

The relationship among partners is a very special one, as potentially each could expose his fellow partners to unlimited liability. The law determines that it is a relationship of utmost good faith. This means that intending partners should make a full disclosure of any relevant facts or circumstances. The relationship is often described as being more difficult than marriage – after all in marriage there is only one spouse to contend with! Wherever possible, potential disputes should be avoided by a carefully drafted partnership agreement. Unless such an agreement is drawn up, the rules of the Partnership Act 1890 will apply. The Act provides that partners are entitled to share equally in the capital and profits of the business. That certainly may not accord with the wishes of partners who have contributed capital in differing proportions. A formal agreement allows for detailed rules about capital and profit sharing, and matters like the running of the firm, engaging in other business activities, retirement of partners and dissolution of the partnership.

Forming a partnership introduces others who will share liability for debts of the business, but that liability continues to be unlimited. All the partners are fully liable for all the debts, and again, creditors are not restricted to assets connected with the business. Even the advantages of a partnership (e.g. privacy with regard to the conduct of business affairs and accounting) may not outweigh the drawback of unlimited liability. An alternative which can avoid this is the formation of a company for the running of the business.

Company incorporation

The process of forming a company is referred to as incorporation. This is a very significant legal step, because the business then becomes known in law as a corporation. This is to distinguish it from groups or bodies which are not incorporated, and which are usually referred to as unincorporated associations. These latter groups have no separate legal existence or personality distinct from their members. So, the Mid-Tyne Angling Club is simply an identifying name for a group of people who have come together with a common interest. There is no legal person called Mid-Tyne Angling Club, i.e. it is not a separate legal entity or personality.

The phrase 'legal personality' means a person or organization which is recognized by the law as having legal rights and owing legal duties. All human beings in this country have a legal personality, although the law may impose restrictions on their capacity to act in certain ways, (e.g. any child under eighteen years of age has legal personality but lacks capacity to make a will, or own land). The law can also confer legal personality on an organization which has become incorporated.

Incorporation can take place by three methods:

1 *By Royal Charter* – this was the method used to create the original trading companies (East India Trading Company and Hudson Bay Trading Company) but it was more important as the means of creating many royal boroughs and institutes.
2 *By Act of Parliament* – many of the public corporations and local authorities have been established in this way, e.g. British Coal, British Airways.
3 *By virtue of the provisions of the Companies Act 1985 as amended* – under the rules of the Act, a number of different types of company can be created. For trading and general business purposes, a company limited by shares is by far the most common. The amount of shares owned by each of the members of the company then fixes the extent of his or her possible liability.

The fact that a company has a separate legal personality, quite distinct from its members (i.e. its shareholders) confers a number of advantages. These include a power to sue and be sued in its own name, the right to own and transfer property, the possibility of perpetual succession unless the company is wound up and the possibility to enjoy limited liability.

If a company gets into financial difficulties, creditors will be paid out of the assets of the company. If those assets are insufficient, the individual shareholders do not become personally liable. The extent of the liability of a shareholder is the amount he owes for his shares. Once he has paid for his shares he risks losing that amount if the company founders but no more. Limited liability is the most appealing feature of forming a company. However, a price must be paid for this advantage; limited liability companies are under statutory duties to publish certain financial details in the interests of both their creditors and the general public.

Creating a company

The actual creation or formation of a company is achieved by drawing up appropriate documents (including a Memorandum of Association and Articles of Association) which are submitted to the Registrar of Companies. The Memorandum is the company's most important document as it sets out the objects for which it is formed and thereby determines or limits the powers of the company. Although the Memorandum can subsequently be altered, this is by no means easy and it is preferable to make sure that it is correct from the outset. The Memorandum is a public document which may be inspected by persons interested in the company.

The Memorandum must set out:

1 *The name of the company* – the company will have a registered name but it may choose to trade under a different name, its business name. Business names are regulated by statute (Business Names Act 1985). The name chosen for the company must not be the same or too similar to that of an existing registered company, nor likely to give the public a false idea of the purpose for which the company is being formed. Any word such as bank, government or trust in a company name could be very misleading. In the case of a limited company, the word 'limited' must usually be the last word of the company's name. Where the company is a public company, it must include the words 'public limited company' or the abbreviation 'plc' in its name.

2 *The location of the registered office* – all that is required in the Memorandum is a statement as to which country the company is located in. This establishes the 'domicile' of the company, which may be important in choosing which system of law should apply. For example, Scots law applies to companies registered in Scotland. The company must also file a separate note of the address of its registered office.

3 *The objects clause* – this clause sets out exactly what business the company can engage in. Traditionally, these clauses have been very long, with the draftsman attempting to cover every possible activity which a company may wish to undertake. This was justified because of the *ultra vires* principle, whereby any activity undertaken by the company outside the scope of its objects clause was *ultra vires* (beyond its power) and void.

Two recent changes made by the Companies Act 1989 have greatly altered the position. Now, a company's memorandum may simply state that the object of the company is to carry on business as a general commercial company. If an objects clause of this new type is used, then the company can indeed carry on any trade or business whatsoever, and it would have power to do anything incidental or conducive to the carrying on of any trade or business by it. All existing companies could change their more limited objects clauses to incorporate this new wider provision. However, one cautionary note should be added. Although this simple new wide-ranging clause is now perfectly acceptable in theory, in practice companies may still opt for much more specific and detailed objects clauses, because institutions such as the banks, who may be considering loans to the company, will be keen to establish in which fields their money will be risked.

So far as the *ultra vires* rules are concerned, for any third party who deals with a company, the position set out in the 1989 Act is that the validity of acts done by the company shall not be called into question on grounds of lack of capacity by the company. Moreover, the third party is not bound to enquire as to whether the activity is permitted by the company's memorandum. Of course, the directors of the company which thus acted beyond the scope of its objects clause would be in breach of duty, and a shareholder could seek to bring proceedings to restrain any proposed activity which would be beyond the scope of the company's capacity.

4 *Member's liability* – where a company is limited by shares this clause simply states that liability is limited. In some more unusual cases, a company may be limited by guarantee, in which case the Memorandum must state the amount which each member has guaranteed to contribute if the company has to be wound up. Exceptionally it is still possible to create a company with unlimited liability but this is rare.

5 *Capital* – this clause states what the company's share capital is to be, and its division into shares of a fixed amount. This amount is the authorized capital of the company. For private companies there is no legal maximum or minimum fixed but a public company must have an authorized capital of at least £50,000. The figure chosen merely establishes the authorized amount which the company may issue as shares, although it may choose to issue only a portion of the authorized capital.

6 *The association clause* – by this clause the subscribers declare that they agree to take the number of shares noted against their name.

By law, this forms the minimum contents of the Memorandum. The Articles of Association set out the internal rules and regulations for the running of the company, covering matters such as meetings, rights of shareholders, voting, appointment of directors, duties of directors and payment of dividends. There is a model set of Articles provided by the 1985 Act, called Table A, and it is very common for companies to adopt Table A with suitable amendments.

The articles are particularly significant for the members, as they are effectively the terms of the contract between the member and the company. Any violation of the rules contained in the Articles is a breach of contract. Members can sue to enforce their rights in ordinary breach of contract claims.

Public and private companies

It is common to think of private companies as small family firms, and public companies as the large business giants, and in practice this is often the case. A company which starts as a private company may expand its activities to such an extent that it decides to 'go public' and seek the change of status to a public company. There are a number of simple differences between public and private companies. For example:

1 A private company can be formed with one member and one director whereas a public company needs two members and at least two directors. See note at end of chapter.
2 A private company need not obtain from the Registrar a certificate entitling it to do business whereas a public company must have such a certificate, and commits offences by trading without it.
3 A public company must satisfy the minimum capital requirement (i.e. its issued share capital must be at least £50,000). This rule does not apply to private companies.
4 A private company may exercise its borrowing powers immediately upon incorporation. A public company may not do so until it has alloted shares up to its authorized minimum capital of £50,000.
5 Private companies may still restrict the transferability of their shares, if they choose. They would usually choose to do so in cases where they want to preserve control of the company within a family. Restrictions normally operate by giving the directors the power to refuse to register transfers. Where such a discretion is given by the Articles, it places immense power in the hands of the directors, but they must act throughout in good faith. Public companies may not restrict the transferability of their shares if they want to be quoted on the Stock Exchange.
6 Private companies may be freed from some of the rigours of the rules about publication of accounts, depending on their size. For

these purposes size is determined by factors like turnover, assets and numbers of employees.

Management of a company

Companies are run by a board of directors, which is normally appointed at its annual general meeting. With private companies it is a common requirement that directors must have a certain holding of shares in the company. Where it is a small family business this is usually not difficult as directors may be members of the family. The company, being an artificial person, must act through human agents, and that normally means through the board of directors. The board will often delegate duties to paid employees, who may also be directors. Once appointed, a director is under duties to show good faith to the company and to take reasonable care. This means, for example, that a director is obliged to reveal any personal interest in a contract made by the company, particularly if he stands to profit by the contract.

The directors, or those acting under delegated powers, must act within the powers granted to them by the documents creating the company. In particular, the objects clause in the Memorandum of Association sets out the purposes for which the company was formed and the limits within which it can do business.

The first directors of the company will be those nominated to the Registrar of Companies on the formation of the company. Thereafter, the articles should provide for the appointment and retirement of directors. If shareholders are dissatisfied with one of the directors, they may choose not to re-elect him or, alternatively, they may seek his removal from office by passing an ordinary resolution. However, so dramatic a step is surrounded by numerous safeguards. The member proposing the resolution must give twenty-eight days' notice to the company before the relevant meeting; the company must then give other shareholders twenty-one clear days' notice of it; the company must notify the director concerned and he has the right to circulate a written statement setting out his point of view; the ordinary resolution must be passed by a simple majority of those voting. The director concerned may well have enhanced voting rights which allow him to beat off such a threat. If a resolution to remove a director is successful, it should be remembered that he may also have a contract of service with the company, and substantial compensation may need to be paid to him.

A board of directors is likely to have a Chairman, who will preside at directors' meetings. The board may also decide to appoint a Managing

Director, who will often have a contract of service with the company, and to whom much of the day to day management of the company is delegated. A company is legally obliged to have a company secretary, who may also be a director.

How a company divides its management between directors and shareholders is largely a question to be determined by the Articles. In practice, the rights of members to participate in management, particularly in large public companies, is extremely limited. A member does have some say by virtue of his voting power at meetings, but he may be one of a very large number of shareholders, holding very few shares, and his opinion may carry no weight. The wishes of the majority should normally take effect but not where the impact of that is to perpetrate a fraud on the minority.

The law has been alert to the problem of protecting minority shareholders and provides a number of means which can help to secure some protection. One example is shareholders' agreements, whereby the members enter into a separate contract with each other regulating how certain aspects of the company will be run. A statutory example of protection is the right of a minority shareholder to apply to the court if the company's affairs are being carried on in a way which is unfairly prejudicial to the members or some part of them. Alternatively, a member could petition the court to wind up the company on the grounds that such a winding up would be just and equitable. These rules could be a particularly effective sanction in small private companies with, say, only half a dozen members, where the rights of one of them are being seriously prejudiced.

It may be that what concerns members most is the income that their shares will produce. A declaration of a dividend is usually made at the company's Annual General Meeting. Every company must hold such a meeting in each calendar year, not more than fifteen months apart, although a private company can dispense with the AGM, along with some other procedural requirements, if the company has passed an elective resolution. Any kind of business may be transacted at an AGM as long as appropriate notices have been served and appropriate resolutions are passed. If business needs to be transacted before another AGM is likely to be held, an Extraordinary General Meeting may be convened, either by the directors, or by members who together hold one-tenth of the company's paid-up share capital. Resolutions may be ordinary, special or extraordinary. The differences between them relate to the length of notice to be given, the size of the majority required and the need, in some cases, for subsequent registration of the resolution. The 1989 Act introduces a new type of resolution, an elective resolution, which may only be passed if everyone entitled to vote does so in favour of the resolution.

The Chairman of the board will usually chair company meetings. He must process the business in accordance with the rules contained in the Articles, and in accordance with the general rules of the common law. Voting on resolutions can be on a show of hands but sometimes the more formal procedure of a poll may be demanded. The Articles usually provide for situations when members can insist on a poll. The advantage of a poll is that proxies can vote too.

Company finance

In order to provide adequate resources to allow the company to develop and flourish it may be necessary for a company to borrow money or look at other methods of funding such as government grants, as well as taking account of the capital raised by the issue of shares. The advantages of share capital are that it does not have to be repaid (unless redeemable shares have been issued) and therefore has a permanence about it, and dividends are only payable if the company makes a profit.

The size of the company's share capital may well determine how much it can borrow, especially in the early stages of a newly formed and developing company. Lenders will be reluctant to advance more than the members of the business have themselves risked. Where a company needs sizeable amounts of money it may be possible to attract funds under government sponsored schemes like the Business Expansion Scheme or Investment Funds. A person who invests his money in a company under the Business Expansion Scheme enjoys tax concessions, especially if he pays tax at a higher rate.

A company can sometimes improve its financial position by siting itself where it may benefit from central and local government or EEC grants and loans. The schemes will vary from time to time, reflecting the political stance of the government of the day. Examples of schemes have included Enterprise Zones and Regional Development areas. Sometimes, the incentive to set up in a particular place will consist of very favourable rents and rate-free periods.

Where a company needs to borrow money it may turn first to its own shareholders, who can provide loans called debentures. Like all forms of borrowing the issue of debentures is only possible if the directors are empowered to borrow by the Memorandum and Articles. It is possible to give debenture holders greater security by creating a charge over the assets of the company.

All lenders of money worry about how they will be repaid if the company gets into difficulties. A lender may be prepared to make the loan only on condition that he is given some security in the event of the company being unable to pay. If that happens the lender can then realize his security i.e. sell it, to repay himself what he is owed.

One type of security is the fixed charge, which is a mortgage over a specific thing like a building. Inevitably the company cannot dispose of, or realize, the property which is subject to the charge without the consent of the lender. This could be particularly restricting in a developing business, so the company may prefer to borrow against the security of a floating charge. This is a mortgage of assets of the company, such as the stock in trade, which changes from time to time. It is in the contemplation of both lender and borrower that the company should be free to carry on its normal business, dealing as usual with those assets. The company continues to have the flexible use of whatever is comprised in the charge. However, once the borrower defaults the floating charge is said to crystallize. This means that it becomes a charge attached to the assets available at the moment of crystallization and the company is henceforth prevented from dealing with those assets. The parties may have specified a number of circumstances which will give rise to crystallization.

Floating charges are a device by which a company can raise money without affecting its ability to trade. However, they could have the effect of misleading persons who are thinking of dealing with the company, because the company might look in a more affluent and healthy financial position than is truly the case. The problem is overcome to some extent by requiring company charges to be registered with the Registrar of Companies within twenty-one days of being made. The company is legally obliged to effect the registration, but it would be sensible for the lender to do so too, because a charge which is not registered in time is void. Once the charge is registered it is part of the information which is available for public inspection. Any person about to deal with a company, and needing information about its financial viability, could make a search. That would reveal any charges registered, the date the charges were created, the person in whose favour the charge is made and the property over which the charge operates.

Raising money by the device of the floating charge is a procedure only available to companies. This gives a company an advantage over a trading partnership, which could borrow only against the security of fixed assets.

Winding up of companies

A company is an artificial body and theoretically it could continue forever with a constantly changing membership as shares are bought and sold. If a company is to be dissolved the process is referred to as 'winding up' or 'liquidation'. The directors may decide on this course, in which case the dissolution is voluntary, or it may occur by order of the court.

If the winding up is brought about because a company has financial problems the rules of liquidation are now contained in the Insolvency Act 1986. A government White Paper, published in 1984, stated that the role of insolvency legislation is:

1. To establish effective and straightforward procedures for dealing with and settling the affairs of insolvents in the interests of their creditors.
2. To provide a statutory framework to encourage companies to pay careful attention to their financial circumstances so as to recognize difficulties at an early stage and before the interests of the creditors are seriously prejudiced.
3. To deter and penalize irresponsible behaviour and malpractice on the part of those who manage a company's affairs.
4. To ensure that those who act in cases of insolvency are competent to do so and conduct themselves in a proper manner.
5. To facilitate the reorganization of companies in difficulties to minimize unnecessary loss to creditors and to the economy when insolvency occurs.

To achieve these aims a new breed of professionals, called insolvency practitioners, has been created. They are authorized to act in liquidations and must provide a bond or security of £250,000 for each appointment as a practitioner as security in respect of losses caused by fraud or dishonesty.

When a company is in financial difficulties a full liquidation process may be unnecessary. It may be possible to save the company by using voluntary arrangements. The proposal for such a voluntary arrangement can be made by a director. The proposal will be for a composition in satisfaction of the company's debts or a scheme of arrangement of the company's affairs. The insolvency practitioner will then be appointed to act as a supervisor, implementing the scheme. The value of such a voluntary arrangement is that it can be organized with the minimum of involvement of the High Court, as it will be approved at meetings of the creditors and the company.

Where a company is likely to be unable to pay its debts another alternative procedure is an Administration Order. The court will make such an order if it believes that by doing so it can secure the survival of the company or that it would give time for a voluntary arrangement to come into being. There will be a court hearing. If an order is made it directs that, for the time being, the affairs, business and property of the company are to be managed by an administrator. While the order is in force there can be no order for the liquidation of the company. The administrator has three months to draw up a statement of his proposals, which are put to meetings of the creditors and the company.

Where liquidation of the company is necessary, this may be a member's voluntary liquidation, which can occur where the company is solvent; a creditor's voluntary liquidation, which occurs where the company is insolvent and where the wishes and interests of the creditors will prevail; or a compulsory liquidation, which arises by order of the court, following the presentation of a petition seeking such an order.

For a compulsory liquidation, a winding-up petition may be based on a number of grounds, including the ground that the company is unable to pay its debts. This is deemed to be the case where a creditor who is owed more than £750 has demanded his money and has remained unsatisfied for three weeks. He is said to have made a statutory demand. Other evidence accepted by the court that the company is unable to pay its debts may be dishonouring a bill of exchange, or proof that the company's assets are less than its liabilities, or proof that a creditor entitled to execution has been unsuccessful. This means that a creditor was granted a warrant of execution to levy distress against the goods of the company, but his warrant has proved ineffective because there are insufficient goods to satisfy his demands.

Proceedings for compulsory liquidation are commenced in the County Court or the High Court, depending on the amount of the company's share capital. Once a petition has been presented any disposition of company property is void unless approved by the court. This discretion is important because it means that the company can continue to trade, but it allows control over activities which may be detrimental to the interests of some of the creditors.

If the court decides to make a winding-up order, certain immediate consequences follow:

- A liquidator is appointed.
- All actions against the company must be stayed.
- A first meeting of creditors may be convened.
- The director's duties and powers cease.

- The business of the company can carry on only to the extent that it is beneficial to the liquidation.

The liquidator must take control of all the company's property. He must call in all the assets, realize them, and distribute the proceeds to the appropriate creditors. Some property of the company may not be available to the liquidator, e.g. where the contract by which the company was acquiring the property provides that the company's interest in the property will end in the event of liquidation, or where a company is purchasing goods under a contract containing a valid retention of title clause. The liquidator will also be concerned to discover whether transactions immediately prior to the liquidation were undertaken to defeat the rights of creditors, or some groups of them. If so, the transaction may be void. Examples include cases where a company transfers property for no consideration, or for a consideration significantly less than the real value; or a transaction involves a preference, where a company puts one of its creditors into a more favourable position than the creditor would be under the liquidation.

Once the liquidator has realized all the assets he then pays the costs and expenses of the liquidation. He must next pay off the creditors, who must have submitted formal proof of their debt. Secured creditors will usually realize their security to settle what is owed to them but they may prove any outstanding balance not covered by the security. Fixed chargeholders take separately, but those holding floating charges rank after preferential creditors. This tends to emphasize the superior security provided by a fixed charge. The creditors rank in the following order:

1 *Preferential creditors* – this group rank equally among themselves after the expenses of the winding up. Examples of preferential creditors include:
   (a) PAYE, referable to the previous twelve months.
   (b) VAT referable to the previous six months.
   (c) National insurance contributions due for the previous twelve months.
   (d) Employees' remuneration due for the previous four months, not exceeding £800 (this figure can be adjusted from time to time by the Secretary of State).
2 *Amounts secured by floating charges.*
3 *Unsecured creditors* – all other creditors rank equally amongst themselves.

If there are surplus funds left over after distributing to the creditors these are distributed to the members of the company according to their rights and interests.

Where a company has been wound up this may reflect seriously on the competence and integrity of the directors. In that case, the courts have extensive powers to disqualify directors for specific periods. In some cases the disqualification may be based on activities of the director which could involve criminal liability but there is no overall requirement that the behaviour must infringe specific criminal law rules. One example where a director could be disqualified is for 'unfitness' to be concerned with the management of a company .

Once the process of winding up has finished, the company is removed from the Register and is dissolved three months later.

Single member private limited companies

A directive of the European Commission, passed in 1989, required member states to provide mechanisms for forming a company having only one member. In this country, that directive related to companies limited by shares or by guarantee. Regulations made in 1992 now permit the formation of a company by one person. All existing rules applying to private limited companies now extend to these single member companies.

5

The law of contract

The nature of a contract

In business and commercial life the law of contract underpins a huge range of activites as diverse as engaging staff, buying supplies, arranging insurance, leasing premises, raising loans and buying shares. These transactions all involve making a contract which must comply with the general principles governing all contracts. The modern law of contract has largely evolved from case law, although major pieces of legislation such as the Unfair Contract Terms Act 1977 have radically altered some of the rules. There is no complete code of rules governing the making of a contract. Where a dispute arises between the parties, litigation or arbitration may be necessary to determine the rights of the parties. The courts have generally taken a 'laissez faire' approach to the making of contracts, indicating that the parties are free to make the contract of their choice, with which the courts will rarely interfere.

However, the role of the courts ought not to be overemphasized. Of all the millions of contracts entered into daily only a very small proportion ever involve litigation before the courts. On the whole the parties carry out their obligations and by that means discharge their contract. Even where problems do arise some compromise or settlement may be reached between the parties without the need to take action in the courts.

Contracts are based on the idea of a bargain. Each side must put something into the bargain. A contract may be defined as 'an agreement which is binding on the parties'. Within that definition lies the first problem. The ultimate bargain arrived at between the parties may not have been 'agreed' in the commonly understood sense of that word. There are a number of reasons why this is so. First, English law adopts an objective approach to decide whether there is an agreement between the parties. It is pointless to ask an individual who is now in dispute whether he intended to make a contract. Rather, the courts ask the question; 'Would a reasonable man think that there was an agreement between the parties?' Second, 'agreement' may not seem to be present where one party has little or no choice in the terms he is obliged to accept. There are many contracting situations where, because of the superior bargaining strength of one of the parties, the other has little or no choice but to accept the terms. Third, there are many cases where little or no negotiation takes place between the parties (e.g. sales in a self-service store) but the 'agreement' between the parties is largely created by implied terms (e.g. terms implied into contracts for the sale of goods by virtue of the Sale of Goods Act 1979).

To return to our definition of a contract (an agreement which is binding on the parties) it is always possible for an injured party who is suffering from some breach of contract to sue to enforce it. Enforcing the contract means seeking an appropriate remedy from the court. In most cases that remedy will be an award of damages. It is only in very limited types of case that a court would order one party specifically to perform his obligations. If the court does grant such an order it is called specific performance.

Formation of the contract – agreement

It has become customary in English law to regard agreement as consisting of offer and acceptance. This is only a method of analysing the bargaining process. It is not necessary for the parties to use these words. An offer exists when one party effectively declares his readiness to be bound by a set of terms without any further negotiation. In other words, the offeror (the person making the offer) has reached a stage where he is content for the other party to say he accepts, and at that point a valid, binding contract will come into existence. If the parties are not at such an advanced stage in their negotiations, no offer may yet exist. Often, what appears to be an offer may only be some kind of enticement to bargain, which the law calls an invitation to treat.

The courts have had to consider numerous instances where the argument turned on whether a definite offer was made. In the case of goods on display in a shop window, or goods on supermarket shelves, the court would normally find that such a display was an invitation to treat – the shopper must come forward and make an offer to buy. Similarly in auction sales, when an auctioneer calls for bids, he is not 'offering' to sell the goods, but rather inviting potential buyers to put forward their offers.

The problem regarding offers can also arise in relation to price lists or catalogues which are sent to customers. These are usually construed to be a mere indication of the goods available and their likely selling price. Any other interpretation could give rise to serious practical difficulties. Consider a case where a firm sends out a price list of second-hand earthmoving plant which it has available for sale. The circular may go to dozens of companies who could be interested to buy. If there is only one model of a particular type on the list and several purchasers come forward saying that they wish to buy it the supplier could find himself in breach of contract over and over again as he is unable to supply them all with the equipment. It is better to construe the price list as a mere invitation to bargain, a 'display' of what the firm has to sell, and think of the approaches by potential buyers as the offers, which the seller can accept or reject as he chooses.

Where a case arises which causes doubt the court approaches the issue of whether any offer has been made by using an objective test of the intention of the person concerned. Would a reasonable man, knowing all the circumstances of the case, think that an offer has been made? The court takes into account the words used. In one case, where a letter said 'I may be prepared to sell . . .', the writer could not yet be said to be ready to be bound in contract without further negotiation. Even if the word 'offer' is used, that is not conclusive in law that an offer exists.

If it is clear that the intention is to make a firm offer it may be made orally or in writing. It comes as something of a surprise to many people to discover that English law has very few rules about formality or written evidence in contracts. Apart from exceptional cases such as contracts for the sale of land, transactions worth hundreds of thousands of pounds can be binding even when made orally. Of course there may be sound reasons for making a contract in a more formal way, as subsequent disputes may be capable of being resolved simply by reference to a written agreement.

An offer can be made to an individual, when only that person can accept. The offer may be made to a group and, in certain cases, it may be made to the world at large, for example, in cases offering a reward to a finder. The offer is only effective when it has been communicated to the other party, the offeree.

Counter offers

The offeree has a number of options open to him. He may wish to accept the offer, in which case he must exactly accept the precise offer which was made to him. If he seeks to bargain further or to introduce new terms he is not accepting, rather he is making a counter offer. The law provides that the effect of a counter offer is to terminate the original offer. So if the counter offer is not acceptable the person who made it cannot try to revive the original offer. He must make a fresh offer. This situation can be of vital importance where the person making the final offer may be fixing the terms of the deal. The offer – counter offer situation arises frequently in business where each side in the negotiations may be trying to make the contract on the basis of their own 'pro forma' standard documents. This gives rise to what is known as the battle of the forms. A classic example can be seen in *Butler v. Ex Cello Machine Tool Co.* where the seller offered to sell a machine tool for £75,000, which would be delivered in ten months' time. The seller's offer was made on a printed form containing various conditions, one of which was a price variation clause allowing for an increase in certain circumstances. The buyers placed an order, using their own standard printed order form. On the bottom of the buyers' form was a tear-off slip, which the seller was asked to return to the buyers. The seller duly returned the slip. It contained the words 'we accept your order on the terms and conditions thereon'. When the tool was ultimately delivered, the seller sent a bill for over £78,000. Its claim for the increased amount was unsuccessful. Its original quotation was an offer. The buyers had made a counter offer which did not contain a price variation clause. The seller had accepted the counter offer by returning the tear-off slip.

Of course, not every communication from an offeree will necessarily amount to a counter offer. Once an offer has been made, the offeree may simply wish to seek further explanation of its terms to satisfy himself that he is getting the best deal possible. So there may be a fine distinction between the cases. For example X offers to sell Y his car for £3,000. If Y replies 'Will you take £2,700?' he is still negotiating and making a counter offer. But if Y replies 'Is that cash or is there any possibility of HP?', he may simply be seeking to clarify the terms of the offer.

Acceptance

Not only must the offeree accept the exact and precise offer made to him; he must also communicate his acceptance to the other party.

Normally, the offeror must actually receive the acceptance. This tends not to create problems where the parties are face to face, or using some form of instantaneous communication such as the telephone or a telex machine. In all of these cases, the contract is made when the acceptance is received by the offeror.

The situation is somewhat different if the parties are negotiating at arm's length. This will usually be a case where they have been dealing with each other by post. The offeror can lay down rules about how the offeree has to accept but in most cases he will not have done so. Then it may be appropriate to use what is called the 'postal rule'. When this applies, an acceptance made by letter is effective, and the contract comes into existence, when the letter is posted. Inevitably this means that a legally binding contract exists for some time before the offeror actually learns that he is legally bound. That can sometimes be inconvenient, but an offeror could overcome the problem by stipulating as part of his offer that he will only be bound on *receipt* of the acceptance. In *Holwell Securities v. Hughes*, the offeree wanted to accept an offer under the terms of which he had to accept by a notice in writing by a fixed date. He posted a letter of acceptance but it did not arrive until after the fixed date. The court held that the phrase 'notice in writing' indicated that the offeror must actually receive the acceptance by that date. Merely posting a letter was no use in those circumstances. The 'postal rule' is merely a rule of convenience and the courts would not apply it in cases where it was manifestly absurd to use the post (e.g. a case where the offeror had indicated that he needed a very speedy acceptance). Nor will the rule be applied if the offeror has prescribed some other method of acceptance.

The offeror can lay down quite elaborate rules for acceptance. The courts may then be faced with the problem that an offeree has purported to accept but has not followed the precise method stipulated. Generally the courts will allow an acceptance by any method which is equally as effective as the one which was prescribed. It is a question of construing whether the offeror was making a suggestion, or whether that method and no other must be used. For instance, if the offeror says 'Drop me a line to tell me if you want to buy', does that mean he is insisting on acceptance only by letter?

The rule that acceptance must be communicated means that silence cannot usually amount to acceptance. This is important because otherwise an offeror could impose obligations on the offeree: 'I offer to sell you my car for £3,000. If I do not hear from you before 9 pm I will assume that you have accepted'. If this were permitted, it would force the offeree into a bargain.

Termination of offers

An offer can be said to be terminated by its acceptance because at that moment there is a legally binding contract. There are, however, a number of ways in which an offer may be brought to an end. The offer itself may have been made subject to a time limit. If there is no acceptance before that time, the offer will automatically lapse. Where there is no fixed time limit the offer will remain open for acceptance for a reasonable time. What is reasonable will vary with the circumstances and the nature of the subject matter.

Once he has made an offer, the offeror may change his mind and wish to withdraw it. Revocation is always possible before an offer has been accepted. To be effective, the fact of revocation must be communicated to the offeree; in postal cases this means the letter must actually be received and not merely posted. This can cause conflict where the parties are negotiating at arm's length using the post where the postal rule of acceptance is relevant. The letter revoking the offer may be posted first, but if it arrives *after* a letter of acceptance has been posted it will be too late to withdraw.

Where an offer has been made for a fixed time the question may arise whether it can be revoked during that time. In reliance on the offer remaining open, the offeree could be going to considerable trouble to put himself into a position where he can accept, e.g. by raising finance. If the offeror revokes before the fixed time is up the offeree may have no legal basis for complaint. Effectively this fixed offer situation is really two distinct offers: 'I offer to sell you my car for £3,000, and I offer to keep the offer open for seven days.' If the offeree wishes to enjoy the benefit of the seven days time limit he must accept the second offer and 'buy' it by giving some consideration for it. This is what happens when a person buys an option but in most everyday situations it would not occur to people to offer to 'buy' time in this way. The rules of consideration are considered later.

Certainty of the agreement

Even though the parties think they have reached agreement the courts may find that there is no legally binding contract because of lack of certainty. Naturally the courts do not want to disappoint people's expectations so they approach any question of vagueness or uncertainty with the idea in mind that 'a thing is certain if it is capable of being made certain'. An agreement which appears vague at first sight may be capable of being given a sensible meaning, either because the parties have dealt together before or because there is some custom of the trade or business. In some cases the parties themselves may be aware that their agreement is vague or incomplete and they may provide for future resolution of these aspects by including an arbitration clause. Sometimes it is possible to fill in missing terms by reference to a statute. For example, if the parties have failed to agree on a price and goods have now been supplied under the contract, s.8 of the Sale of Goods Act 1979 provides that where no price has been fixed and no method has been agreed for determining the price, the buyer must pay a reasonable price.

The agreement

Once the processes of offer and acceptance have culminated in the formation of the agreement the terms have become fixed between the parties. That has been the object of their negotiations – to determine the extent of their obligations to each other. Once agreement is reached it cannot be altered by one of the parties without the consent of the other. If any alterations or variations to the agreement are to take place then the relevant rules of consideration should also be borne in mind.

There can be great difficulty in establishing what has become a term of contract. A distinction has to be drawn between statements made which go to the core and essence of the transaction and those statements which merely influence the decision to enter into the contract. The former are terms, where a breach gives rise to a right to claim for damages for breach of contract. The latter are representations, where rights may be available if the statement made was untrue and in some way induced the contract.

Consideration

English contract law is based on the idea of a bargain. There must be some exchange between the parties. One party must be doing something or giving something in consideration of what the other is doing or giving. Consideration can be defined simply as the 'price' each party pays for the right to enforce the other party's promise. It is important to remember that consideration will not always be money although it is quite usual for money to be the consideration provided by one of the parties. If I sell you my car for £3,000 my consideration is giving you the car and your consideration is giving me the £3,000. If we make that agreement and provide for the exchange to take place at the end of the month then our agreement is still a binding contract *now* because our exchange *now* is an exchange of promises. In consideration of you promising to pay me £3,000 at the end of the month I promise to hand over my car to you then.

This type of consideration, based on promises which are to be performed in the future, is called executory consideration. It is surprising how often people fail to realize that there is a legally binding obligation in such a case. Take the example of a shopper who orders a chair in a furniture store, delivery to be in six weeks' time, when the customer will be invoiced for the chair and will then send payment. Later the same day, having placed the order, the customer wants to back out. That would amount to breach of contract. The customer might say that the furniture store had not even had time to send the order off, but the store has lost the profit on the sale and no doubt has been put to some trouble and expense in dealing with the customer.

In order to provide a valid consideration, the law insists that it must not be past. *Re McArdle* illustrates the principle. Members of a family were all entitled to a share in a house on the death of their mother. While the mother was still alive one of the sons and his wife went to live with her. The son's wife made several improvements and alterations to the property for her own comfort and convenience. When the mother died the other members of the family promised to pay the son's wife for all the work that had been done. When they failed to pay she could not enforce their promise as she had given no valid consideration for it. The work she had done was in the past and had not been done in consideration of their promise to pay. The promise and the act were independent of each other. To be a valid consideration there has to be an inter-dependence of promise and act.

There are some situations which look like past consideration but which are not true instances. Take the case of a person requesting someone to perform services for him, in circumstances where he must be expecting that he will have to pay, for example where he asks an accountant to deal with his tax affairs. Once the service is performed any subsequent promise to pay in such a case is enforceable because by asking for the service there is an implied promise to pay a reasonable amount.

Although contracts are based on the idea of a bargain the law does not insist that the two sides of the bargain must be equal. This is often expressed as a rule that consideration need not be adequate. It is up to the parties to make the contract of their choice. One person, for good reasons of his own, may be prepared to pay far more for goods than they are really worth. Generally, contract law allows great freedom to the parties in making their bargain and each is supposed to look after his own interests and strike the keenest bargain possible. One of the maxims of the law is *caveat emptor* which means 'let the buyer beware'. Where the two sides of the bargain are seriously unequal that may be evidence of fraud, duress, mistake, or incapacity, in which case there may be some basis on which the court could set aside the contract. However, if the person complaining has simply made a bad bargain the law will not intervene.

While not insisting that the two sides of the bargain must be equal the courts will insist that each side of the bargain must have some value in the eyes of the law. This may be so even where the consideration consists of something intrinsically worthless. This point is illustrated in *Chappell & Co. v. Nestlé & Co.* where the chocolate company was running a promotional scheme. In return for three chocolate wrappers and a postal order for 1/6d (7½p) the chocolate company would send the customer a pop record. Chappell's were entitled to receive royalties on the selling price of all sales of the record. The question arose here whether the record was sold for 1/6d or whether the wrappers formed part of the consideration. Even though the chocolate company simply threw the wrappers away the House of Lords found that they did form part of the consideration.

Duties already owed

It seems obvious that you are not providing any consideration where you simply behave in a way in which you were legally obliged to act. This view is accepted by the courts in relation to both public and contractual duties owed. Take the case of *Collins v. Godefory* where a witness was under a public duty to attend and give evidence because he had been ordered to do so by the issue of a subpoena. The witness was promised money by the defendant in the case if he came and gave evidence. Once he had given his evidence was it possible to say he had done so in consideration of the defendant's promise to pay? The court held that the witness gave no consideration for the promise as he was already under a duty to give evidence. He was doing nothing more than his duty.

In those cases where it can be shown that a person does more than his public duty requires, then he may have provided consideration for a promise to pay. This can be seen to be the basis of the decision in *Glassbrook Bros v. Glamorgan CC*, where a mine owner requested police to provide a twenty-four hour guard over his property. Although the police were under a duty to prevent crime and protect property they were not required to mount a continuous guard. The mine owner promised to pay for the service and it was held that the local authority could recover the money from him as the police were doing more than their public duty. They had, therefore, given consideration for his promise to pay.

The same sort of situation can arise where a duty is owed under a contract. Suppose you have asked me to supply 100 sausage rolls for your village fair and you have paid me for them. Under the terms of our agreement I am due to deliver the sausage rolls at 12 noon on 1 June. I fail to arrive at 12 noon and when you contact me I sound reluctant to make a delivery. You then promise to pay an extra £10 and, in consequence, I make the delivery. What consideration do I give for your promise to pay £10? I was already contractually bound to deliver the goods so I am doing nothing more than performing an existing duty. In an example such as this, the courts have traditionally found that I could not enforce your promise to pay £10 as I had given no further consideration for it. An old case, *Stilk v. Myrick*, has long represented that view of the court. There, sailors were promised extra pay for sailing a ship back to port when some of the crew had deserted. Later, the shipowner refused to pay, as the sailors were already obliged by their contract to sail the ship back, and had therefore simply performed the duty they already owed under the contract.

Such a decision may be entirely correct as a pure application of the law, and no doubt it prevents situations of near blackmail arising where a person could apply pressure to the other contracting party to pay more, simply by threatening not to perform. But inevitably that 'pure' approach may fail to recognize the true pressures of business life and the market place. In a more modern shipping case, the *Atlantic Baron*, the court was able to enforce a promise of extra payment for the building of a ship by finding that the shipyard did more than they originally promised, and had therefore given consideration for the extra payment.

Some judges have gone so far as to say that *any* promise to perform a pre-existing duty should be good consideration, because it is a benefit to the person receiving performance. When the argument is carried to these lengths, in our sausage roll example, I could recover the extra £10. The benefit to you would be that you receive performance with none of the aggravation of needing to sue for breach. The detriment to me is that I give up the possibility of being in breach of contract – and there is no doubt that this can be a very valuable right. Having agreed to sell sausage rolls to you for a fixed price, I may now find that I can sell the same goods so advantageously elsewhere that it actually pays me to be in breach with you and suffer the consequences.

Much of this thinking has been further developed by the Court of Appeal decision in *Williams v. Roffey*. The facts there were that building contractors had subcontracted the carpentry work on a block of 127 flats to the plaintiff, for a price of £20,000. The plaintiff was entitled to interim payments under the contract. After he had completed nine of the flats, and done preliminary work on the others, he had already received over £16,000. He then realized that his price for the work was far too low. The building contractors began to get anxious, as they would become liable under penalty clauses in their main contract if the work was not completed on time. They knew that the plaintiff was having difficulty because he had underpriced the job. They suggested a meeting with him, at which they agreed to pay him £10,000 more, on condition that the work was completed on time. In view of the extra payment, the plaintiff went on with the work. He completed eight further flats but then stopped work. When he tried to recover what he alleged was due to him under the contract, the building contractors refused to pay any part of the extra £10,000, on the grounds that the carpenter had given no extra consideration for it. In the Court of Appeal, the carpenter was held to be entitled to the extra money, because the building contractor had obtained a benefit, in that they were avoiding the penalty under another contract.

Of course, an important feature of the case was that there was no suggestion of 'blackmail' by the carpenter. Indeed, he had been approached in the first instance by the builder who had made the suggestion to pay extra. Moreover, if the court had been determined to look for something in the carpenter's behaviour which could amount to performing more than his mere contractual duty, no doubt they could have 'found' it. For example, it could be said that the carpenter's agreement to undertake the work in a different sequence could be a detriment to him and a benefit to the builders.

It is difficult to make a judgement yet as to the ultimate effect of the Williams case on the development of this area of law, but it is clear that the courts have found it easier to discover a valid consideration where performance of a duty owed to a third party is involved. Consider the following example. A is contractually bound to supply the engines for a ship being built by B. The engines are due to be delivered by 1 January. C has a contract with B to charter the ship from 1 June, immediately on its completion. It begins to look as if A is going to be late in making delivery of the engines. C is extremely anxious that the ship should be finished on time as he stands to lose an enormous sum of money if his charter of the ship is delayed. C approaches A and promises him £10,000 if he will get the engines to B by 1 January. If A does deliver the engines to B on time, can he then claim the £10,000 from C? The courts have been inclined to take a more relaxed view in this type of three-party situation, and in a number of cases (e.g. *New Zealand Shipping Co. v. A. M. Satterthwaite & Co.*) they have allowed A to sue for the £10,000. The reasoning seems to be that there is a detriment or burden to A because, by binding himself in a contract to C as well as to B, he lays himself open to two actions for breach of contract if he fails to deliver the engines. Moreover, it might have been more to A's liking to be in breach of contract with B and face whatever consequences that might have brought. Now he foregoes that possibility because of his agreement with C.

In such cases it would be important to check that there was nothing in the transaction which was contrary to public policy. Latterly the courts have shown themselves willing to control these situations by developing rules of economic duress (see page 109).

Part payment of a debt

Problems can arise in finding valid consideration where one party seeks to perform his contract by doing something less than he is contractually obliged to do. Take the example of A who owes £100 under a contract to B which is due for payment now. A may be in financial difficulties, so much so that B agrees to accept £75 in full settlement. What consideration has A given for B's promise to forego £25? The courts would say that the £25 was still due and B could sue to recover it despite his promise. It would be quite different if, at B's request, A had paid £75 at an earlier date. That would be a detriment to A and a benefit to B and A would then be providing consideration for B's forebearance.

The difficulty with the proposition advanced above is that, in practice, people dealing together do frequently make concessions to each other within the framework of their legally binding agreement. These concessions or variations are usually informal and may work perfectly well so long as the parties remain on good terms. If their relationship becomes strained one of them may wish to revert to the strict terms of their original agreement. The other will want to know if the variation can be regarded as binding. Traditionally, in order to enforce the variation, it has to be 'paid for' by some extra consideration. There is sometimes a way round this problem, however, by the application of the equitable principle of promissory estoppel.

Equitable principles operate to abate the rigours of the operation of the common law. Undoubtedly, there can be grave hardship where one contracting party has relied on some concession by the other, only to see it abruptly withdrawn. That was the state of affairs in *Central London Property Trust v. High Trees House*. The contract in that case related to a lease of a block of flats in London where the rent was to be £2,500 per annum. Because of the war it became extremely difficult to find tenants for the flats, so the landlord agreed to reduce the rent by half. When the war ended the landlord wanted to revert to the original rent and was also keen to know whether he could recover the arrears of rent. Effectively, what the landlord had done was to make a promise, within the framework of an existing contract, where the tenant gave no consideration for the promise, but where obviously the tenant had relied on it. On the authority of earlier cases the High Court found a principle of equity whereby if A leads B to suppose that A will not enforce his strict rights under the contract, then A cannot go back on his promise where it would be inequitable. Applying this principle in the High Trees case, the court would not allow the landlord to recover the arrears of rent. But the court also found that the end of the war acted as notice that the promise (made only because of the war) was no longer operative.

At first sight this case may seem to strike at the very core of the doctrine of consideration. Further examination shows that not to be the case. A number of points must be borne in mind. It was made clear in cases subsequent to the High Trees case that this equitable principle of promissory estoppel can only be pleaded by way of defence. In the colourful language of judges it is a shield and not a sword. There must be circumstances which show it would be inequitable for the promisor to go back on his word, at least without giving notice of his intention to revert to the original terms of the contract. Of course there may be cases where once a concession has been made it is impossible to revert to the original terms e.g. if the concession related to an extension of time for delivery. This raises important questions about the scope of promissory estoppel, as it seems that the effect of the promise or concession should be intended only to suspend legal rights and not to extinguish them completely. This fits well with the idea that the promisor can revert to the original terms by giving reasonable notice. Certainly if the scope of the rule is seen to be thus limited it gives rise to fewer legal difficulties.

It must be stressed that the development of promissory estoppel is based on equity. 'He who seeks equity must do equity' and 'he who comes to equity must come with clean hands'. It follows from these maxims that a person seeking to take advantage of promissory estoppel must himself have behaved properly. This aspect was critical to the decision in *D & C Builders v. Rees*, where two jobbing builders were owed money by Mr and Mrs Rees. Mrs Rees knew they were in a grievous financial state and she persuaded them to accept less than they were owed in full settlement. So desperate was their plight that they accepted the smaller sum but determined to see what they could do later to recover the rest. Mrs Rees claimed that they had promised to accept the smaller sum, that she had acted in reliance on their promise and that it would be inequitable to allow them to go back on their word. The court had no hesitation in finding that her own behaviour was inequitable as it verged on intimidation so they refused to allow her to take advantage of the equitable principle.

Although promissory estoppel can be seen as an important safeguard against some of the harsher aspects of the common law rules it must be remembered that consideration is still essential for the creation of a valid contract, although there are limited circumstances where it may not be necessary to support some modification within the contract.

Only parties to a contract may sue to enforce it

It is a general rule of contract that in order to enforce a contract you must have given consideration, i.e. you must be a party to the contract. You have no right to sue merely because you were the intended beneficiary of a contract. In *Tweddle v. Atkinson*, where a young couple were about to be married, their respective fathers promised *each other* that each would pay a sum of money to the son. One father did not pay. The son attempted to sue him but failed as he was not a party to the contract. The defaulting father would have to be sued by the other father.

The importance of this rule is that it creates the notion of *privity of contract*. Although the rule can be inconvenient at times, it remains a fundamental part of contract law. It can have significant repercussions in the law relating to exclusion clauses (see page 95).

Intention

If both parties to a contract provide valid consideration that may be taken to indicate that they intend to make a legally binding arrangement. However, in English contract law, the courts look for a separate element of intention. Traditionally the courts have viewed the cases as falling into two categories: commercial arrangements where the parties are presumed to have intended to create a contract; and family, domestic and social arrangements where the parties are presumed not to intend to create a contract. The presumptions are sensible in most cases. Take the example of a parent who gives pocket money to a child or two friends who arrange to meet for supper. In either case, one party would never expect to be sued for breach of contract if such arrangements collapsed.

The problem facing the courts arises where one party to an

arrangement is seeking to rebut the normal presumption. In family arrangements he may want to prove that a legally binding agreement was indeed intended. He must then satisfy the court that this was so. Many of these cases turn on problems between husband and wife on the breakdown of marriage. In one case, *Merrit v. Merritt*, relations between the parties were so strained that they met in the neutral territory of a car park to discuss the disposal of their matrimonial home. Once they had arrived at a decision the wife insisted that the husband should jot down the main points on a piece of paper. The court found here that the parties did intend to create a legally binding contract, influenced no doubt by the rather formal steps that the wife had insisted on, and also by the fact that the relationship between the parties was breaking down, so that they were no longer closely bound in a family or domestic situation.

Where one party seeks to rebut the presumption, the court may also be influenced by the degree of certainty and detail with which the parties made their arrangements. In *Jones v. Padavatton*, the details of the arrangements with regard to a house were so sketchy that it was impossible to believe that the parties could have intended to make a legally binding agreement. In other cases it seems that as the family relationships become less close, so the strength of the presumption diminishes. In *Parker v. Clark*, where the parties were uncle and nephew, there was found to be an intention to create a legally binding agreement, but no doubt the amount at stake, together with the fact that the terms of the agreement obliged the nephew to give up his own home to come to look after the uncle, also influenced the court. There seemed to be little difficulty in establishing the appropriate intention in *Simpkins v. Pays*, where three people living together regularly participated in a competition. When one of them was ultimately the winner, she was obliged to share her winnings with the other two.

In a commercial situation it is considerably more difficult to rebut the presumption that a contract was intended. The courts will require convincing evidence that the parties intended otherwise. In *Rose and Frank Co. v. Compton Bros* the parties indicated in a written document that their arrangement was a 'gentleman's agreement, binding in honour only'. In the face of such convincing evidence the court was obliged to find that there was no contract.

Sometimes the parties want to lift their transaction outside the framework of contract law because they do not want the trouble and expense of litigation. This can frequently be seen as a condition in competitions and is one explanation of the football pools cases, where a punter who believes he has a winning line cannot sue to enforce his win. Provided the parties have made their intentions sufficiently clear, the court will accept their decision.

Form of the contract

As a general rule, English contract law does not require contracts to be made in any particular form. In practice many contracts are oral, which often leads people to assume that it would be difficult to sue to enforce such a contract. Certainly there may be problems of proof but in the end it may be a matter of judging which party the court can more reliably believe. Where a contract is put into writing to provide the necessary proof, the writing can in turn prove to be problematical. There may be a dispute about its precise meaning, or it may fail to spell out the entire contract between the parties.

Exceptionally the law sometimes requires contracts to be made in writing (e.g. hire purchase contracts). In these cases it is usually necessary to look at the appropriate statute to see the precise form the writing must take and also to determine what will happen if the parties have not complied with the rules. Rules about the precise form of the contract can be a useful device for protecting consumers. In the Hire Purchase Acts certain information contained in HP agreements needs to be displayed in prominent boxes, by which means it is hoped to draw those aspects more particularly to the attention of the consumer.

Despite the freedom about the form of the contract there has long been special treatment for contracts for the sale of land. As early as 1677, rules were laid down that contracts for the sale of land must be evidenced by a note or memorandum in writing, otherwise the contract would be unenforceable. These rules were re-enacted in more or less the same form by s.40 of the Law of Property Act 1925. Evidence in writing was meant as a safeguard against fraud but ironically the need for such evidence has sometimes been the very means by which people could perpetrate a fraud. The rules were regarded as unsatisfactory and, when the Law Commission took a detailed look at the workings of s.40, it reported: 'As a result of judicial attempts to prevent the statute being used as an instrument of fraud, it is virtually impossible to discover with acceptable certainty, prior to proceedings, whether a contract will be found to be enforceable . . . s.40 would appear ripe for reform'.

That reform has now taken place with the repeal of s.40 by the Law of Property (Miscellaneous Provisions) Act 1989. Section 2 of the new Act provides that contracts for the sale or other disposition of interests in land can only be made in writing, and must incorporate all the terms expressly agreed between the parties. The contract must be signed by both parties, or their agents. Incorporation of the express terms may be by reference to another document, and the rules also make provision for the methods employed by solicitors engaged on behalf of the parties, where it is common to 'exchange contracts'. Sales of land by public auction are excluded from these rules, as are short leases (see page 288).

Although these new rules seem designed to promote greater certainty, in that it should be easier to determine when a legally enforceable contract for the sale of land exists, one major problem remains to be tackled. It is still the case that people who are unaware of these rules may 'shake hands' on an oral agreement for the sale of land, and the purchaser may then incur expenditure on the land concerned, only to find that the vendor refuses to complete the transaction. Take the example of an oral agreement for the sale of a house, where the parties agree that the prospective purchaser can have a key for the property, and may undertake work on the property in advance of the legal transfer to him. If the purchaser were to install central heating and undertake extensive decorating work, it would be disastrous for him if the vendor then refused to transfer the house to him. In such a case, the vendor would seem to receive improper protection from the new rules. The aggrieved purchaser would be unable to sue for breach of contract because the s.2 rules state that a contract for the sale of land *can only be made in writing*. Lack of writing, therefore, means that there is no contract.

Problems of the sort outlined above were previously solved by the use of discretionary equitable relief, in the form of the equitable doctrine of part performance. In the absence of the necessary note or memorandum in writing, this equitable principle used to permit specific performance of the contract if the behaviour of the parties was in itself sufficient to indicate the existence of a contract. Unfortunately, there is no scope now for the application of this discretion, as it depends for its efficacy on the fact that lack of writing under the old s.40 rules merely rendered a valid contract unenforceable. It did not deny the very existence of a contract, which is the result achieved under the s.2 provisions of the new Act. Clearly, the equitable doctrine of part performance has been repealed along with s.40.

The Law Commission did consider the difficulties that the new rules would create, and in particular they examined the plight of a would-be purchaser who had expended money and effort on the property. They considered that the present law contained a sufficient armoury of weapons by which to assist such a person, such as an action for restitution, or suing on a *quantum meruit* if work had been undertaken on a property. It will take time for cases to come before the courts to show whether the Law Commission's faith in these solutions is justified.

One interesting application of the new rules had already come before the courts in *Spiro v. Glencrown Properties Ltd*. This case involved an option to purchase land, where the vendor and prospective purchaser had both signed an agreement creating the option, which conformed in all respects with s.2. Later, when the would-be purchaser served a signed notice to exercise the option, the vendor attempted to argue that the documents already signed merely amounted to an irrevocable offer and that it was the notice exercising the option which brought about a contract for the sale of an interest in land. As this notice was signed only by the purchaser, it did not satisfy the rules of s.2. The court dismissed that argument, preferring instead to construe the original agreement as the contract, conditional upon the exercise of the option. When the purchaser served the signed notice, he was thereby fulfilling the condition. It was the original agreement which needed to satisfy the s.2 rules, and as it clearly did, the vendor was bound.

Subject to contract

It is common in dealings relating to land for the parties to be anxious to take expert advice before finally committing themselves to a binding contract. Negotiations are often carried on 'subject to contract'. Those words will usually be interpreted by the courts to mean that the parties have not yet concluded a contract. The very words deny the existence of a present contract. Of course, much depends on the precise words used, as the role of the court is to seek to construe what the parties must have intended. In one case the parties declared their agreement to be a 'provisional agreement', but the effect of that phrase was quite different and the court held that their agreement was binding from the start (*Branca v. Cobarro*). It seems to follow from the cases that where the parties use the phrase 'subject to contract' in a written document, that document is not a contract, even if it complies in all other respects with s.2.

Agreements which are subject to contract may be contrasted with conditional contracts. For example, a contract for the sale of land may be subject to the condition that planning permission is obtained by a particular date. If the condition is not satisfied, then the parties are not bound by the contract.

The contents of the contract

Once it is clear that a contract has been agreed, it may then be necessary to establish precisely the obligations which each party has undertaken. Of course, this analysis will not be necessary if everything goes well between the parties and each performs the contract to the entire satisfaction of the other. However, the rules are important not only in problem or dispute solving, but also in contract planning, particularly in a business situation. The more the parties understand the rules, the better they are able to use them to ensure that subsequent problems do not occur, or where they do, that they can be solved with a minimum of expense, delay and irritation.

Where a contract consists of buying a cup of coffee from a machine, or paying the fare on a bus, it may be difficult to imagine what the contents of such a contract are. But all contracts have terms. These may be express or implied. The express terms will be those specifically negotiated between the parties, either orally or in writing. Remember that the parties are largely free to fix their own terms. Contract law still adheres generally to the principle of freedom of contract. The parties must strike their own bargain. In effect, this imposes a serious restriction on the extent to which terms can be implied in a contract.

Implied terms may be incorporated from an Act of Parliament. One of the commonest examples is the set of implied terms incorporated into sale of goods transactions. Without any need for specific negotiation by the parties, terms are implied relating to fitness for purpose of the goods, and their merchantable quality. The courts have a limited role to play in judicially implying terms, but great care is needed to ensure that judges are not seen to be making the agreement for the parties. It is usually said that terms can only be implied by the judges if it is necessary to give 'business efficacy' to the arrangements made by the parties. This may occur where the parties have already undertaken a substantial part of the contract and they come up against some problem or difficulty which they did not provide for. In such cases a judge can imply a term where it is clear that, had the parties put their minds to the problem at the appropriate time, both would have agreed to the term proposed.

Terms are the obligations under a contract but inevitably they will not be of equal importance. Take, for example, a term about the time of a delivery of goods. This may be of paramount importance where goods are perishable or where it is clear that they will be of little commercial use unless delivered on time. In other cases the time for delivery may be far less significant. The law has to find some way to grade or classify terms, because it must seek to offer remedies appropriate to the significance of the term which has been broken. Traditionally, the law classified terms as major terms, which it called conditions, and minor terms, which it called warranties. The terminology is unfortunate, as both of those words have sundry other meanings and uses in the law of contract. Conditions are said to be those obligations which are of the essence, which go to the very core and heart of the contract. Breach of a condition always gives the injured party the right to sue for damages. Over and above that, the injured party may also have the right to repudiate the contract. This means that he is entitled to regard himself as no longer bound by the contract. A warranty is of more peripheral importance, often said to be merely collateral to the main purpose of the contract. An injured party can sue for damages for breach of warranty but has no right to repudiate the contract.

The problem with this traditional approach to classifying terms is that it requires a decision to be made about the status of the term at the time when the contract was made. In effect this emphasizes the importance of the intention of the parties, as it should be clear from what was said and written, and from all the surrounding circumstances, whether they intended a particular term to be a condition or warranty. This approach creates business certainty, as it would be possible to say at the outset what remedies would be available for particular breaches. But it takes little account of the realities of business life, where the term breached may be only a warranty, but the consequences could turn out to be very serious. This would leave the victim of the breach in a situation where he would be forced to go on with performance of the contract and be limited to a claim for damages.

Consider as an example the problem which arose in *Hong Kong Fir Shipping Co. Ltd v. Kawasaki Kisen Kaisha Ltd*. One of the terms in a contract to charter a ship for two years provided that the ship should be in every way fitted for ordinary cargo service. When the ship was delivered at the beginning of the two-year charter period it was unseaworthy and when it made its first voyage, repairs were needed which took four months to complete. Did the breach amount to a breach of condition? If it did, the charterer could repudiate liability under the contract and thus rid himself of the burden of this unsatisfactory vessel over the remainder of the two-year period *and* sue for damages. If it was only a breach of warranty his claim would be limited to damages and he would be left with the vessel for the remainder of the charter period.

In the Hong Kong Fir case, the Court of Appeal chose a novel solution, preferring instead to say that some terms are 'intermediate' or 'innominate'. Such terms defy initial classification as conditions or warranties and the parties must wait until the scale of the resulting breach is known, at which point the term can be suitably classified. Obviously this produces a much less certain situation between the parties during the currency of the contract.

This development in judicial thinking did not in fact assist the charterer in the Hong Kong Fir case. The term in issue was held to be only a warranty, so the charterer had no right to repudiate the contract.

As he had wrongfully repudiated, he was obliged to pay damages to the shipowner.

The same problem of classifying a term was before the Court of Appeal in another charter party case, *The Mihalis Angelos*, where the term related to the date when the ship would be ready to load. In the event, it was nearly a month after that date before the ship was ready. The term was held to be a condition. The court made the point that it was influenced by earlier cases holding similar clauses to be conditions and it was vital to remember that 'one of the important elements of the law is predictability'. One important point emerging from this case is that it may already be established, either by statute or an earlier precedent, that a particular type of clause falls into the category of a condition. Then there is no scope for attempting to classify it as an innominate term. But where a term could be either a condition or a warranty, and it is not clear which the parties intended, and where the range of possible breaches and the scale of the resulting harm is very wide, then the court may classify it as innominate. This classification is theoretically made at the time when the contract is made, but the remedy available for breach of the term will only be established once the scale of that breach is known.

The appropriateness of this 'wait and see' approach is well demonstrated by the facts in the Hong Kong Fir case. There, the clause in question required that the ship was to be 'in every way fitted for cargo service'. These words were so wide in potential interpretation that there could be breaches ranging from failure to supply the vessel with a proper number of anchors, or to put on board medical supplies, to totally defective engines at the other end of the scale.

It is clear that the courts remain anxious about creating business uncertainty. The House of Lords has confirmed that the traditional division into conditions and warranties is still acceptable in many cases (*Bunge Corporation v. Tradex Export SA*). They have, however, left the way open to use the more flexible Hong Kong Fir approach in a limited number of cases where this may be justified. Clearly the problem can be largely overcome by the parties being sufficiently specific at the outset about the result of each type of breach. This is a classic example of how rules of contract law can have a planning function.

Exclusion and limitation clauses

Although many terms in contracts will be concerned with creating obligations, it has become increasingly common for the parties to seek to limit or exclude liability which would otherwise arise under the contract. Exclusion and limitation clauses are frequently encountered in contracts involving car parks, dry cleaners, film processing, package holiday deals, furniture removal and insurance. They may take forms such as 'No liability accepted for any damage caused by flooding' or 'In the event of a breach, our liability is restricted to ten times the contract price'.

Exclusion clauses fall into two main categories: those negotiated between parties of equal bargaining strength, where such clauses can be seen to facilitate business activity and to allow the parties to allocate risks between themselves and organize which of them should insure; and those imposed on a weaker party by a stronger party who may enjoy a monopolistic position, where the weaker party has no option but to accept. Clauses falling into the latter category are, not unnaturally, unpopular with the courts. In the past the judges strove to protect victims of harsh exclusion clauses, but some of the techniques developed by them to afford protection produced unfortunate results in business cases, sometimes upsetting the very carefully negotiated arrangements between the parties. The courts have never had a power at common law simply to strike out exclusion or limitation clauses on the grounds that they are unfair or unreasonable because contract law proceeds on the assumption that the parties are free to fix the terms of their bargain for themselves.

Judicial dislike of exclusion clauses has always meant that courts will only allow reliance on such a clause if it has effectively become part of the contract. The incorporation of the clause into the contract can occur:

1 By one party signing a document containing the clause, of which he would then be deemed to have notice whether he had read it or not.
2 By one party giving reasonable notice of the clause to the other. Reasonable notice may be given in a contractual document but not every piece of paper passing between the parties can be so classified. Generally, contractual documents are those pieces of paper on which a reasonable man would expect to find conditions printed. This would usually include documents such as airline tickets or standard company order forms. It would not include things such as mere receipts for the payment of money. In any case, such a receipt would often be handed over too late, as it is vital that reasonable notice be given before, or at the time when, the contract is made. Once the parties have finished negotiating and finalized their deal it is too late for one of them to try to change it or add to it unilaterally. Where no contractual document passes between the parties the exclusion clause can be incorporated by the reasonable display of a notice at an appropriate place, so that its contents could have been read before the contract was made.

3   By virtue of a 'course of dealings' between the parties. Where two people have dealt with each other regularly, always using the same terms, the court may be prepared to say that adequate notice has been given of any exclusion clause. This principle is only likely to be applied between business parties who have regular commercial dealings.

In cases where the court accepts that the exclusion clause is properly incorporated and forms part of the contract, the judge will next consider what the words of the clause mean, and whether the wording covers the actual facts which have arisen. In the construction or interpretation of the clause the standard policy of the courts is to construe any ambiguity strictly against the interest of the party seeking to rely on the clause. In one case involving motor insurance, a clause in the contract provided that the insurers need not pay out on a claim under the contract if the vehicle was carrying 'an excessive load'. In fact, it was a five seater car carrying six persons, but the judge held that the word 'load' did not include people.

Prior to 1977, when the Unfair Contract Terms Act was passed, the judges had to use these construction and interpretation techniques boldly in order to protect victims of harsh clauses. In consequence, there was a very distorted result in many of the cases. Such striving on the part of the judges is much less necessary now since the 1977 Act was passed, and they have been urged by the House of Lords to give the relevant words in an exclusion clause their plain and natural meaning. Nevertheless, the two common law rules relating to incorporation and construction are still relevant, though the provisions of the 1977 Act now govern the majority of problem cases in this area.

Unfair Contract Terms Act 1977

The scope of this Act is wider than its name would suggest as it deals with notices limiting liability in the law of tort as well as exclusion clauses in the law of contract. One advantage of introducing statutory control of exclusion clauses is that a flexible approach can be taken, making a distinction between the type of clause negotiated between businessmen on an equal footing, and other clauses 'imposed' by businessmen on consumers. The Act does not apply to all contracts – insurance contracts for example are outside its scope. Although the Act is fairly general in its coverage there are other specific Acts of Parliament which also prohibit particular types of exclusion clause.

The main sections of the Act only apply in 'business liability' situations. This phrase is defined as meaning liability for breach of duties arising from things done in the course of business (s.1(3)). So, except in sale of goods transactions which are considered separately, (see page 139) the Act does not extend to what may be regarded as private transactions. Once the Act is shown to apply, it has two different techniques for dealing with offending exclusion clauses. Some are rendered totally ineffective however they are drafted. Others can survive and take effect if they satisfy the requirement of reasonableness set out in the Act.

Sections 2 and 3 are the most important general sections of the Act. Section 2 deals with liability for negligence, which is defined by s.1 of the Act as:

1 Breach of a contractual obligation to take reasonable care or exercise reasonable skill in the performance of the contract; or
2 breach of common law duty to take reasonable care or exercise reasonable skill; or
3 breach of the common duty of care imposed by the Occupiers' Liability Act 1957.

If negligence results in death or personal injury, then it is not possible to exclude or limit liability either by an exclusion clause in a contract or by a notice displayed generally. So, where a contract contains a term that 'no liability is accepted for death or injury howsoever caused', then if death or injury is caused by negligence, the relevant clause will be rendered totally ineffective.

Section 2 also provides that where loss or damage (other than death or personal injury) is caused by negligence, an exclusion clause *can* take effect *if* it satisfies the requirement of reasonableness. This requirement will be analysed in conjunction with s.3.

Section 3 covers two different situations. First, where one party deals as a consumer, and second where one party deals on the other's written standard terms of business. Both of these situations require further explanation. Dealing as a consumer means that one party is not making the contract in the course of a business but the other party is. A classic example would be taking your car to be serviced. You deal as a consumer, while the garage deals in the course of a business. Or it could be having your house rewired. You deal as a consumer and the electrician deals in the course of a business. What constitutes 'written standard terms of business' is not entirely clear as the phrase is not defined by the Act.

In either of these two situations s.3 provides that the other party, having committed a breach, cannot exclude or restrict his liability for that breach unless the exclusion clause satisfies the requirement of reasonableness. An example will help to illustrate the working of the section. Imagine that you take a film to be processed by a high street photographic shop. On the counter is displayed a clear notice to customers that in the event of the film being lost or spoilt, liability is limited to a replacement film. When you go to collect your photographs, they cannot be found and the shop tries to rely on the exclusion clause. It will only be able to do so if it can prove to the court that the clause satisfies the requirement of reasonableness. This means that, in relation to contract terms, it must be a fair and reasonable term to include, given what the parties knew or ought to have known at the time when they made the contract. The burden of showing that it was fair and reasonable is on the person now seeking to rely on the clause. The answer will obviously depend very much on the circumstances of each case, but a judge approaching the problem is given some limited help by the Act itself. It is clear from s.11 that there are cases where the judge must consider whether it was possible for the parties to have protected themselves by insurance. For instance, there are situations where a businessman may be prepared to quote two prices for a job. A high price will reflect the fact that he is prepared to accept all liability. A low price may be conditional upon him excluding some liability. Naturally, when I consider his prices I may well choose the low price if I can easily and cheaply insure against the risks myself or if I already have insurance which covers those risks.

The Act itself provides guidelines which will sometimes be relevant when determining reasonableness. The court may take into account:

1 The relative bargaining strength of the parties. One factor which might be important here is whether the goods or services could have been obtained easily elsewhere.
2 Did the customer receive some inducement to agree to the term? The example quoted above where a customer chooses the low price is an instance of an inducement. The court will also keep in mind whether a similar contract made with someone else would have been likely to contain an identical term.
3 Did the customer know, or ought he reasonably to have known, of the existence and extent of the term? Bearing in mind the importance of showing that the term has been properly incorporated into the contract this guideline seems to suggest that there is one standard for deciding on incorporation but a different one for deciding reasonableness.
4 Where the term excludes liability only if a party fails to comply with some condition, was it reasonable to assume when the contract was made that it would be practicable to fulfil the condition? An example might arise where a clause states: 'No liability for any breakages in transit which are not reported to us within five minutes of delivery.' Even when the contract was made it would be clear that such a condition was not practicable.

5 Where the contract relates to a sale of goods which have been specially manufactured, processed or adapted to the buyer's requirements, that may give the seller wider scope to limit his liability.

There are relatively few decided cases to clarify the requirement of reasonableness. That may be a good thing, as it is important to remember that these cases turn on their own individual facts. *George Mitchell v. Finney Lock Seeds Ltd* makes a useful illustration of the rules. The farmer wanted to plant Dutch winter cabbage in a field of some sixty acres. He was sure that he could make a handsome profit on the vegetable as it would be ready to harvest when home-grown vegetables were in short supply. He obtained the seed from Finney's at a cost of £192. The terms of the contract provided that liability was limited. In the event of the seed being defective, the sellers would refund only the price of the seed. They also excluded 'all liability for any loss or damage arising from the use of any seeds supplied by us and for any consequential loss or damage arising out of such use, or for any other loss or damage whatsoever.'

Whatever the seeds supplied might have been, they were not Dutch winter cabbage. The crop failed and the farmer sued for £63,000. The company was prepared to refund £192. The case turned on whether Finney's could rely on the clause. They could only do so if the clause was a fair and reasonable one. In the House of Lords the judges were conscious that this was the first interpretation of the 'fair and reasonable' provision. They made clear that arriving at a decision as to what is fair and reasonable is not like the exercise of discretion, but the court must put a number of factors on the scales and see on which side the balance comes down. 'There will sometimes be room for a legitimate difference of judicial opinion . . . and the appellate court should treat the original decision with the utmost respect.' When looking at the facts of the present case, the judges took into account that the farmer had been a regular customer of the seed firm for many years and had had plenty of opportunities to read the exclusion clause. It was not a difficult clause to understand. Similar clauses would be found in every seed firm's contracts, and had never caused farmers as a body to protest through the National Farmers Union. So far as the question of insurance was concerned, the seed firm could have insured against the risk of crop failure without needing to significantly raise the price of seed. The court was also influenced by the fact that the seed firm usually sought to negotiate settlements in cases like this, rather than rely on the strict terms of the exclusion clause. This latter fact was regarded by the court as particularly significant as they thought it pointed to the seed firm themselves recognizing that their own clause was unreasonable. Taking all of these factors into account, the House of Lords held that the clause was not fair and reasonable.

The 1977 Act has been of great importance and value in that the flood of complex, and often irreconcilable, decisions on exclusion clauses has virtually stopped. It is clear that it will be easier to justify a clause which merely limits liability, as opposed to one which seeks to exclude liability completely. Other relevant aspects of the Act are considered separately in relation to misrepresentation, sale of goods and occupier's liability.

Misrepresentation and mistake

The basic idea of misrepresentation is a very simple one to grasp. One party makes a false statement which has induced the other to enter the contract. This area of law is regarded as complex, largely due to the number of overlaps with other areas of law, so that it is seldom possible to view a misrepresentation in isolation. The overlaps occur either when misrepresentation induces a mistake; or when the statement made, depending on its relative importance, may be classed as a representation or a term; or because the law of contract and the law of tort have separately developed ideas about negligent statements.

In complex contractual negotiations, both parties will make many statements to each other about the subject matter. We have already considered the problem of classifying terms of a contract. Terms may result from statements made which now form part of the contractual obligations. Alternatively, the statement may be of less importance and may not become part of the contract. Nevertheless, that statement may weigh with one party in influencing him to make the contract. He is induced by the representation. Some statements may simply have been designed to create a good bargaining atmosphere and these may give rise to no liability at all. These examples are sometimes referred to as 'tradesmen's puffs'. In one case, where a seller described a car as 'a great little goer', no liability could arise from such a claim.

The court must determine into which category a statement falls. When deciding between terms and representations they are influenced by factors such as:

- How early in negotiations was the statement made?
- Was an oral statement subsequently included in a later written contract?
- Did the person making the statement have special knowledge on the subject matter of the contract?

The answers to these questions will not necessarily produce a definite solution. Each case will turn on its own facts.

Where a false statement is classed as a representation it is possible for it to be made innocently, negligently or fraudulently, depending on the maker's state of mind. The classification is important because each type of misrepresentation gives rise to different remedies. It is also important to distinguish terms from representations because of the different remedies available. Non-compliance with a term results in breach of the contract. Misrepresentation is a false statement standing outside the contract and not forming part of the obligations and does not, therefore, involve a breach of contract. The difference in the remedies available used to be so great that the distinction between a term and a representation assumed immense importance. Since the changes in the law introduced by the Misrepresentation Act 1967, it is now far less significant.

Figure 2 shows how any particular statement can be graded. Take an example like 'The car has only done 20,000 miles', or 'This lorry can carry thirty tons'.

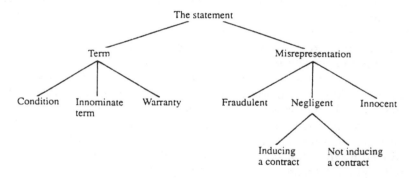

Figure 2   *The grading of terms and misrepresentations*

Misrepresentation is defined as a false statement of fact, made by one party to the other, before or at the time of contracting, which is one of the causes which induces the contract. Within the definition, a number of points of difficulty occur.

1   There must be a statement, but the courts have been prepared to imply a statement from behaviour. 'A nod or a wink, or a shake of the head or a smile' from the purchaser, intended to induce the vendor to believe the existence of a non-existing fact, would be sufficient.

2 The statement must be of fact. This excludes statements of intention, statements of opinion and statements of law. Statements of opinion give rise to particular problems, as they can be made by someone who has more knowledge and information in the particular bargaining situation. This makes the statement seem especially authoritative. It seems as if the maker of the statement knows facts which justify him in holding that opinion. In one case a vendor, trying to sell property which was let, described the tenant as 'a most desirable tenant'. The vendor knew that the tenant was in arrears with his rent and earlier rent had only been extracted with considerable difficulty. In these circumstances, knowing the facts as he did, the court held that the vendor could not reasonably hold such an opinion. They regarded his statement as actionable misrepresentation (*Smith v. Land and House Property Corporation*). In that case, the vendor had superior knowledge. He alone knew the problems about the tenant. Compare the situation in *Bisset v. Wilkinson*, where a vendor was selling land in New Zealand. In response to an enquiry from the purchaser, the vendor said that the land could carry 2000 sheep. The vendor had never kept sheep nor had the land ever been used for that purpose, so he had no facts on which to base this opinion. It was only an opinion and not actionable as misrepresentation.

3 Since there must usually be a false statement, it follows that keeping silent cannot usually constitute misrepresentation. If I want to sell my second-hand car, and I say nothing at all about its condition or performance, it will be up to a prospective purchaser to enquire specifically about anything that concerns or worries him. This is part of a general principle in contract of 'let the buyer beware' (*caveat emptor*). I would be under no duty to point out faults and defects. Of course, if I am asked specific questions then I should answer honestly, otherwise my statements could amount to misrepresentation. There are some situations, however, where the law imposes a positive duty to speak. In these cases silence can constitute misrepresentation. These situations are:

(a) Where one party tells a half truth but remains silent about some fact which distorts the positive statement made. An example is where a prospective purchaser asked about the farms on an estate, enquiring whether all the farms were let. The vendor replied that they were (a positive statement) but failed to go on to say that all the tenant farmers had given notice to quit, (silence distorting the positive representation) (*Dimmock v. Hallett*).

(b) Where the relationship between the parties is fiduciary (a relationship of trust) e.g. solicitor and client, or banker and customer. If one then makes a contract with the other, there is a positive duty to disclose known material facts. This rule is significant when considering undue influence.

(c) Where the contract is one involving utmost good faith. The commonest example of such a contract is insurance. Here the facts are all within the knowledge of one party. On the strength of what he tells the other, that other must decide whether or not to accept the risk and how much to charge as a premium. The law imposes a positive duty in these cases to reveal all material facts. Insurance contracts are discussed in greater detail on page 152.

4 The false statement must actually help to induce the contract. There will be many occasions when false statements are made, but they do not act as an inducement. This may be because the other party does not hear the statement, or does not believe it, or is not in any way influenced by it, but makes his decision to enter the contract for quite separate reasons.

Where all the relevant features of misrepresentation can be proved, the injured party will want to know if he can escape from the contract, and also whether he has any claim for damages. The answer depends largely on the type of misrepresentation involved. Was it a statement made fraudulently, negligently, or innocently? Once the maker's frame of mind is established, the remedies can be considered

Fraudulent misrepresentation

This is defined as a false statement made knowingly, without belief in its truth, or recklessly, careless whether it is true or false. In the very limited number of cases where this can be proved, the behaviour of the person making the false statement would amount to the tort of deceit. Unfortunately, it is very difficult to convince a court that a statement falls into this category. As one judge commented: 'A charge of fraud is such a terrible thing to bring against a man that it cannot be maintained unless it is shown that he has a wicked mind'.

Negligent misrepresentation

This category of misrepresentation was introduced by the Misrepresentation Act 1967, and is one of the factors which has effected a significant improvement in this area of the law. A statement is negligent if it was made in circumstances where the maker had no reasonable grounds for believing it to be true. He may have thought that he was making a true statement but the court will evaluate whether the basis for him so believing was reasonable. This may be a difficult question to resolve, and much may be at stake. An important commercial case, *Howard Marine and Dredging Co. Ltd v. Ogden & Sons Ltd* will illustrate the problem.

In order to tender for a large excavation contract with a water authority, Ogdens needed to hire or buy barges to move spoil out to sea. They approched Howard Marine and had extensive discussions and negotiations with Howard's manager. He thought that two particular barges would be suitable for Ogden's needs, and he told Ogdens that each barge could carry 1600 tonnes. He was basing his statement on his recollection of information contained in *Lloyds Register*. In the company's files about the barges, the manager could have discovered that the *Lloyds Register* figure was wrong. The actual carrying capacity of each barge was nearer 1000 tonnes. Ogdens were successful in getting the excavation contract, and they immediately hired the two barges from Howards. As soon as they began using them they realized the vast discrepancy between the manager's statement and the true carrying capacity. Ogden's refused to pay any further hire charges for the barges and wanted to claim damages from Howards. Obviously the court had to classify the statement made. First, they had to consider whether it was of such importance that it could be regarded as a term of the contract. Given the fact that the statement was made quite early in the negotiations, and not subsequently included in the charterparty (the written hire contract), the court found that it was not a term. What kind of misrepresentation was it? By a majority verdict only, the Court of Appeal found that this was a negligent misrepresentation. One of the judges thought that it was reasonable for the manager to rely on *Lloyds Register*, which was normally famous for its detail and accuracy; but the other two judges thought that the manager should have referred to, and checked in, the authoritative file on each barge.

Liability for negligent misrepresentation under s.2 of the 1967 Act can only arise if the statement causes the other party to enter into a contract. There is considerable overlap in this area with the law of tort, and negligent statements are a problem area. Just before the 1967 Act was passed the courts had an opportunity to consider what liability, if any, should arise from negligent statements. In the case of *Hedley Byrne v. Heller* the facts involved a company who wanted assurances about the financial stability of a prospective customer. They asked the customer if it was in order to approach the customer's bank for a credit reference. The bank negligently informed the company that the customer was financially sound. Subsequently, the customer went into liquidation, and the company suffered losses in excess of £15,000. As they were unlikely to get their money from the customer, the company was anxious to sue the bank. On these facts, it is clear that the bank's statement had not induced any contract between the bank and the company, so s.2 of the 1967 Act could not cover such a case. That section requires that a person must have entered a contract. Could the company recover the damages against the bank in the tort of negligence? The law on negligence had not developed at that stage to cover such mis-statements causing pure financial loss, but the House of Lords in the Hedley Byrne case showed themselves ready, in appropriate circumstances, to develop the law and award damages.

In the mid-1960s, this created a sudden new wealth of possibilities where a person was a victim of a negligent statement. It has become common to refer to negligent statements which result in a contract as negligent misrepresentations, and those which do not result in a contract as negligent mis-statements. While it may be to a plaintiff's advantage to have more possible causes of action, these twin developments in respect of negligent statements make the law more complex than it need be. See page 229 for further treatment of negligent misstatements in the law of tort.

Innocent misrepresentation

Since the 1967 Act, this is limited to those cases where the maker of the statement honestly believed that his statement was true *and* he had reasonable grounds for believing it to be true.

Remedies for misrepresentation

Once it is clear which type of misrepresentation is involved, the next problem is to establish the appropriate remedy. All types of misrepresentation have the same effect on the contract, which is to make the contract voidable at the option of the party misled. This means that the injured party can seek to rescind the contract, to have it set aside as if it had never existed. The parties would then be restored to their original positions. This is known as the remedy of rescission, and in some circumstances it can be achieved on a self-help basis without any court assistance or intervention. Inevitably such help will often be required, especially where the party at fault will not cooperate. There will be times when a court is unable to order rescission. The remedy is discretionary and will not be available in the following circumstances:

1 Where the court is unable to restore the parties to their original position. This could be the case where goods sold have been

consumed, or inextricably mixed with other goods.
2 Where the injured party has affirmed the contract. This means that the injured party is fully aware of the false statement but has shown himself prepared to go on with the contract.
3 Where the injured party has delayed too long in seeking rescission. There is no precisely defined time limit, but a long delay once all the facts are known may in itself be an affirmation of the contract.
4 Where third party rights have intervened.
5 Where the court decides to exercise its discretion under the 1967 Act to award damages in lieu of rescission.

Damages

When considering the availability of damages for misrepresentation, it must be remembered that the false statement in question may have induced the contract, but it does not form part of the contract obligations. If the statement was regarded as sufficiently important by the parties to form part of the contractual obligations, it will be classified as a term. If it then turns out to be false the appropriate remedies for breach of contract will be available.

As misrepresentations stand outside the contract, it follows that there can be no damages available in the law of contract. Where the statement is fraudulent, damages have always been available in the tort of deceit. Prior to the recognition of negligent misrepresentation, a misrepresentation made without fraud was innocent, for which no damages were available. Now, if a plaintiff can prove negligent misrepresentation under s.2, the court has power to award damages. It seems clear that these damages will be assessed on a tort basis rather than a contract basis. *Royscot Trust Ltd v. Maidenhead Honda Centre Ltd* is a recent decision showing a most favourable application of s.2 in the assessment of damages. Where the claim is for negligent misstatement within the Hedley Byrne principle, damages for the tort of negligence are available. The difference between tort damages and contract damages is significant, as it could result in quite different calculations. Tort damages are designed to put the plaintiff into the position he would have been in but for the tort; contract damages are designed to put the plaintiff into the position he would have been in if the contract had been properly performed.

Damages in lieu of rescission

Even if a false statement has been made, the remedy of rescission is sometimes far too extreme to produce a fair result. If I sell you my car, and tell you that it has done 30,000 miles, and you subsequently discover it has done 31,000 miles, it can often be quite satisfactory to award damages in recognition of the false statement. It may seem to give undue protection to the purchaser to allow him to back out altogether, although this will often be a question of degree. Under s.2(2) of the 1967 Act, the court has a discretion to award damages in lieu of rescission for negligent and innocent misrepresentation.

Limiting or excluding liability for misrepresentation

In the same way as it is possible to limit or exclude liability for obligations which would otherwise arise under a contract (see page 96) it is also possible to limit or exclude liability for misrepresentations which induce a contract. The relevant rules are contained in s.3 of the Misrepresentation Act 1967 which has been substantially amended by the Unfair Contract Terms Act 1977. Section 3 permits exclusion of liability for false statements (negligent or innocent, but not fraudulent), provided that the clause satisfies the requirement of reasonableness (see page 99).

Mistake

As a broad rule, English law admits few mistakes as affecting the validity of a contract. Often, when one party says: 'I made a mistake', what he really means is: 'I failed to watch out to make sure I was making a good bargain'. In such cases, the maxim *'caveat emptor'* will apply – let the buyer beware.

In some instances where the parties are mistaken, no contract ever comes into existence between them. The extent of their misunderstanding is so great that it prevents them reaching true agreement. In one case the parties believed they had an agreement relating to a cargo of cotton coming to this country on a ship called *The Peerless* which was sailing from Bombay. By coincidence there were two ships, both called *The Peerless*, both carrying cotton and sailing from Bombay. As it was impossible to say which ship the parties had in mind, it was held that they had never reached a true agreement (*Raffles v. Wichelhaus*).

Sometimes, the state of confusion is caused by one person seeking deliberately to lead the other into making a mistake. This can occur where one person misrepresents his identity. Such fraudulent behaviour would be likely to give rise to remedies for misrepresentation. But in cases where it can be shown that the identity of the person was crucial, the mistake about identity may be so important that there is no real agreement. The contract would be void.

Other factors affecting validity and enforceability

A party seeking to enforce a contract may be unable to do so in cases where it was induced by duress or undue influence, or in cases where there is some illegality. These issues will be considered separately.

Duress

It is hard to imagine that any contracts are made nowadays as a result of threats of physical violence. The old rules about duress would only permit relief if the contract came about because of threatened or actual violence. Originally the result of such behaviour was that the contract was void i.e. there was no contract at all. Subsequently, the rule was modified. Now, if such facts occur, the injured party could seek to have the contract set aside. The modern effect of duress is thus to render the contract voidable.

These rules are of such limited significance that they would not be worth mentioning if it were not for the more recent trend in the cases towards establishing principles relating to economic duress. The cases which have developed this theme tend to be complex, but the usual scenario is that one party threatens the other with breach in order to gain some new concessions from the other. For example, one party sees that he has not negotiated a particularly good price for building a ship, so he threatens to abandon the contract unless the price is renegotiated. This will put the other party into an impossible position if he has already contracted to charter the new ship. Unless he agrees to the higher price he will find himself in turn in breach with his charterer. He could resort to litigation against the shipbuilder, but that is costly, uncertain and very time-consuming. The potential loss of business and goodwill may be so great that he reluctantly agrees to pay the higher price. At a later date, would a court allow him to recover the extra sum paid?

The courts have not yet developed a particularly coherent set of principles to solve such a problem. There must be more than mere commercial pressure. To establish economic duress, it is necessary to show that there has been such 'coercion of will' that there is no true consent to the new deal. Factors which the court will take into account were set out in a case, *Pao On v. Lau Yiu Long*, 1979:

1. Did the victim protest?
2. Did the victim have any other course of action open to him, e.g. could he have sued?
3. Did the victim receive independent advice?
4. Did the victim take immediate steps to try to avoid the new deal?

This developing area is one which is likely to cause considerable difficulty for the courts, as these situations may arise between businessmen who are of apparently equal bargaining strength, where the courts have a traditional reluctance to interfere with the contract as agreed between the parties. However, the argument has succeeded in a recent case, *Atlas Express Ltd v. Kafco Ltd*, 1989.

Undue influence

Rules of undue influence were developed in equity because of the narrowness and harshness of the old common law rules about duress. Where there is behaviour falling short of such open violence or threats of harm, equity may grant relief if a contract has been brought about by undue influence. There is no precise definition of this phrase, but it usually occurs where there is a dominant/subservient relationship between two people and the dominant party seeks to take advantage of, or exploit, the other.

Undue influence is automatically presumed to exist in some relationships, e.g. solicitor and client, or parent and child. In other instances it is not presumed but may be proved to exist. Where a contract is highly favourable to the dominant party it may be avoidable on grounds of undue influence. The courts will look to see whether the dominant party took advantage of the weaker party, or abused his own position. If so, the contract may be set aside where the transaction is to the manifest disadvantage of the weaker party.

Some of the cases in this area involve the big banks, where they have obtained signatures on guarantees and mortgages from relatives or debtors, often visiting them in their homes and giving them no chance to receive independent advice. The court must weigh very carefully whether such a transaction was only agreed to because of the bank bringing undue influence to bear. *National Westminster Bank Ltd v. Morgan* and *Midland Bank Ltd v. Shepherd* are recent useful illustrations of these rules in practice.

Illegal contracts

The phrase 'illegal contracts' is used here to cover those situations where contracts are either void or illegal at common law or by statute. A wide range of situations is thus encompassed, for which the law offers numerous solutions. Sometimes the illegality involves an immoral contract, such as a contract to commit a crime, or to pervert the course of justice. On some occasions Acts of Parliament designate particular types of contract as unlawful, for example gaming and wagering contracts. In these cases the results of making such a contract will frequently be governed by the Act.

There is a general power available to the courts to interfere with contracts where it can be seen that such contracts are contrary to public policy. Such a potentially wide-ranging doctrine must be applied with caution, because the judges must not forget that the parties are largely free to fix their own bargain. Inevitably the definition of 'public policy' must remain somewhat flexible as it will need to shift and change to reflect the changing needs of society. One judge remarked that public policy was 'a very unruly horse and once you get astride it, you can never know where it will carry you'.

Where contracts involve some considerable element of moral wrong-doing, it has been usual to describe them as illegal at common law. Examples would be contracts promoting sexual immorality. In these cases, neither party will be able to enforce the contract. But if one party is the 'less guilty' of the two, he may be able to recover money he has paid or property he has transferred. This will particularly be the case where one party can be seen to have clearly 'repented'.

Where less moral wrongdoing is involved, for example in contracts in restraint of trade, the approach of the courts is to declare void either the contract as a whole or its offending parts. A contract in restraint of trade is one whereby one party promises or agrees to restrict his individual freedom to contract. These clauses are frequently found in employer–employee contracts or in contracts for the sale of a business. They are regarded as important because so many disputes in this area are litigated in the courts.

It is useful first of all to establish why such clauses are regarded as contrary to public policy. Surely, one might argue, if the two parties agree, that should be the end of the matter. That would be to take a pure freedom of contract approach. However, viewed from the point of view of the public interest, it is seen as wrong and contrary to public policy that a person's skill and experience should be rendered effectively useless and sterile by a contract. The courts have, therefore, been active in controlling those contracts in restraint of trade which they regard as being unreasonable. The modern rule is that such an agreement will only be enforced if it is reasonable in the interests of the public and reasonable in the interests of the parties.

When judging what is reasonable, each case will tend to turn on its own individual facts, so precedents are of little value other than as examples. The courts have always made clear that it is not appropriate to incorporate a restraint of trade in every contract of employment. The employer cannot include such a clause merely to prevent competition. He may only restrain an employee when he has some legitimate business interest to protect. That interest is usually his trade secrets or his business connections i.e. his customers. Take the case of an employee who is engaged as a travelling salesman within a twenty-five mile radius of London. It may be permissible to include a clause in his contract that when he leaves his present employment he will not solicit any of the employer's customers within that twenty-five mile radius for a period of one year. It would clearly not be reasonable to restrict the salesman throughout the UK, as he only poses a threat to the employer within a very small area.

When evaluating the reasonableness of a restraint clause the courts will pay close attention to factors such as the area of the restraint and the time limit involved. The restraint should be no wider and no longer than is strictly necessary to protect the employer's business interests.

It will be clear that there is no precise science about clauses in restraint of trade. An employer who wishes to put such a clause into an employee's contract will have to tread carefully in his drafting to make sure that it goes no further in the restrictions it imposes than will be found acceptable by the courts. If the employer 'gets it wrong' in his drafting, he will pay a high price as the clause will be held to be void and will be struck from the contract. The only help the courts will give is to construe a restraint of trade clause in the light of the circumstances existing when the parties made their contract, in order to determine what they must have intended. By interpretation of the clause, it may be possible to limit in scope what otherwise appears to be too far-reaching. An example of this approach can be seen in *Home Counties Dairies Ltd v. Skilton*, where a milk roundsman, once he left his present employers, was to be restrained from selling milk or dairy produce for a period of six months. The phrase 'dairy produce' was so wide that, on a literal interpretation, the man could not even go to work in a supermarket selling yoghurt. The court held that the phrase had to be construed in the light of what the parties must have intended, and should therefore be more confined in its meaning.

The courts tend to concentrate on the issue of reasonableness as between the parties, but it should be borne in mind that rules of public policy are in issue, and public interest is always a factor which could sway a court's decision. This can clearly be seen in the case, *Pharmaceutical Society of Great Britain v. Dickson*. The society intended to change the rules for its members, many of whom were owners of small chemists' shops. As a result they would be severely restricted by their professional rules from dealing in non-traditional lines. For example, their shops would have to stop selling goods like paperback books or cassettes. For many small chemists these were high profit lines which effectively kept the business going. One chemist sought a declaration that such a change in the rules would be an unreasonable restraint of trade. The House of Lords clearly thought so and considered public interest to be a paramount consideration. Such a change in the rules could well have caused many small chemists' shops to close and thus deprived large sections of the public of an important amenity and public service.

Over the years the courts have been vigilant in finding restraint of trade clauses hidden in some unusual situations. In one case, the rules of a pension fund were held to operate as an unreasonable restraint (*Bull v. Pitney Bowes*). The principles have also been applied in cases involving licences for racehorse trainers (*Nagle v. Fielden*) and the transfer system for footballers (*Eastham v. Newcastle United*). These cases clearly demonstrate that the courts will not allow the parties to achieve through a back door what they could not do by a direct method.

Where the courts are prepared to interfere and find a restraint unreasonable it is void and struck from the contract. Where it is only one clause of a much larger contract, the remainder of the contract may be perfectly valid and enforceable. In some cases, especially borderline situations, the inevitable effect of declaring a restraint to be void is that the employer is left with no protection for his legitimate business interests. An example of this can be seen in *Commercial Plastics Ltd v. Vincent*, where a research chemist had been working on PVC but only on a limited aspect of its application. The restraint in his contract was drafted to cover too general a field of work on PVC and so the whole restraint was held to be void. The judges felt sure that the employer had genuine interests to protect, and with a properly drafted clause would have secured such protection.

This rigorous approach of the courts stems from the principle that the parties must make their own terms and cannot expect the judges to do this for them. In one limited circumstance the judges may be able to give some assistance. Where the restraint is couched as a series of alternatives, some of which are valid but some of which are too wide and thus unreasonable, the court may be able to sever parts which are excessive. Consider a restraint for 'twelve months within a ten mile radius of our shops at Newcastle or York'. If the employee has never worked in the York shop he probably poses no threat to the employer there and that aspect of the restraint could therefore be too wide and unreasonable. Using an approach known as 'the blue pencil test' the courts will simply cross out the words 'or York' and leave a valid restraint behind. Of course, this will be no help in a great many cases where the restraint consists of one overwide restriction e.g. a restraint for 'twelve months within 100 miles of our shop in York'. The judges have no power here to cut this down in scope if it should turn out to be too wide to be reasonable.

It is obvious that the rules about restraint are not clear or precise. The main value of putting a restraint clause into a contract is its deterrent effect. It acts as an early warning to an employee that he will not get away with interfering with his employer's interest. But even if no restraint is imposed, or a restraint has been struck out as unreasonable, it should be remembered that employees owe duties to their employers at common law under their contracts of service (see page 178).

Discharge of the contract

Discharge by performance

In the normal course of events, both parties to a contract expect that the contract will be properly carried out. Where both parties perform their obligations under the contract exactly and precisely as promised, then the contract is discharged by its complete performance. It follows that if either party does not exactly and precisely perform, there is a breach of contract. Contractual obligations arise by agreement, and it is right that both parties should carry out their obligations in full. A necessary result of this rule is that a person who only partly performs his obligations cannot be paid. This is clearly demonstrated in a case involving the installation of central heating. The job was done and the customer had not paid. But the heating failed to give the promised temperatures in the house and the boiler emitted vile fumes. As the installer had not exactly and precisely performed, it was held that he was not entitled to be paid (*Bolton v. Mahadeva*).

That case does demonstrate how harshly the rule of exact and precise performance can operate. Not surprisingly it is subject to a number of exceptions.

1 A party may be paid for partial performance where he has substantially performed the contract. So, if one party has performed exept for some trivial details, he will be entitled to be paid, less the value of the other party's counter-claim in respect of those trivial breaches.
2 A party may be paid where the other party is prepared to accept part performance. In these cases, however, the other party must have a genuine choice whether to accept part performance or not. So where I am due to deliver ten tons of coal and I arrive with only eight tons, you may agree to accept eight tons. But in the central heating case, the plaintiff had no such choice. The work was only partly done but it could not be undone. Of course, in those cases

where the other party does agree to accept part performance, his obligation to pay will be correspondingly adjusted.
3 A party who has partially performed may be paid as much as the work or performance is worth in cases where the other party will not permit him to complete his performance. In this case, of course, the person who will not permit performance is in breach, and the party who has partly performed may either sue for damages in respect of that breach, or claim a reasonable sum for the value of the work he has already done.
4 Where a contract is divisible, a party who had performed some parts may be paid for them. So, in a contract of employment with salary paid monthly, each complete month could be regarded as a divisible contract. In building contracts, any agreement about progress payments will make it a divisible contract.

Although the rules about precise performance are quite strict, there are, nevertheless, some defences which are regarded as acceptable excuses for non-performance or partial performance. For instance, the parties may have agreed to some lesser performance. The validity of their agreement will be the subject of rules about discharge by agreement (see page 116).

Or again, the parties may be unable to perform their contract as originally contemplated because of some external factor or event outside their control. The validity of such an excuse is considered under discharge by frustration (see page 117).

The parties themselves may have taken account of the need to change the rules about performance e.g. in a building contract more time may be needed if a contractor unearths antiquities, and their contract may make provision for altered performance.

The time for performance of the contract

It is often said that 'time is of the essence' in a contract, meaning that the time stipulated for performance is a major term, and any breach of that term would entitle the other to repudiate the contract. This will often be true, but each case will turn on its own facts. Have the parties expressly provided for a time for performance? If not, then the contract must be performed in a reasonable time. If a precise time has been fixed, the surrounding circumstances may make it clear that it must be adhered to. If I order a wedding dress to be completed in time for my wedding, time is clearly of the essence. But if I order a new armchair and you undertake to deliver it next week, it may not be so disastrous if you do not make delivery on time. Inevitably, in that situation, the time may come where the patience of the customer starts to wear thin. Then, although time was not originally of the essence in the contract, the customer can make it so by serving notice on the seller. The customer may indicate that, unless he has the chair by the end of the week, he regards the contract as at an end.

Tender of performance

Sometimes it is only possible to perform the obligations under a contract with the cooperation of the other party. Take the example of a sale of goods where the seller agrees to deliver. He needs the other party to be there to receive the goods. If he offers to perform, i.e. tenders performance, by taking the right amount of goods of the correct description and quality, and the other party will not accept them, then the seller can argue that he has done all he can, and he can then consider himself discharged from his obligations. In *Startup v. McDonald*, a delivery of oil was due to be made before 31 March. The seller arrived with the oil on 31 March, which happened to be a Saturday. It was very late in the evening, and the buyer refused to accept the delivery. This was held to be a valid tender of performance so the seller could recover damages for the non-acceptance of the goods.

Where the obligation under the contract is to pay money, there is only a valid tender if money of the exact amount is offered, made up in such a way as to constitute 'legal tender'. Of course it is quite common for the parties to have agreed expressly to payment in some other way e.g. by cheque. If no such agreement exists, the other party is entitled to expect money.

Discharge by agreement

What is created by agreement can be discharged by agreement. Although this statement is largely true, discharging a contract by agreement depends on the operation of the rules of consideration. Effectively, if the parties agree to discharge a contract they are making a new contract. That new contract will be supported by valid consideration if each party is giving up something of value by discharging the first contract. If I agree to sell you goods for £10, and I have not yet delivered and you have not yet paid, then we can discharge that contract by a new agreement. My consideration for the new agreement will be giving up the right to receive £10 and your consideration will be giving up the right to receive the goods.

Obviously, it is not so straightforward to achieve a discharge by agreement if one party has already performed. If I have already delivered the goods, my part of the bargain is complete. It would be possible to discharge our agreement by drawing up a deed, but this would be absurd to contemplate except in commercial cases involving very large sums of money. In these cases, the rules about waiver play an important part. In the example already given, I may agree to forego payment of the £10 due. What would prevent me from changing my mind at a later stage and suing for the money? It could be argued that I am prevented from going back on my earlier promise because of the defence of promissory estoppel (see page 86). As this is an entirely discretionary principle, the other party may feel less than secure in relying on it. I would also be prevented from suing for my £10 if we have made a valid variation of our original contract.

Discharge by frustration

When two parties make a contract their natural expectation is that both of them will perform it. However, in some cases, one party may find it impossible to perform his part of the bargain due to events quite beyond his control. The law of contract basically adopts a harsh principle that if contractual obligations are undertaken, then any non-performance for whatever reason amounts to breach.

However, since the nineteenth century, the courts have developed rules relating to 'frustration' or subsequent impossibility, in an attempt to achieve justice between the parties when things go wrong in an unforeseen way. If one party chooses not to perform because he now regards the contract as a bad deal he will be in breach. If the parties themselves have provided an express term to cover some particular event, then the terms of their agreement will take effect. The rules developed by the courts provide that if there has been a frustrating event it will have the effect of terminating the contract and discharging the parties. The resulting financial arrangements between the parties will then be dealt with, when appropriate, by the Law Reform (Frustrated Contracts) Act 1943.

The first problem is the difficulty of establishing which events should be regarded as frustrating events. Take as an example the closure of the Suez Canal. If my ship is locked in the canal as a result of the closure, then any contract I have made to hire the ship to you would surely be frustrated. But if I had a contract to transport your goods, and the closure of the canal simply means that I must use a longer and more costly route, then this is by no means certain to be found to frustrate the contract. In such a case my major complaint would be that I am not going to make a profit on our contract. This factor alone will not influence the court. Contracts inevitably involve risk, and the parties must look out for their own interests. *Davis Contractors Ltd v. Fareham UDC* is a good example of this point. In July 1946 the contractors had agreed to build seventy-eight houses for £92,000, the work to be completed within eight months. Owing to quite unexpected circumstances, and through no fault of either the contractor or the council, there was a serious labour shortage and necessary building materials were extremely difficult to obtain. As a result the job took almost two years to complete, and the contractor incurred extra expense of £17,500. He wanted to argue that the very long delay had frustrated the original contract, and as that would avoid the original contract, he could then claim payment for a reasonable sum. The House of Lords rejected the contractor's argument. One of the judges in the case said: 'The possibility of enough labour and materials not being available was before their eyes and could have been made the subject of special contractual stipulation. Frustration is not to be lightly invoked as the dissolvent of a contract.'

Examples of situations in which the court is prepared to find frustration include:

1 *A forecast event fails to take place.* A classic example was the cancellation of the coronation of Edward VII. Many people had made contracts to hire rooms to view the coronation procession. In one such case, *Krell v. Henry*, the contract for the hire of a room overlooking Pall Mall was held to be frustrated by the cancellation of the coronation. It is evident in a case like this that although the hirer could still use the room on the day in question, it is impossible for the parties to achieve the substantial object of their contract. There is an interesting contrast to be made with the case of *Herne Bay Steamship Co. v. Hutton*, where the contract indirectly arose out of the coronation activities. The Royal Naval Review was to take place at Spithead, and the plaintiff chartered a boat to cruise around the assembled foreign fleets in the harbour and to see the naval review. Because of the King's illness the review had to be cancelled. But the contract for the charter of the boat was held not to be frustrated, because it was still possible to enjoy a day's sailing, and see all the foreign fleets. In other words, the substantial object of the contract had not been destroyed.
2 *The subject matter of the contract is physically destroyed.* In *Taylor v. Caldwell* the contract provided for the hire of a music hall which burned down before the date for performance of the contract. The fire was accidental and the contract was held to be frustrated. A similar rule for goods which perish after the contract is made but before the date for delivery is contained in s.7 of the Sale of Goods

Act 1979. The contract would be frustrated provided neither party was at fault. The same principle also applies where the person who was to perform the contract dies, or becomes too ill to perform. This rule is only appropriate where the contract calls for the services of one particular person. So, if I hire a famous concert pianist to give a recital, then death or illness of the pianist could frustrate the contract. Where there is an ordinary on-going contract of employment, short-term illnesses would not usually frustrate the contract. There will inevitably be borderline cases, where a long-term illness rendering the employee unfit to perform his former duties may amount to frustration. This sort of example should not give rise to too many difficulties when it is remembered that such an employee's contract could be terminated by giving appropriate notice (see page 188).

3 *Subsequent intervention makes performance of the contract illegal.* This situation can occur when parties have made a perfectly valid contract, but before they perform it some change in the law makes that type of contract unlawful. So, in *Metropolitan Water Board v. Dick Kerr & Co.*, the defendants had undertaken to build a reservoir for the Water Board. They were already at work when they were ordered to stop by the government, which directed them to build a factory instead. The court held that the contract with the Water Board was frustrated. It is true that the defendants could have resumed work on the reservoir very much later, but because of changing costs of materials and labour this would have then been a substantially different task from the one originally undertaken. Obviously the reservoir example can be compared with the council house case (Davis) discussed earlier. In the one case the rules of frustration operated and the defendant was released from what had become a bad bargain. In the other case, Davis Contractors were held to their original contract. How can such a distinction be justified? It seems that in most cases the courts will abide by the traditional view that you must perform as you agreed, but there will be exceptional situations where the increased cost of performance is so great that it can be said to make the contract into something radically different from what was contemplated by the parties. No doubt this allows the courts to find frustration in extreme cases, but it does demonstrate how difficult it is to pin down the scope of the doctrine with any precision.

4 *Frustration affecting contracts concerning land.* It used to be thought that the rules of frustration would not operate in the case of leases or contracts for the sale of land. Two important cases emphasize how significant the rules could be in such instances. In *National*

Carriers v. Panalpina Ltd a warehouse was let on a ten year lease. With five years of the lease still to run, the street giving access to the warehouse was lawfully closed off by the local authority because there was a listed building in a dangerous condition. It could only be demolished with the consent of the appropriate minister, which took almost two years to obtain. During all that time the tenant was unable to use the premises for warehousing as no lorries could gain access. He stopped paying the rent and claimed frustration of the lease. The House of Lords rejected his claim. One judge said: 'Under the bargain the land has passed from the lessor to the lessee with all its advantages and disadvantages. Why should justice require that a useless site be returned to the lessor rather than remain the property of the lessee?' In *Amalgamated Investment Co. Ltd v. John Walker & Sons Ltd* the facts involved the purchase of a piece of land which the purchaser intended to redevelop. He contracted to pay £1.7 million for the land. On the day after he signed the contract, officials at the Department of the Environment listed a building which was presently on the site as a building of special interest. This made it extremely difficult to go ahead with development plans, and the site without development potential was worth only £250,000. Despite the scale of the financial disaster the purchaser was held to be bound by the contract. Although it was now apparent that he had made a bad bargain, the contract was not frustrated.

From the cases and examples we have looked at, it is possible to state only a very broad principle. Frustration occurs when performance of the contract is impossible or when the parties can no longer achieve the substantial object they had in mind. It is essential that the supervening event should be outside the control of either of the parties. It it comes about through some act or election of one party, or as a result of negligence, then it cannot be frustration. Both of these points are well illustrated in *Lauritzen A. S. v. Wijsmuller B. V.* (the *Super Servant Two*), where the defendants had contracted to transport a drilling rig from the Japanese shipyard where it was being built to Rotterdam. The contract provided that the carriage would be undertaken between June 20 and August 20, and would be effected using either *Super Servant One* or *Super Servant Two*, both of which were large transportation units. In fact, it was always the intention of the defendants to use *Super Servant Two* to fulfil the contract but, unfortunately, it sank in January. In the meantime, the defendant had made other contracts to use *Super Servant One*, so that vessel was not available to be substituted. The defendants argued that the sinking of the *Super Servant Two* had frustrated the contract. The Court of Appeal held that this was not a supervening event such as to frustrate the contract, because the defendants could have performed without breach by using *Super Servant One*, and it was by their own election that they chose not to do so.

Inevitably, frustration often occurs after the parties have embarked on performance. Their valid contract becomes void at the moment of the frustrating event, and the parties are therefore discharged from further obligations under it. This can produce a very untidy and unfair situation, as it means that obligations which fell due before the frustrating event are still owed. Take the example of a contract to import and sell £10,000 worth of a particular type of toy, delivery to take place in November, with £2,000 payable on the making of the contract, and the remaining £8,000 to be paid within one month of delivery. The contract is made in June and £2,000 is duly paid. In October the government bans the import of the toy in question. At that point the contract is frustrated. But the buyer has paid £2,000 for which he has received nothing. The early approach of the courts to sorting out these problems was to decide that the loss lay where it fell. In an attempt to solve the financial plight of the two parties to a frustrated contract, the Law Reform (Frustrated Contracts) Act 1943 was passed.

The Act makes no attempt to define what frustration is. It simply deals with the consequences. Its two main objectives are to allow for the recovery of money paid and for compensation for partial performance. These will be considered separately.

1 *Recovery of money paid* – the Act provides that all sums paid shall be recoverable and all sums due cease to be due. Applying this to the sale of toys example, the £2,000 deposit could be recovered and the £8,000 would cease to be due. Of course, this may not be particularly fair to the seller, who may have incurred considerable expense in going about the performance of the contract. The Act recognizes this by providing that where expenses have been incurred before the frustrating event, and those expenses were 'in, or for, the purpose of performing the contract', the court may allow a party to recover in respect of those expenses, to the extent that the court thinks is just and equitable. In the toy example, the seller may have made phone calls and sent telexes and may have had transport expenses and insurance costs. The judge may say that he can keep £500 of the £2,000 deposit. Now the buyer is £500 out of pocket. That is an inevitable price which has to be paid when a transaction has gone wrong. The rules of frustration only operate where neither party is at fault, so inevitably, the courts are trying to make the best

of a bad situation and no solution will ever be perfect. They can only aim to achieve as fair a compromise as possible.

2 *Recovery of compensation for partial performance* – we have already seen that a frustrated contract will have commenced life as a valid contract, under which the parties may have started to perform their obligations. Suppose I have contracted to build you a house for £60,000 payable on completion. If I have completed three-quarters of the work when the government interferes and puts a ban on private building work, can I recover the value of the work I have done? The Act provides that if one party obtains a valuable benefit, by reason of work done in the performance of the contract by the other, the court may allow the party who has performed to recover a sum which does not exceed the value of the benefit conferred, and which is just in all the circumstances of the case.

The scale of disaster when such frustrating events occur can be seen in *BP Exploration Co. Ltd v. Hunt*. Hunt owned an oil concession in Libya. He did not know if oil would be found there or, if so, how much. He wanted, somehow, to finance the exploration of the oil field but to hold on to a share of any profits if large quantities of oil were found. He contracted with BP to share the field. BP would bear all the exploration costs and, if oil was found, Hunt would repay them from his share of the oil revenues. The exploration work cost a staggering $87 million. A large amount of oil was found and extraction had only just begun when the Libyan government confiscated the land and nationalized the concession. BP had done work in performance. Had that work conferred a valuable benefit on Hunt? Merely exploring for oil gave Hunt no valuable benefit, but its discovery greatly increased the value of his concession before it was nationalized. The judge calculated Hunt's valuable benefit to be the value of the oil already removed and the compensation he would receive from the Libyan government. This formed the upper limit of what he could have awarded to BP. He then had to consider what figure, up to that upper limit, he thought was just in all the circumstances. Although the judge thought that the 'valuable benefit' to Hunt was over $80 million, he awarded $35 million to BP, as it had been an agreed basis of the contract that it would bear the exploration costs. The exploration might have found no oil at all, and in that case it would have had to bear those costs in total with no claim at all against Hunt.

The 1943 Act may not apply if the parties themselves have included a term to cover the results of a frustrating event. It will be a question of construing or interpreting their term, to discover if they intended it to operate in such extreme cases. The rules of the Act have no application to contracts of insurance, so if I insure against sickness, pay a year's premium and then die after the first month, there can be no recovery of part of the premium. In some sale of goods situations the appropriate rules are found in s.7 of the Sale of Goods Act 1979, rather than the 1943 Act.

Discharge by breach

A breach of contract may occur because one party does not perform his obligations at all, or performs then late, or performs them in an unsatisfactory way. One of the major difficulties when looking at discharge by breach of contract is trying to establish exactly which breaches have the effect of discharging. Where such a major breach can be identified, the injured party may nonetheless want to go ahead with the contract, and simply ignore the breach, or more likely, seek compensation for it.

All breaches of contract, however large or small, give the injured party the right to sue for damages. In some cases it may be possible to exercise that right through set-off. Where a painter contracts to decorate the outside of your house for £500, to be paid on completion of the work, he may commit breaches by not burning off where specified, or by applying only one undercoat instead of two. In such a case, the injured party could solve the problem by deducting something from the £500 due. In some cases a contract may even provide a schedule of damages for specified breaches, thereby anticipating those which are likely to occur and providing a solution. Such an example is the case where a builder undertakes to complete a job by 30 June, and if he fails to do so, to pay £100 per day (agreed or liquidated damages) in respect of each day's delay thereafter.

A much more difficult question still remains – when does a breach put an end to the contract itself, or permit the innocent party to choose to treat the contract as at an end? And what exactly is meant when we say the contract is 'at an end'? As a general rule, where the party at fault is in breach of a major term (a condition) or in breach of an innominate term where the effects are serious, the innocent party can treat that behaviour as repudiatory, and regard himself as discharged from further performance. Where the breach is of a minor term (a warranty) the injured party is restricted to an action for damages and must continued with performance of the contract.

The various categories of terms have already been examined (see page 92). Let us assume that when the breach occurs it is clear which type of term has been broken. Take the case where you want to ship goods under refrigeration. You enquire whether the ship has refrigeration plant and are assured that it has. On that basis you contract to ship ten tons of soft fruit under refrigerated conditions. When you arrive to load the fruit, you discover that the ship had no refrigeration equipment. This is likely to be a breach of a major term, a condition. You now have a choice. You can treat it as a repudiatory breach and regard yourself as discharged from further performance, i.e. the obligation to pay the shipping charges. You can also sue for damages in such a case. No doubt you would take this course of action if you were able to find alternative shipping facilities immediately, or if you could sell the fruit in a local market. Where that is not possible, you may decide that you have no option but to ship your fruit in the unrefrigerated ship, thereby choosing to affirm the contract, but of course, you can still sue for damages for the breach. As you have affirmed, you have to pay the shipping charges, but if your fruit deteriorates as a result of the breach you can be compensated by the damages.

This example assumes that, even though the breach was major, you were left with a choice of courses of action, either to affirm and go on, or consider the contract at an end. Sometimes that will not be the case, as the very nature of the breach itself may put an end to the contract. Take the case where you engage a central heating contractor to install a system for you for £2,000, the work to commence on 1 June. If the contractor informs you in May that he has found more lucrative contracts and does not intend to come to do your job, he is effectively repudiating the contract in advance. You could now wait until June to see if he comes. If he fails to turn up then, this is a case of breach by non-performance. Or you could sue immediately, making clear that you are choosing to treat the contract as at an end. Where these situations occur in practice, the difficulty may be that the injured party needs to respond quickly to minimize the harm and inconvenience, and he may therefore act precipitately without the benefit of proper advice. Yet, once it is clear that the other party does not intend to perform it might be dangerous to wait and see if he will change his mind, because in the meantime the injured party could lose his rights through some supervening frustration.

Enforcing the contract – remedies for breach

The law provides remedies for breach of contract, but the extent to which purely legal remedies can be truly effective is limited. A remedy is necessary because things have gone wrong. An ideal solution would be for the parties themselves to plan in advance for situations which may arise. They can then impose solutions which are satisfactory to both of them. This contract planning role is very significant in the business and commercial world. The standard form building contract is an example where many contingencies are provided for, without the need to go to court. One advantage of such a system is that it creates a climate within which the parties can stay on reasonably good terms, which is vital if they need to do business with each other frequently. Sometimes the contract planning role can also avoid those situations where court action would probably be useless. There are business situations where the customer is asked to pay a deposit. Not only does this create a situation where the customer is more likely to perform but it also provides a ready small sum of compensation against his non-performance. An example would be 'same-day' photographic processing, where deposits are common.

One problem of seeking to enforce a contract through the courts is that the usual remedy for breach is damages. The courts are prepared to order the other party to carry out his obligations, but only in a very limited number of cases. If they will grant a remedy, it is called specific performance. Such an order is not readily available because the court may then need to keep the defendant under constant supervision, and it has neither the desire nor the means to do so. Moreover, in many cases, money will be adequate to compensate the plaintiff. The types of contract where an order of specific performance may be granted usually relate to sales of land and leases. No two pieces of land are identical, and it is thought that money cannot adequately compensate. That is quite a difficult idea to accept when the 'land' in question may be a semi-detached house on an estate surrounded by dozens of properties which are seemingly identical! The argument about the unique quality of the property can also be used, although less frequently, in relation to sale of goods. If the goods are in some way unique, or not readily obtainable, a court may be prepared to grant an order for specific performance. Indeed they have a discretion to do so under s.52 of the Sale of Goods Act 1979.

As the power to award specific performance is equitable and discretionary, a number of points will be borne in mind by the courts. They will not grant an order if it would operate very harshly on the defendant. Nor will they grant it where a plaintiff has delayed too long in seeking relief – delay defeats equity. This rule is not precise, and what counts as excessive delay will depend on the facts of each case. The order will not be granted if constant supervision would be required, nor in cases involving personal service, as it is thought that this would be tantamount to slavery. So if I contract with a famous painter to paint my portrait and he backs out of the contract I will not get an order of specific performance. It used to be suggested that an order of specific performance would not be granted in cases involving breach of contract to pay money, as damages would surely be an adequate remedy. This may not be so, and the problem was highlighted in the case of *Beswick v. Beswick*, where a coalman sold his business to his nephew, who was to pay for it by paying the uncle £5 per week for his life and thereafter the money was to be paid to the coalman's widow. After the death of the coalman the payments to the widow were not made. She sued as the personal representative of her late husband and was awarded specific performance of the contract. This was the only way to achieve justice in this case.

There is another form of equitable relief, the injunction, which can be useful in a limited number of contract situations. This is particularly the case where a party to a contract has promised *not* to do a particular thing. A common example is a clause in restraint of trade, where an employee or a seller of a business promises not to compete. Where a breach of such a clause is proved the court may grant an injunction, but it should be remembered that the relief is entirely discretionary.

Awarding damages

In theory, an award of damages can be made in respect of every breach of contract, however large or small. The aim of the court when awarding damages is to put the injured party into the position he would have been in if the contract had been properly performed. This compensates him for loss of expectation. Somehow, the court has to arrive at an appropriate figure for the damages. This is described as the measure of damages. The figure can only be worked out once it is established for which harm the plaintiff can recover. He can only recover in respect of losses which 'arise naturally, according to the usual course of things; and losses which must have been in the contemplation of the parties when they made the contract'. This is the rule of remoteness of damage which was laid down in the case of *Hadley v. Baxendale*. The plaintiff's losses must also have been caused by the breach of contract. The two aspects of the remoteness rule can be seen at work in the case of *H. Parsons Ltd v. Uttley Ingham & Co. Ltd*, where the defendants manufactured and installed an animal feed dispensing hopper for Parsons, who was a pig breeder with a class 1 pedigree herd. The installation was carried out after the hopper had been transported by road. During the journey the driver had jammed the hopper ventilator with a rag to stop it rattling and he did not remember to remove it before the hopper was erected. Pig food was poured into the hopper and at first all seemed well. Inevitably, because the ventilator was jammed shut, the pig food eventually turned mouldy. Of course, there was a breach of contract here, and Parsons could undoubtedly claim damages in respect of the mouldy pig food. Unfortunately, the pigs had eaten the food and developed a disease which killed off a large proportion of the herd. In consequence, not only had Parsons lost more than 250 pigs, but he had also sustained the loss of a breeding season. Should he be allowed to recover in respect of the loss of the pigs? The scale of the problem can be clearly seen in this case where the defendants accepted liability with regard to the pig food and were prepared to pay damages of £1,802, but Parsons was trying to claim about £30,000. Parsons' claim was successful. The defendants must have contemplated that if they installed a defective hopper which made the food mouldy, the pigs could become ill. It was matterless that the degree of harm (the death of the pigs) was greater, as long as the type of harm (illness) could have been contemplated.

It is clear that the remoteness rules operate very flexibly, but the more the defendant knows about the plaintiff and his requirements, the more he is likely to be able to contemplate by way of potential harm.

Once the courts are clear about the harm or loss which falls within the remoteness rule they must then assess an appropriate measure of damages. Sometimes this will be quite straightforward, as in cases for the sale of goods where the seller has failed to supply. The buyer must now look elsewhere for the same goods and may only be able to buy at a higher price. The measure of damages is the difference between the two prices. Or a buyer may refuse to accept delivery of goods properly tendered under a contract. There, the measure may well be the loss of profit on the deal. Or a holidaymaker may cancel a booking for a country cottage late in the day, and the damages will be the loss of profit on the booking.

Where there has been a breach the injured party cannot just sit back and wallow in his misery. The law imposes on him a duty to take reasonable steps to mitigate his losses. A plaintiff cannot recover for any harm which could have been prevented by taking reasonable steps in mitigation. If a buyer refuses to take delivery of goods under a contract, the seller must take reasonable steps to find another buyer. In the holiday cottage example above, he must take reasonable steps to relet the property. An ironic result of taking reasonable steps to mitigate may be that the plaintiff actually increases his losses. In the holiday cottage example, he may place advertisements in newspapers and still not manage to relet the property. He can then include the cost of the advertisements in his claim for damages.

Of course, the plaintiff is the person who has been wronged so there is a limit to what he can be expected to do under the duty to mitigate.

As the defendant has committed the breach, he is hardly in a strong position to 'call the tune'. This is well illustrated in *Pilkington v. Wood*, where a solicitor gave incorrect advice to a client, in consequence of which the client found that his house had a defective legal title and was virtually unsaleable. When the solicitor was sued for damages he argued that the plaintiff had failed in his duty to mitigate as he could have solved the problem of the defective title by suing the person who had originally sold him the house. This was hardly asking the plaintiff to take a reasonable step, as such litigation would have been expensive and the outcome by no means predictable. The plaintiff was right not to have embarked on such a costly and uncertain procedure.

It used to be a feature of damages for breach of contract that nothing would be awarded for distress, injury to feelings, or disappointment. One justification for this rule may have been that it is very difficult to quantify such losses. The rule has been subject to an exception over recent years in cases where the plaintiff was contractually entitled to expect pleasure and enjoyment, not pain and distress! However, the sums awarded for distress and disappointment have not tended to be large. *Jarvis v. Swans Tours Ltd* shows the sort of case where an award for disappointment may be appropriate. The plaintiff had booked a holiday for himself and his family on the strength of a brochure which claimed that he would be joining a house party and special resident host. He would enjoy a welcome party and special yodelling evening, plenty of gemutlichkeit (good atmosphere) and, generally, he would be 'in for a great time'. In fact, by the second week of his holiday there were no other guests at all and in every other respect the holiday failed to live up to the brochure claims. Even the yodeller was a local man singing in his working clothes. When awarding him damages for disappointment the judge said: 'Mr Jarvis has only a fortnight's holiday each year. He books it far ahead and looks forward to it all that time. He ought to be compensated for the loss of it.' The court awarded him £125.

It has already been emphasized that by effective contract planning, the parties can provide in their contract for the sums to be paid in the event of specified breaches, a common example being late performance. Such contractually determined damages are called liquidated damages. The advantages of such a clause are clear. Problems can be resolved quickly and privately, without resort to the courts and, hopefully, in a manner which will preserve good relations between the parties. The parties also know in advance the risk and cost involved in particular breaches. On the other hand, such clauses can cause problems, particularly where the party agreeing to the clause lacks bargaining strength. In that case the result of the clause might be unfavourable to him. Either part may miscalculate the real effects of a specified breach and thus be unhappy to be regulated by what turns out to be an unfair clause. Against that, of course, must be set the fact that such clauses are freely negotiated, at least in theory!

The courts themselves are alert to the possibility that such a liquidated damages clause could operate harshly. If there is a dispute about the clause it is open to the court to consider whether it represents a genuine attempt by the parties to estimate their losses, in which case the clause will be enforced even if the estimate is wrong; or a clause inserted by one party to frighten or terrorize the other into performance. In the latter case the clause is classed as a penalty clause and is unenforceable. It is matterless what name the parties give to their clause. The court will determine which category it falls into, using a number of fairly flexible tests:

1 Does the clause provide for the same amount to be paid in respect of different breaches, some of which may be serious while others are trivial? If so, it will be hard to show that the clause was inserted as a genuine pre-estimate.
2 Does the clause provide for damage for breaches involving nonpayment of money at a level far higher than the sum unpaid? If you are due to pay £25 per week and I insert a clause saying that for every late payment you must pay me £100 that looks like a penalty.
3 Does the sum provided for by the clause appear to be harsh and unconscionable, given the breaches which are likely to occur?

Suing on a quantum meruit

Quantum meruit means 'as much as it is worth'. Generally, the rules about *quantum meruit* have no particular significance in the law of contract, because the contract itself should fix the amount of any payment due or the method of calculating it. In some cases, however, even the law of contract provides for the payment of a reasonable price. This could happen where minors have been supplied with necessaries. Another example is where goods have been sold and delivered, and accepted, but no price has been fixed. (s.8, Sale of Goods Act 1979).

There are times when the parties may think that they have a valid contract. They may have acted in reliance on that belief in performing obligations, only to find that the contract has not taken effect. In such cases if one party confers a benefit on the other, in circumstances where both believe that the benefit will be paid for, then the courts will allow the performer to recover a reasonable sum under the *quantum meruit* rules. This turned out to be a useful solution in *British Steel*

Corporation v. Cleveland Bridge and Engineering Co. Ltd, where the parties planned to enter into a contract but it had not been finalized between them. In ancipation of that contract one party began to perform. When a contract failed to materialize the performer was able to be compensated for the work done on a *quantum meruit* basis.

Limitation periods

The remedy of damages, as the principal relief for breach of contract, must be sought in time, otherwise any claim is statute barred by the Limitation Act 1980. The imposition of time limits is sensible because evidence eventually becomes stale and unreliable and a party at fault must ultimately have some certainty about whether he will be sued or not. In claims for breach of contract the time limit is basically fixed at six years from the date when the cause of action accrued. That usually means six years from the date of the breach of contract. The six year rule is extended to twelve years where the contract was made by deed under seal.

These rules are fairly rigid except where a plaintiff fails to discover that he has a cause of action because of the defendant's fraud. In these cases time only begins to run when the plaintiff discovers the fraud, or when he could with reasonable diligence have discovered it. 'Fraud' for these purposes could include cases where a defendant has deliberately concealed relevant facts from the plaintiff. This is what happened in *Applegate v. Moss* where a builder constructed a house on foundations which did not comply with the contract terms. The house was finished in 1957 but it was 1965 before the defects began to manifest themselves. The house was declared unsafe for habitation. Although eight years had elapsed, and normally the plaintiffs would have been statute barred, the time was extended to run from the date when they learnt that the foundations were defective because of the builder's concealment of facts relevant to the plaintiff's claim – he had covered the foundations up before they could discover how inadequate they were.

These strict time limits do not apply to equitable remedies such as specific performance. But in exercising any discretionary equitable relief, the court works on the maxim that delay defeats equity. It will be a question of fact in each case whether a plaintiff has unduly delayed.

6
Special contracts

Agency

The agency relationship lies at the heart of commercial and business life. Whether it is the sales assistant or the bus conductor, the travel agent or the auctioneer, all are acting as agents. The relationship of principal and agent arises whenever the principal authorizes an agent to make contracts on his behalf. Generally, there will be a contract between the principal and the agent, with the latter being paid for his services, but it is possible for the agent to act gratuitously.

Agency is usually created by express or implied agreement, but agreement may be lacking in cases of agency of necessity, or where the agent binds the principal under his apparent authority (see page 132). The usual rule is that the agent may only bind the principal in contracts which are within the scope of his authority.

Where the agency has been expressly created, whether orally or in writing, the scope of the authority may have been clearly established. More often, however, it is a question of implying whatever authority is customary in the circumstances. So, a sales assistant in a department store would have authority to sell the merchandise in the store, but is unlikely to be authorized to sell the shop premises.

Where an agent is appointed for a particular purpose, he will also have authority to carry out any necessary incidental acts. Where the agency arises as part of a trade, profession or business, the agent has implied authority to perform such acts as are usual in that trade, profession or business. As a general rule, an estate agent has no implied authority to take a deposit in respect of the property to be sold, nor has he the right to sign a contract of sale.

Agency of necessity

Agency of necessity arises where someone acts to protect the interests of another, without any prior authority. In one example, where goods were rescued from a stranded ship, the rescuer then stored them in a warehouse to prevent further deterioration. He was able to recover the costs of warehousing from the owner of the goods. Similarly, where the owner of a horse failed to collect it promptly from a railway station, the railway company could recover the costs of feeding and stabling the animal.

Of course, in cases such as these, an agency of necessity will only be appropriate if the agent acts reasonably, and in good faith, in circumstances where it was impossible to obtain the owner's instructions. The actions of the agent must be for the benefit of the owner and not for the convenience of the agent. The courts are naturally anxious about imposing liabilities in such cases. Indeed, one judge remarked that 'liabilities are not to be forced upon people behind their backs'.

Apparent authority

It is clear that an agent can always bind the principal if he makes a contract of a type for which he has actual authority. Additionally, the acts of an agent will bind the principal if they are within the scope of the agent's apparent authority. Simply, this means that an agent may have exceeded his real authority, but in the eyes of the third party with whom he is negotiating, there was nothing to raise any doubts about the agent's authority to act and to bind the principal. In other words, there was the appearance of authority. This is sometimes regarded as an example of the doctrine of estoppel.

There are a number of ways in which such apparent authority can arise. For example, the principal may allow the agent to create an impression that he has wider authority than he actually possesses. Take the case of a firm of solicitors which has five partners. All partners are agents of each other for all purposes, by virtue of their relationship of partnership. They decide to take in a new young partner, whose name appears alongside the original five on the notepaper. Because the new partner is young and inexperienced, the others tell him that for the first year, he cannot buy equipment for the firm to a value greater than £100. Within the first month, the new solicitor places an order on firm's notepaper for a computer for the firm costing £800. Are the partners bound by this contract?

The answer is likely to be yes, because the computer salesman would know that he was dealing with a partner, and could therefore presume that the partner had authority to bind the firm. The young partner was acting within his apparent authority, i.e. as it would appear to a third party. If the other partners wanted to ensure that their young partner did not bind them in such a contract, they would have needed to make known the restrictions on his usual authority.

Apparent authority may also exist where a person allows someone to act as if he were an agent. In this instance, unlike the case above, the 'agent' has never been appointed as such and has never had any actual authority. In one example, a debtor came into an accounting office and paid his debt to a person he found in the office there. That person was a friend of the creditor, but was not and never had been an agent, but it would appear to a person coming into the office that he worked there and had authority to receive the money.

Another example of apparent authority occurs where a person has acted as an agent, but has had his authority totally withdrawn, e.g. where he has been dismissed. Once this happens, the principal should make this clear to third parties, and if he has not done so, the former agent may still be able to bind the principal on the basis of apparent authority.

The law is inclined to hold the principal liable for the acts of the 'agent' in these cases of apparent authority for reasons of equity and business confidence.

Architects and engineers as agents

Although architects and engineers are frequently appointed under standard conditions of engagement which lay down in detail the duties to be carried out, the conditions rarely set out the extent of the agent's authority. The scope of their agency may thus depend upon implied or apparent authority.

Ratification

In some cases where an agent has no authority to act but has done so, it is possible that the principal will subsequently ratify the transaction, and thus become liable to the third party. Ratification may be express, or it can be implied from the behaviour of the principal. However, there can be no implied ratification without full knowledge of the unauthorized act. If ratification is possible, then it puts all the parties into the same position they would have been in if the agent had been duly authorized to act all along. An undisclosed principal cannot ratify the act of an agent.

Powers of attorney

Wherever an agent needs to be given authority to execute deeds, e.g. where he is given power to sign a conveyance on the transfer of land, then the agent's authority must be conferred by deed. The deed conferring such authority is called a power of attorney.

Duties of the parties

Where the agency relationship is created by a contract, the duties of the parties will be rather similar to the common law duties between an employer and an employee. The chief duties of the principal are to pay the agent the agreed commission or remuneration, and to indemnify him against liabilities reasonably incurred when carrying out his duties.

Paying agreed commission or remuneration

Problems about when payment is due have arisen mainly in relation to estate agents' fees. This is not surprising when the various phrases used in their contracts are examined. Some contracts provide for the fee to be paid if the agent introduces a purchaser. Or the contract might provide for payment if the agent secures an offer. Sometimes the fee is only due on completion of sale. There are many variations on these themes.

The courts had always been inclined to construe these commission clauses very strictly (*contra proferentem*), because it was thought that estate agents enjoyed some advantage over their clients, by being able to impose standard terms on them. Despite the passing of the Estate Agents Act 1979, this aspect of the law, about when commission is payable, was not clarified. S.18 merely requires that the agent should give the client particulars of the circumstances in which the client will become liable to pay remuneration to the agent. Inevitably, these cases will continue to come before the courts, requiring them to construe in each case when the commission is payable.

Indemnity

The principal must indemnify the agent against all liabilities and expenses reasonably incurred in the execution of his authority. An agent would not however be entitled to an indemnity in respect of illegal transactions, or in any situation where liability is caused by his own breach of duty.

Lien

An agent has a lien (a right to hold on to goods belonging to another) over property belonging to his principal until the principal settles all outstanding dues, such as remuneration and expenses, with the agent. Often, this will be a particular lien, where it will relate only to the property in respect of which the dues have been incurred. In some cases the law grants an agent a general lien over all property of the principal. This could be so in the case of solicitors and bankers.

Duties of the agent

The agent is bound to carry out his duties with due care and skill, and the standard of care required will vary. If the agent holds himself out as being skilled in some particular field, then he must carry out the agency duties with the skill of a reasonable practitioner in that field.

The agent cannot normally delegate performance of the duties, unless the contract with his principal expressly or impliedly provides for delegation.

Whilst undertaking his duties, the agent must not put himself into a position where his own interests conflict with the duty he owes to his principal. So, for example, if he is asked to sell goods on behalf of his principal, he should not buy them himself. He would also be in breach if he agreed to act as agent for the third party too.

Strict rules apply if the agent has compromised himself by making a secret profit from the transaction. In some cases, the secret profit may amount to a bribe. Not only must the agent hand over the secret profit to the principal, but also he is liable to be dismissed for such behaviour. He would also lose the right to any commission earned on that transaction, and in some cases, could be prosecuted under the criminal law.

Relationship between principal and third party

Once the agent has created a contract between the principal and the third party, his role is effectively over. His task has been to act as an intermediary, but in the normal course of events the agent is not a party to the ultimate contract. That contract is made between the principal and the third party.

The rights of the principal to enforce the contract against the third party now depend on whether the existence of the principal was known to the third party when he was making the contract. If it was, the principal is referred to as a *disclosed* principal. If not, he is an *undisclosed* principal. If the principal was disclosed, he could then be either a *named* or an *unnamed* principal.

Generally, a disclosed principal may sue the third party. The difficulty in allowing an undisclosed principal to sue the third party is that it raises issues in relation to privity of contract. That principle basically enshrines the rule that only the parties to a contract can sue to enforce it, and a stranger to the contract cannot sue. It is accepted that in allowing an undisclosed principal to sue, effectively this is an exception to the privity rule. However, in order to be fair to the third party, an undisclosed principal cannot sue him if the identity of the principal is material to the contract, and the third party believed he was contracting with the agent. And if the undisclosed principal sues the third party, the latter can raise any defences that he would have been able to plead against the agent. Of course, where an undisclosed principal does decide to sue, his identity will then become known.

As for the liability of the principal to the third party, he is fully liable on the contract whether his position was disclosed or undisclosed. If the principal gives the agent money to settle with the third party, but the agent fails to do so, the principal is still liable to the third party.

Relationship between the agent and third party

As a general rule, once the agent has brought about a contract between the principal and the third party, his role is complete, and he should not be liable on the concluded contract, nor should he be able to sue to enforce it. His role is one of communication between the parties. However, in some exceptional situations, the agent can incur liability. This is particularly the case where he acts for an undisclosed principal. In these circumstances, the third party will have thought he was contracting with the agent himself, and it would seem very unfair if he was deprived of the right to enforce against the agent, when a principal suddenly appears on the scene.

The position of the third party is enhanced in these cases, as he may elect whether he will sue the agent or the principal. Once he has made his election, however, the result is to discharge the other from liability.

Termination of the agency relationship

In many cases, the agency relationship exists within the framework of a contract, and where this is so, any method of terminating a contract will terminate the agency relationship. So, for example, the relationship may terminate because the contract is frustrated, or because it has been completely performed, or because it is discharged by a repudiatory breach. It may also be possible to discharge the relationship by either party giving notice. Sometimes, the contract itself may provide for a particular period of notice, but where none is specified, it may be terminated by giving reasonable notice.

In two instances, however, the agent's authority may be irrevocable. The first case is where the agent's authority is coupled with an interest. An example of this could occur where the principal entrusts goods to an agent for sale. The agent advances money to the principal, who authorizes the agent to repay himself out of monies received for the goods. If the principal then tried to withdraw the agent's authority to sell, the agent would have no way to recover the money he had loaned to the principal. In such a case, his authority is said to be coupled with an interest, and would effectively be irrevocable.

In the second situation, statutory rules under the Enduring Powers of Attorney Act 1985 provide that once a power of attorney has been drawn up in appropriate form, then even if the principal subsequently becomes incapable, e.g. because of mental illness, the authority of the attorney under the power of attorney is not terminated.

The Estate Agent's Act 1979

The Act empowers the Director General of Fair Trading to prohibit an unfit person from undertaking certain or all estate agency work, as defined by the Act, or to warn him that future breaches of obligations imposed by the Act will result in prohibition. Before exercising his powers under the Act, the Director General must instigate an inquiry into the person's fitness. He may instigate such an inquiry if the person has:

- been convicted of an offence involving fraud, dishonesty or violence, an offence under the Estate Agent's Act or other specified offences;
- discriminated, sexually or racially, in the course of estate agency work;
- failed to comply with obligations imposed by the Act, or
- carried on a specified undesirable practice.

A person may appeal against the Director General's decision to issue a prohibition or warning order within twenty days of being notified of it. Appeal is made to the Secretary of State for Trade and Industry. There is a further right of appeal to the High Court, on a point of law.

The Act provides that a person who is bankrupt shall not engage in estate agency work except as the employee of another.

The Act requires estate agents to inform clients of the circumstances in which they will become liable to pay remuneration or make other payments (e.g. advertising costs and expenses). Clients must also be informed what the amount of remuneration or other payments will be or what the method of calculation will be.

The Act also requires estate agents to disclose personal interests that they have (or will have following the proposed transaction) in land, before entering into negotiation for its acquisition or disposal.

The Act lays down requirements in relation to clients' money held by an estate agent. Such money is held on trust for the person entitled to it. It cannot be used to pay off the estate agent's creditors if he becomes bankrupt. The Act requires an estate agent to keep properly audited accounts.

Clients' money must be kept in a special client account. Payments out of the account can only be made to persons and for purposes specified by the Act. For example, where an estate agent holds money belonging to the purchaser, he is entitled to make authorized payments to the vendor.

Property Misdescription Act 1991

S.1 of the Act creates an offence of making a false or misleading statement about a prescribed matter in the course of estate agency or property development business, otherwise than in providing conveyancing services. The Act provides that a statement is misleading if what a reasonable person may be expected to infer from it, or from any omission from it, is false. A statement may be made orally, in writing, by conduct, in pictures or by any other method of signifying meaning.

S.2 provides a defence where a person charged with the offence under s.1 has taken all reasonable steps and exercised all due diligence to avoid committing the offence.

Regulations made under the Act specify the matters where misdescription in the course of estate agency business or property development business will constitute an offence. The schedule lists more than thirty specified matters. Common examples include misdescriptions of location, view, fixtures, size, proximity to services or facilities, history or age of the property, amount of service charges, easements and covenants.

Sale of goods

A contract for the sale of goods is defined by the Sale of Goods Act 1979 as one where: 'The seller transfers or agrees to transfer the property in goods to the buyer for a money consideration called the price.' The Act covers a wide range of transactions from selling loaves of bread to buying cars. It must be read in conjunction with all the basic rules of contract, as the Act is not a complete code in itself. Its significance lies in the fact that it sets out:

- The terms to be implied into a sale of goods transaction.
- The extent to which those implied terms can be excluded, subject to the Unfair Contract Terms Act 1977.
- The rules for determining when ownership and risk pass from the seller to the buyer.
- The remedies for the buyer and seller where breaches of contract have occurred.

The Act does not apply to hire purchase transactions, nor to contracts of exchange (e.g. trading stamps for goods), contracts for work and materials (e.g. installing central heating), or contracts for the supply of services, where separate statutory provisions apply.

Implied terms

It is quite common for people to enter into contracts to buy goods without ever speaking to the seller e.g. when buying from a vending machine, or in a self-service store. If asked what were the terms of their contract, they might reply 'none'. However, such transactions are covered by the Act, and the following terms are automatically included for the buyer's protection.

1 *Quality of the goods*

Section 14 provides that goods sold in the course of a business must be of merchantable quality (i.e. fit to be sold), bearing in mind the description applied to them, the price paid, and other relevant circumstances. This condition does not apply if:

(a) defects were specifically drawn to the buyer's attention before the contract was made. If a buyer buys a washing machine at a reduced price, having been told that the reduction is because the cabinet is dented, he cannot subsequently try to reject the washing machine because of this damage.
(b) The buyer examined the goods before the contract was made and the defects ought to have been revealed by that examination.

One of the practical problems of applying the rule about merchantable quality is the question of how long the goods should remain merchantable. How long should a pair of shoes last? If they leak at the first wearing, clearly they would seem not to be of merchantable quality. But what if they leak after six weeks? How often did the buyer wear them in that time? In what weather conditions? Whether goods are merchantable is a question of fact and, in disputes, it would need to be answered on the idividual circumstances.

2 Fitness for purpose

Section 14 further provides that where goods are sold in the course of a business, and the buyer makes known the purpose for which he requires the goods, then the goods must be reasonably fit for their purpose. What must a buyer do to 'make known the purpose' for which he wants the goods? In most cases, where goods have one common purpose, and that is what the buyer wants to use them for, then he need do nothing in particular, as he is regarded as impliedly making known the purpose. It is only where he wants to use goods for some unusual purpose that he must make that purpose known in order to benefit from this condition. The protection of this rule about fitness for purpose will not apply if the buyer did not rely on the seller's skill and judgment. This might be shown by the buyer specifically asking for goods by their brand name. In such a case, the inference is that the buyer knows that such goods will satisfy his purpose, i.e. he is relying on his own skill and judgment.

One obvious advantage of these clear duties imposed on the seller of goods is that the buyer has rights against someone whom he can identify and locate. In practice, the seller may try to indicate that he is not liable and that the buyer will have to take the matter up with the manufacturer. The buyer could possibly have some legal action against the manufacturer as well by virtue of the tort of negligence or under the Consumer Protection Act 1987, but he should not be fobbed off by the seller in this way. The seller is under a strict liability to sell goods which are of merchantable quality and reasonably fit for their purpose.

In cases covered by s.14 these conditions are important consumer protections, and they limit the operation of the usual principal of *caveat emptor* (let the buyer beware). That principle is still significant in cases not covered by s.14 e.g. private sales of secondhand goods, because s.14 specifically states that in circumstances outside the rules set out above, there is *no* implied condition or warranty about the quality or fitness for any particular purpose of goods supplied under a contract of sale.

3 Sales by description

Section 13 provides that if goods are sold by description there is an implied condition that they must correspond with that description. There is obviously a sale by description in any case in which the buyer has not seen the goods, but they have been described to him, e.g. goods selected from a catalogue or by reference to a price list. The Act specifically states that there can still be a sale by description even where the goods are displayed and selected by the buyer. The importance of the protection under this section is that it gives a right to a buyer to reject the goods even if they are, in themselves, entirely satisfactory. For example, a buyer who orders a five gallon can of white spirit can reject the goods if the seller supplies five separate one gallon cans, even though the white spirit itself is perfect.

4 Implied terms about title

A person buying goods is naturally anxious to become the proper legal owner of them. Section 12 provides that there is an implied condition that the seller has the right to sell the goods, or will have when the time comes to pass ownership. This condition can never be excluded.

5 Sale by sample

The mere fact that a buyer sees a sample during the course of negotiations does not necessarily make the transaction a sale by sample. What is necessary is that the sale should be made by reference to that sample. The parties may well indicate this, e.g. when the buyer orders goods 'equal to sample'. In such a case s.15 then implies conditions:

(a) That the goods supplied will correspond with the sample.
(b) That the buyer should have an opportunity to compare the goods supplied with the sample.
(c) That the goods will be free from any defect rendering them unmerchantable which would not be apparent on a reasonable examination of the sample.

6 Implied terms about payment and delivery

Although the parties may not have negotiated specifically about any of the previous terms, it is likely that they will have thought about payment and delivery. This is reflected by s.10 which applies 'unless a different intention appears'.

Section 10 states that stipulation about time of payment are not 'of the essence', which means that any agreement as to when money is due does not go to the root of the contract. In practice, a buyer who fails to make payment on time can be sued for damages for non-payment, but his behaviour will not allow the whole contract to be repudiated.

As to the time of delivery, s.29 makes it clear that if no time is fixed, delivery must be within a reasonable time if the seller is obliged to send the goods. It is important to note that there is no general rule that a seller must send the goods. Whether it is for the seller to send or the buyer to collect is a question to be determined by the contract. In other words, what was expressly or impliedly agreed between the parties? In many cases, the operation of these rules will be straightforward. Suppose that a buyer purchases a deep-freeze cabinet from a shop which promises, as part of the bargain, 'free delivery within twenty miles'. The buyer lives within the twenty mile area, and agrees a day when the goods will be sent by the seller. In this case, by the terms of the contract, the seller is bound to send the goods. If he fails to send them on the appointed day the provision about the delivery date may be a major term, the breach of which would entitle the buyer to repudiate the contract. Alternatively the failure may be a breach of a minor term, merely entitling him to sue for damages. The solution will depend on the weight attached to the term by the parties themselves. Their contract may make it clear that delivery dates are vital and must be complied with.

Unless otherwise agreed between the parties, delivery of the goods and payment of the price are concurrent conditions (s.28). It is very common practice to exclude the operation of this rule, e.g. in cases where goods are bought on account and a bill is rendered at the end of the month.

Special contracts 143

Exclusion of implied terms

Notwithstanding the extensive 'writing-in' of terms into contracts for the sale of goods, the Sale of Goods Act (SGA) provides by s.55 that any or all of these implied terms may be excluded, subject to the Unfair Contract Terms Act 1977 (UCTA). When considering the relevance of UCTA to sale of goods transactions, a distinction must be drawn between those buyers who 'deal as consumer' and those who do not. The UCTA identifies dealing as a consumer occurring when:

1. A person dealing as a consumer *does not* make the contract in the course of business.
2. The seller *does* make the contract in the course of a business.
3. The sale is of goods of a type ordinarily supplied for private use or consumption (UCTA 1977, s.12).

So, a housewife buying a loaf of bread in a baker's shop deals as a consumer. A contractor buying supplies from a builders' merchant is an example of a buyer not dealing as a consumer.

One party dealing as consumer (UCTA 1977, s.6)

The rule in such cases is that there can be no exclusion of any of the terms implied by SGA 1979, ss. 12-15 (i.e. no exclusion of the terms relating to title, sales by description, merchantable quality, reasonable fitness for purpose or sales by sample). Any attempt by a seller to exclude or limit his liability in these areas will be ineffective.

Neither party dealing as consumer (UCTA 1977, s.6)

Although no exclusion or limitation of SGA 1979, s.12 (title to the goods) is allowed, the seller may limit or exclude his liability under SGA 1979, ss. 13-15, if the contract term by which he imposes the limitation satisfies the requirement of reasonableness, i.e. if, given what the parties knew or ought to have known when they made the contract, it was a fair and reasonable term to have included. UCTA 1977 contains guidelines for the application of the reasonableness test, which suggest that a court should consider:

- The relative strength of the bargaining position of each party.
- Whether the buyer received an inducement to agree to the term.
- Whether the buyer knew, or ought reasonably to have known, of the existence of the term.
- Whether the goods were manufactured, processed or adapted to the special order of the buyer.

144 Building and Land Management Law for Students

Passing ownership and risk

A contract for the sale of goods is defined by the Sale of Goods Act 1979 as one where the seller transfers or agrees to transfer the property in goods for a money consideration called the price (s.2). In this definition the word 'property' is used to mean ownership. It is important to realize that ownership can pass at a different time from possession. When making the contract the parties could agree to ownership passing immediately, but no delivery of the goods (i.e. transfer of possession) may take place until a later date. It is important to establish the exact time at which the buyer becomes the owner because:

1 The buyer can only effectively transfer ownership to a third party after he has become the owner.
2 As a general rule, risk passes with the ownership, irrespective of whether the goods have been delivered. This means that if goods should be insured, the buyer should do so, despite the fact that he has not yet received them. If the goods are lost or stolen at this stage, the loss will fall on the buyer. But the seller who remains in possession is under a duty to take reasonable care of the goods.

The parties are free to fix for themselves when ownership will pass but if their contract does not deal with this point, SGA 1979 provides rules to cover the various situations which can arise. The following points should be noted:

1 The time when ownership can pass depends on how the law categorizes the goods in question. If they are unascertained (goods mixed with others and not yet separated or identified as fulfilling the buyer's contract) no property can pass until they become ascertained (goods identified in accordance with the contract after the contract is made). For example, if a builder orders a quantity of timber from a timber merchant who has huge stocks, the goods are unascertained until the merchant selects timber of the correct quantity and specification. Once he makes the selection, the goods become ascertained. Goods may also be specific. These are goods identified and agreed upon by the parties at the time when the contract is made.
2 Ownership of specific or ascertained goods passes at the time agreed by the parties. If they do not make their intention clear, the rules of SGA 1979, s.18 will be applied. These can be summarized as follows:

(a) In the case of a sale of specific goods which are fit for their purpose and ready to be delivered, ownership passes when the contract is made.
(b) When specific goods have to be put into a deliverable state (e.g. they must be repaired, cleaned, tested or weighed), ownership passes when this has been done, and the buyer has notice to that effect.
(c) In a sale of unascertained goods ownership passes once appropriate goods are set aside in fulfilment of the particular contract.

It will be seen that it is better for the parties themselves to fix when ownership will pass, as the operation of these rules leaves a buyer with little control over the timing.

Recently sellers have tried to lay down more precise rules in the contracts, to try to retain ownership of goods until the buyer has paid for them. This can be particularly useful when a buyer becomes insolvent, as the seller can insist that he still owns the goods in question and can take them back. Otherwise he would simply rank as an ordinary creditor in a bankruptcy or liquidation, with little hope of being paid in full. Reserving ownership of goods which are now in the buyer's possession is not without practical problems. One of the first significant cases on this point was *Aluminium Industrie Vaassen BV v. Romalpa Aluminium*. Indeed, it was from this case that 'retention' clauses began to be known as Romalpa clauses. AIV had sold aluminium to Romalpa, on terms that ownership would only vest in Romalpa once the goods had been paid for; that if the foil was used in manufacture, the end product would belong to AIV; and that if this end product was then sold, the proceeds would be due to AIV. This clause was held to be valid and was enforced by the court. Their reasoning was based on the fact that Romalpa were not owners but merely bailees under a bailment of the foil. A bailment is a fiduciary relationship, a relationship of trust which would allow the bailor to recover his goods, and also give him the right to trace the proceeds of sale.

This case appears to hold out important advantages for retention of ownership clauses, but it must be viewed in the context of later decisions, where the court reached a different conclusion. In *Borden UK Ltd v. Scottish Timber Products Ltd*, the material supplied under the contract was resin to be used in the manufacture of chipboard. An 'ownership retention clause' was inserted. When Scottish Timber went into receivership, Borden wanted to trace their resin (which was now in manufactured chipboard) and trace the proceeds of sale of chipboard which contained their product. They failed on both counts. There could be no bailment here as a bailment implies a right of redelivery. Borden were supplying small quantities at short, regular intervals and must have realized that their product was being incorporated immediately into the manufacturing process. If the clause was intended to create some kind of charge over the chipboard, it needed to be registered, and it was not.

The difference between the cases seems to rest on the fact that no mixture took place in Romalpa, whereas that was inevitable in Borden. Romalpa clauses could still be effective, as long as the seller makes clear that the buyer is a bailee, and no mixing occurs.

Remedies

The remedies available for breach depend on which party is in breach, and what type of breach has been committed:

1 *Seller in breach* – where the seller commits a major breach, e.g. by supplying goods which are defective, and not therefore of merchantable quality, the buyer ordinarily has the right to repudiate the contract (put an end to it), usually by rejecting goods. He also has the right to sue for damages. In practice, the buyer may choose not to reject the goods, and indeed in some cases the Act specifies that he cannot reject them. This is likely to happen when the buyer has purchased specific goods (goods which were identified by both parties at the time when the contract was made), where he had a reasonable chance to inspect the goods, and where he has now accepted them. He may have indicated his acceptance by selling part of the consignment of goods to a third party. In these cases the right to reject is lost, and the aggrieved buyer must content himself with an action for damages.

2 *Buyer in breach* – the most likely situation is that the buyer has failed to pay for the goods. The seller's remedy will then depend on whether he is still in possession of the goods. If so, he can exercise his right of lien over the goods (i.e. holding on to them) with a view to ultimately exercising his right to resell them. If delivery has already been made to the buyer, then the seller must sue for the price of the goods.

Supply of goods and services

New provisions were enacted in the Supply of Goods and Services Act 1982 (SGSA), relating to contracts under which ownership in goods will pass, where the contract also provides a service element. The Act also regulates contracts for services only. Obvious examples of contracts falling within the Act are builders who supply materials and do the work and garages who supply parts and undertake the repair. Less obviously, the supply of goods provision may be important where a 'free gift' comes with a contract: 'Buy a luxury fridge/freezer during January and get this electric coffeemaker free.' The coffeemaker is not a 'sale of goods' because there is no money consideration given for it. But it is a supply of goods.

The Act deals separately with the supply of goods and the supply of services. Where the goods are supplied under a work and materials contract, a number of terms are implied:

1 An implied condition that the seller has a right to transfer the ownership of the goods (s.2).
2 Where goods are transferred by description, there is an implied condition that the goods must correspond with their description (s.3).
3 The goods transferred must be of merchantable quality, and reasonably fit for their purpose (s.4). These conditions apply only where the goods transferred are supplied in the course of business.

These conditions are couched in terms very similar to those used in the Sale of Goods Act 1979. Cases interpreting and applying those rules will no doubt be helpful in clarifying the 1982 SGSA provisions. Very similar terms are incorporated by ss. 7-9 in contracts for the hire of goods. These are significant nowadays, as hiring cars, office machinery and heavy plant is very common in business.

As usual, the possibility of excluding these implied terms is limited by the operation of the Unfair Contract Terms Act 1977. Generally in consumer transactions these protective conditions cannot be excluded, but in business transactions an exclusion clause may be acceptable if it satisfies the requirement of reasonableness under the 1977 Act (see page 99).

Supply of services

The 1982 SGSA does not define what is meant by a service, except to make clear that the services provided by professions such as architects and surveyors are included. Obvious examples are activities like car servicing, boiler maintenance, dry cleaning and watch repairing. In any of these cases the customer will feel aggrieved if the job is not done effectively, if it costs too much or if it takes too long. In essence, those are the three areas where the Act now prescribes implied terms.

Exercising reasonable care and skill

Section 13 states that if a supplier of a service is acting in the course of business there is an implied term that he must carry out the service with reasonable care and skill. One interesting feature of this section is that it does not prescribe whether the term is a condition (major term) or a warranty (minor term). The standard laid down is apparently imprecise, but it means that a builder must exercise the care and skill of a reasonably competent builder, a surveyor must exercise the care and skill of a reasonably competent surveyor, and so on. The standard is thus set objectively. It is no use asking whether the builder did his best, or even whether he tried reasonably hard given his limitations! He must be measured against a mythical, reasonably competent builder. The courts will not find this new provision difficult to apply as s.13 merely codifies the existing common law rules, i.e. similar duties would previously have been implied into such contracts by the courts. As soon as the idea of 'reasonable care and skill' is introduced it is clear that there is the possibility of a claim in the tort of negligence too. Usually it is more straightforward to sue in contract, as the relevant duty is owed by virtue of the contract and need not be specifically proved. The singular advantage of codifying the 'reasonable care and skill' requirement in s.13 rather than relying on a judicially implied term, is that it makes the position between the parties much clearer and more certain.

The time for performance

Section 14 provides that there is an implied term that a supplier acting in the course of business, where no time is fixed by the contract, must perform the service within a reasonable time. Establishing what is a reasonable time is a question of fact, which can sometimes be implied from the nature of the contract. If a firm is engaged to come and spray fruit trees when they are in blossom there is only a short span of time within which the contract can be performed.

The cost of the service

Section 15 provides that where the contract does not fix the charge or provide a method of fixing it, the customer must pay a reasonable charge. What is reasonable is a question of fact.

It is possible to exclude the terms implied by these sections, but again, only within the limits provided by the Unfair Contract Terms Act 1977. It should particularly be remembered that where, by negligence, a person causes death or personal injury, he cannot exclude or limit his liability by a contract term (s.2).

Hire purchase, credit sales and conditional sales

It is often the case that a buyer can only afford to purchase goods if payment can be spread over a period. Historically, hire purchase developed at the end of the nineteenth century as a means of allowing people to buy more expensive durable goods like pianos and sewing machines by spreading payments out over a period, and yet achieving a measure of protection for the seller who remained the owner of the goods until the last payment was made. If a buyer fell into arrears with payments, the seller could 'snatch back' his goods. Where a buyer found that he had overcommitted himself financially, he could usually terminate the agreement, but often on terms which were disadvantageous to him. Recognizing these evils, and the potential inequality of bargaining power between the buyer and the seller in such transactions, Parliament passed a series of Hire Purchase Acts, the rules from which are now consolidated in the Consumer Credit Act 1974.

A number of different ways exist for a customer to buy on credit. Traditionally the law provided several types of transaction, which give flexibility about the terms incorporated and which vary in the protections afforded to the customer. The supplier could choose to use:

1. A hire purchase agreement, where the customer hires goods for a period, with the opportunity to exercise an option to purchase at the end of the period. In this type of agreement, the supplier remains the owner of the goods until the option is exercised. The hirer may terminate the hiring agreement without exercising the option, although this would not usually be wise, as it would prove to be a costly way to hire goods. As the goods remain in the ownership of the supplier he has certain rights to seize back the goods if the hirer defaults on payment. In this type of transaction, terms about merchantable quality and reasonable fitness for

purpose are implied in a form similar to the implied terms of the Sale of Goods Act 1979. HP agreements are controlled either by the Consumer Credit Act 1974 or by rules of common law.

2 A credit sale agreement. These extend credit to the buyer of goods e.g. where he arranges to pay by instalments, but the ownership in the goods passes to the buyer immediately. They are governed mainly by the Sale of Goods Act 1979.

3 A conditional sale agreement, where ownership of the goods remains vested in the seller, but the buyer has no right to terminate the agreement. Because credit is extended to the customer, the Consumer Credit Act 1974 may apply.

Of course, these types of contract are not the only ways in which credit may be given. There has been a dramatic rise in the use of credit cards – either those issued by particular stores for use in their own shops, or those issued by credit card companies for use in any accredited outlet of the company. In either of these cases the sum owed can represent many different contracts made by the debtor. Sometimes, by the terms of his agreement with the credit card company, the debtor must repay at regular monthly intervals and may not exceed a fixed credit limit. In other cases the debtor may owe up to a fixed credit limit and may choose whether to settle in full at the end of the month, in which case he pays no interest, or he may pay part of what he owes and run up interest on the balance.

Inevitably, most organizations which are prepared to extend credit do so only for a return. That return is the interest they charge. It may be thought that evil moneylenders are a thing of the past, but 'loan sharks' are their modern equivalent. The law has long believed that rules are needed to protect people from extortionate bargains, and such rules are all the more vital when seen as part of 'consumer protection'. The rules were dramatically altered by the introduction of the Consumer Credit Act 1974. As its name applies, the Act is of special significance to 'consumers', and the protections of the Act range from licensing people and organizations which offer credit to insisting on sufficiently full information being available, so that a customer can establish the cost of the credit. To give the Act 'real teeth' there are criminal as well as civil sanctions.

For the businessman making contracts to purchase supplies or equipment the Act may be of limited significance. Its main provisions only operate where the credit extended is more than £50 but less than £15,000 and the customer is not a corporation. For a company or an individual seeking credit of more than £15,000 the ordinary common law rules apply, the most important of which is *caveat emptor* (let the buyer beware)!

Where the Act does apply there are some very important benefits for consumers, not least that the organization which extends credit, e.g. the finance house, is liable in respect of breaches of the implied terms with regard to merchantable quality and fitness for the purpose. This could be particularly relevant where the actual supplier of goods is not worth proceeding against.

One aspect of the Act is relevant to all credit agreements, with no financial limit. This is the power of the court to examine credit agreements which are extortionate, and seek to do justice between the parties. Section 38 of the Act specifies some factors which can be taken into account when determining whether the credit bargain is extortionate. These include the prevailing interest rate; the degree of risk accepted by the creditor; the creditor's relationship to the debtor; the debtor's age, experience, business capacity, and state of health; and the existence of any financial pressure on the debtor. If the bargain is found to be extortionate the court can adjust the rights and duties between the parties.

Other methods of consumer protection

Many of the protections discussed in this chapter can only be fully effective if a person is prepared to bring a court action to enforce his rights. But the public is often in fear of the cost of litigation and, especially with smaller cases, there can be a feeling that the time, trouble and effort involved is disproportionate to the size of the claim. An interesting trend in consumer protection is towards state enforcement of duties. Examples of this can be seen in rules such as those covering food and drugs and weights and measures, which create numerous offences for which prosecutions can be brought. A further development is the Trade Descriptions Act 1968, which makes it an offence to apply a false trade description to goods. A common example of an offence under this Act is that of the car salesman who falsifies the recorded mileage of a car by interfering with the instrument. If a prosecution is brought against him the court has power not only to fine the salesman, but also to make him pay compensation to the buyer. From the buyer's point of view the advantage of this is that someone else will take the initiative in bringing court proceedings. The drawback, however, is that compensation may not be nearly so accurately assessed as damages would be by a civil court. Moreover, the buyer may not want money, and might prefer to repudiate the contract and hand back the goods.

The individual's best consumer protection comes from dealing with a reputable trader with high standards who wishes to maintain his reputation. In an effort to improve trading standards generally the Office of Fair Trading was established in 1973, under the control of the Director General of Fair Trading. The Office encourages groups of traders to prepare and abide by codes of practice, and generally seeks to discourage unfair trading practices.

Insurance contracts

When a businessman considers the risks inherent in running his business (e.g. that his workmen may injure third parties, the workmen themselves may be injured, his business premises may take fire, a customer may be injured by a defective product) it takes little imagination to realize that potential claims against him could be enormous. It is true that some organizations are sufficiently large to bear their own risks. This tends to be the case only where the insurance premiums charged are so high that these large organizations find it more economical to meet claims, if and when they arise, out of their own funds. In all normal cases a firm would be wise to secure insurance cover, and in some circumstances the law makes insurance compulsory, e.g. employer's liability and motor insurance.

Commercial insurance is subdivided into fire, life, accident and marine. It is common to speak of life assurance, while using the word insurance in relation to the other classes. The difference is said to be that with assurance, the risk (i.e. death) is bound to happen sooner or later, whereas insurance covers risks which are foreseeable but which may never occur, e.g. a fire or an accident. In practice, the difference is not important.

Two vital principles are fundamental to all insurance contracts.

1 Utmost good faith

A person who seeks insurance usually knows all the relevant facts. On the basis of what he reveals the insurers can decide whether to accept the risk (i.e. give insurance cover) and how much to charge by way of premium. Because of this inequality in the bargaining position of the parties the law imposes a duty on a person seeking insurance to disclose all the material facts to the insurer. The duty of utmost good faith (*uberrimae fidei*) only requires disclosure of facts which are material and which are known or ought to have been known by the insured.

(a) *When are facts material?* The answer seems to be when they would influence a prudent insurer as to whether or not he will accept the risk, or as to the amount of the premium. With motor insurance previous motoring convictions are obviously material. In the case of most types of insurance the fact that another insurer has refused the risk is material. In life insurance previous illnesses may be material.

(b) *Knowledge of the facts*. The duty on the insured is quite strict. He must reveal facts he knows or is presumed to know. He cannot avoid this duty by deliberately keeping himself in ignorance. If the facts within the knowledge of the insured actually reduce the risk, or are facts which are common knowledge so that the insurer would be presumed to know them too, then no disclosure is necessary.

These rules about disclosure recognize the possibility that an insured may be silent despite being under a duty to speak. They are an exception to the general contract rule that silence does not usually amount to misrepresentation. This is normally the case because of the principle of *caveat emptor* (let the buyer beware). Usually it is up to the parties to negotiate their own bargain and if one of them requires information, he must ask the other for it. The law does not generally require the other party to volunteer it. Of course, a person seeking insurance may have made positive statements which turn out to be false. In such a case the normal rules of misrepresentation would operate and the insurance company's remedy would depend on the type of misrepresentation, which can be innocent, negligent or fraudulent.

When there has been non-disclosure of a material fact the remedy of the insurer is to avoid the policy. It is irrelevant that the insurer has not asked a specific question on the proposal form. This rule can sometimes operate very harshly in practice. In one case, *Woolcott v. Sun Alliance of London Insurance*, the plaintiff did not complete the proposal form himself. He was buying a house with the aid of a mortgage from a building society. His mortgage application form indicated that the building society would insure the property. The house was subsequently damaged by fire. When the plaintiff sought to claim on the block policy held by the building society the insurers avoided the policy and refused to pay. The grounds for their refusal were that the plaintiff had been convicted for robbery in 1960 and sentenced to twelve years in prison. They regarded this as a material fact which had not been disclosed to them.

2 Insurable interest

If people could insure goods or lives in which they had no legitimate interest the result would be a wager or gambling agreement. It is a necessary ingredient of an insurance contract that the insured must have an insurable interest, i.e. some foreseeable financial loss or liability.

(a) *Life insurance* – it is accepted that everyone has an insurable interest in his own life and a person may insure himself against death or personal accident to an unlimited extent, so long as he can pay the premiums. In certain circumstances a person may have an insurable interest in the life of someone else, e.g. a wife has an unlimited insurable interest in her husband and vice versa; a creditor has an insurable interest in the life of his debtor, up to the amount of the debt. Parties to civil proceedings in the courts have an insurable interest in the life of the judge who is hearing their case, as they stand to lose large sums of money if he dies and the hearing has to begin again in front of a new judge.

(b) *Property insurance* – the owner of property has an unlimited insurable interest up to its full value. In other cases, where a person is not the owner, the law allows insurance to the extent of that person's interest or liability. For example, a building society could insure a building for the amount loaned on mortgage. Where goods are concerned a buyer can insure before he takes delivery, or a warehouseman can insure other people's goods entrusted to him for storage.

(c) *Liability insurance* – where a person can foresee liability, (e.g. arising out of his occupation of premises, or use of a vehicle) he has an insurable interest. The law does not allow him to insure against criminal liability, as that would be contrary to public policy. So a motorist cannot insure against being fined for careless driving.

The insurance contract

Subject to the two principles discussed above, the usual rules about the formation of a contract apply. The following points should be noted:

1 *Agreement* – this usually results from the person seeking insurance completing a proposal form (the offer) which is then accepted by the insurers.
2 *Consideration* – insurance contracts are not usually made by deed, so consideration is necessary. This usually consists of paying a premium in return for the promise of a payment out whenever the risk occurs.
3 *Form of the contract* – as a general rule contracts of insurance do not need to be in writing. An oral contract could be valid and

enforceable. In practice it is usual to record the terms of the agreement in a policy. Should any dispute arise over the risks which are covered the words of the policy must be construed (i.e. interpreted) to see whether they cover the facts of the case. It is usually said that the words of a policy should be read so as to give effect to the intention of the parties. This suggests that both play a part in drawing up the policy, whereas the truth is that it will usually be a standard form contract drawn up by the insurers. Mindful of this, the courts will construe any ambiguity against the insurers. It should also be remembered that the Unfair Contract Terms Act 1977 does not apply to insurance contracts. In theory, therefore, insurers are free to limit or exclude their liability.

4   *Terms of the contract* – one feature of insurance contracts is that all terms, major or minor, tend to be called warranties. This is particularly confusing because of other meanings which the word can carry, e.g. in the sense of a guarantee. In any contract situation it is always a difficult question to determine whether a statement which has been made is so important that it becomes part of the contract (i.e. becomes a term), or is of lesser importance so that it is merely a representation. A person seeking insurance may make a number of statements in answer to questions on the proposal form, e.g. stating that he has no motoring convictions, or that he has never had an operation. Insurers have been in the habit of making such statements form the basis of the contract, by including a note on the proposal form (which is then signed by the insured) to the effect that all the answers will be regarded as warranties. This practice puts the insurers into a most powerful position in relation to the insured. Should there be anything wrongly answered on the form, then, even if the insured has been acting honestly, and even if it is immaterial, an insurer can avoid liability on the whole contract by treating the error as a breach of warranty. This tactic has resulted in one judge commenting: 'I wish I could adequately warn the public against such practices on the part of the insurance offices.' Insurance contracts are not controlled by the Unfair Contract Terms Act 1977, but insurers do now adhere to a 'Statement of Insurance Practice' which resolves some of these difficulties. Although this 'Statement' was issued following consultations and discussions with the Department of Trade it is not legally binding upon insurers. Some changes in the law may be expected, since the Law Commission has already made recommendations with regard to non-disclosure and breaches of warranty. The 'Statement' is limited in its scope to private consumers. This seems in line with the thinking behind the Unfair Contract Terms

Act 1977, in which protection is given mainly to persons dealing as consumers. The insurance industry has been sufficiently worried about customer complaints that it has cooperated to appoint an Insurance Ombudsman, who will look into customers' grievances.

5 *Duration of the policy* – often an insurance policy runs for a fixed period, usually a year. There are cases where insurance relates to a specific event (e.g. bad weather insurance for a country show, or insurance against twins being born) and the protection lapses once the date of the event has passed. Where the policy is of a continuing type e.g. life assurance, the insurer must go on accepting the premiums even though the risk inevitably becomes greater year by year. With other types of policy, e.g. motor insurance, the expectation is that it will be renewed, but neither party is obliged to make a fresh contract. If renewal does occur, then the duty to disclose material facts becomes relevant again, because each renewal creates a fresh contract.

Types of insurance

1 Employers' liability insurance

Although sensible employers have always tended to have insurance cover for protection against claims by injured workmen, problems have arisen in the past when some employers had no such insurance, and were unable to meet claims for damages from their employees because they had become insolvent. Although an employee injured in an accident at work may have claims under the Industrial Injuries Scheme, the law also provides that he can claim damages from a negligent employer. That claim is now protected by the requirement of compulsory insurance imposed by the Employer's Liability (Compulsory Insurance) Act 1969. The following main points about the Act should be noted:

(a) All employers (except local authorities, statutory corporations and nationalized industries) are covered by the rules.
(b) Employers must maintain insurance against liability for injury or disease sustained by employees in the course of their employment.
(c) The level of cover required is £2 million in respect of claims from one or more employees arising from any one occurrence.
(d) The insurance must be made with an authorised insurer. This term is defined by regulations.
(e) The employer's duty is to maintain an 'approved policy'. The effect of this rule is to make it illegal for the insurer to limit his liability in certain ways.

(f) The rules are enforced by criminal sanctions. The Act provides for fines of up to £200 per day for failure to comply.
(g) The insurance certificate must be displayed and, if required, produced to an inspector of the Health and Safety Executive.

Limitations on the cover provided under the Act are:

(a) It only protects employees who work under a contract of service or apprenticeship.
(b) Only those incidents arising 'out of and in the course of the employment' are covered. The same phrase is used in the Industrial Injuries Scheme, where its interpretation and application has not been without difficulties.

2 Public liability insurance

The employers' liability insurance rules only oblige an employer to arrange insurance cover against injury to his employees. He may owe liability to a third party who is injured by some activity of one of his employees, e.g. if one of his labourers negligently hits a pedestrian with a plank of wood. In such a case the principle of vicarious liability means that the pedestrian is likely to bring his action for negligence against the employer. It is not compulsory for an employer to insure against this type of liability (usually called public liability insurance), although it is very common in practice. Despite not being required by law, such insurance is often specified in building contracts. For example, if a contractor is working under the JCT Standard Form Contract he is required to: 'Maintain such insurances as are necessary to cover the liability of the contractor in respect of personal injury or death arising out of the carrying out of the works'.

It is not only employers who need public liability cover. Anyone may cause death, injury or damage while going about his or her daily life. For example, a pedestrian may cause a motorist to swerve and crash, while a property-owner could be liable if his chimney were to fall on a passer-by. However, public liability cover is often provided as part of another type of policy, e.g. under a householder's policy.

3 Insurance against damage to property

This type of insurance usually covers fire damage, but will often extend to damage caused by explosion or lightning. The policy may then be further extended to cover other risks to property caused by, for example, storm, flood or impact. This type of insurance is not compulsory by law, and it is essential that a policy-holder should ensure that he has all the desired protection in his policy. For example, a policy could cover damage caused by fire, yet exclude liability if the fire was caused by explosion.

It is now quite common to extend fire insurance protection to cover consequential loss. Should a fire render a building unusable a firm's business activities could be severely disrupted. Insurance cover can be provided to cover loss of profits and additional expenses necessarily incurred, e.g. renting alternative property.

4 Motor insurance

In the case of motor insurance, certain aspects of insurance cover are compulsory. As a minimum a motorist must have a policy which covers:

- Liability for death or personal injury to third parties, including passengers.
- Hospital charges (up to a fixed maximum) in respect of the treatment of an injured party and emergency treatment fees.

It is very unusual for such limited cover to be taken out by a motorist. He would normally also want cover against:

- Damage caused to property belonging to a third party.
- Fire and theft in relation to his own vehicle.
- Damage to the vehicle and to himself.

The insurance companies do not find motor insurance altogether profitable. Premiums charged naturally vary with the risk, and most insurers will take into account the type of vehicle, its value, the age and experience of the driver, the locality in which he lives, his claims record and any driving convictions. Premiums still tend to be expensive and the high cost may be one of the reasons why some motorists fail to insure. When a third party is injured by an uninsured motorist, or a 'hit and run' driver, the victim may claim from the Motor Insurers' Bureau. Every motor insurer must belong to this organization which operates under an agreement made with the Department of Transport.

Claims under an insurance policy

Once the need to make a claim under a policy arises, the question of construction of the contract becomes important. Is the event which has occurred actually covered by the policy? Naturally the insurers will have based their premium on a fixed amount of cover to be provided, and they will not wish to pay out more than is necessary.

If the loss which has occurred is covered by the policy then, in the case of most types of insurance (except life and personal accident), the principle of indemnity operates. This means that the amount paid out by an insurer should be sufficient to cover the loss sustained, but should not allow the insured person to make a profit. However, even in those cases where the indemnity principle is applicable, the following restrictions on it should be noted:

1. The policy itself may provide for a maximum sum insured. This can provide a complete indemnity if the sum insured is adequate, but a property may have been underinsured, and the policy may contain an average clause which will then operate. The object of such a clause is to 'penalize' the person who has underinsured, since this practice is unfair from the insurer's point of view, preventing him from receiving a proper premium in return for the risk he has to bear. Of course, underinsurance is not necessarily caused by deceitful motives. A person may genuinely fail to realize that the rebuilding of his property will cost much more now because of inflation than when he purchased it some years earlier. Generally, average clauses work on a pro-rata basis. If you underinsure by 50 per cent, your claim will be settled on the basis of 50 per cent of the loss sustained. For example, a house which could cost £40,000 to rebuild at present-day prices is insured for £20,000. If it is totally destroyed by fire, £20,000 is the maximum payable. If damage amounting to £10,000 is caused (one-quarter of the true insurance value) the insured will be paid one quarter of the sum insured, i.e. £5,000.

2. The policy itself may be subject to an excess. It is quite usual for the contract to specify that the insured will bear an agreed proportion of each claim, often the first £15 or £25. In such cases, the insurance does not provide a full indemnity.

3. An insurer may not be wholly liable if the property has been insured with more than one company. In such case the contribution principle will operate, whereby other insurers who are covering the same risk share in the cost of paying out. The indemnity principle is thus preserved, as the total settlement figure will only cover the insured's loss and not allow him to make a profit. As with indemnity, the contribution principle is not appropriate in the case of life insurance. Double insurance consists of covering the same risk by more than one policy. This is not the same as reinsurance. As soon as an insurer has taken on a risk there

is always the likelihood that he may have to pay out. In cases where a single accident may be very costly e.g. the loss of a telecommunications satellite, the insurer may wish to spread the risk. In such a case, he can reinsure part or all of the risk with another insurer. If the risk then occurs, the insured claims against the original insurer in full. The original insurer then claims under his own reinsurance policy.

4   When the risk occurs, an insured person may have a right of action for damages against the person who has injured him or damaged his property, as well as the possibility of claiming under his insurance policy. If a motorist has his vehicle negligently damaged by another motorist, the subrogation principle will apply. If the victim decides to claim from his insurers they may in turn claim against the negligent motorist. The practical effect of this would be that one insurance company would be claiming from another. Given the thousands of claims which they deal with, this process would be expensive and time consuming. It is, therefore, common for insurers to make agreements amongst themselves not to claim from one another under the subrogation rules. In the long run, given the amount of business involved and the administrative savings made, this will usually be a perfectly equitable arrangement. This type of agreement between insurance companies results in the 'knock for knock' arrangements found in motor insurance. The practice does not affect the rights of the insured.

Assignment

A policy of insurance is a chose in action i.e. property which can only be claimed or protected by taking action in the courts. As a type of property it is valuable in itself and the insured person may wish to use it as security to raise money, or may wish to transfer the benefit of the insurance to someone else. The assignment may be governed by the rules of the Policies of Assurance Act 1867 in the case of life policies. It must be in writing, and the insurers must be given notice. Note that it is the benefit of the policy which is assigned. Assignments of other types of insurance are effected under the rules of s.136 of the Law of Property Act 1925. This requires the assignment to be absolute, made in writing and notified in writing to the insurer.

7
Special contracts – building contracts

Control of work under a building contract

Many of the examples used so far to illustrate the working of the law of contract have been simple transactions involving small sums of money where the contract is made and performed within a short period of time. Contrast that with a building contract where the work may be massive in scale, taking months or years to complete and costing millions of pounds. The very scale of the work means that many subcontracts will be involved, as well as the main contract. Despite the complexity of a building contract it is judged by the ordinary general principles of the law of contract. It would be uneconomic to draft a complex new set of terms for every building contract and although this type of contract has not been codified by a statute, standard forms of contract have developed in the construction industry. These have been subject to extensive interpretation by the courts, and amendments to the standard form are issued regularly. Inevitably, the terms of any standard form contract may be seen to lean against the interests of one of the parties, unless representatives of both sides played a part in drawing up the original version.

A building contract is one whereby the contractor or builder undertakes work for an employer or building owner. Building works are normally controlled by JCT contracts, issued by a Joint Contracts Tribunal made up of bodies such as building trades employers, the architects' association and local authorities. JCT contracts come in a number of forms, for use with a local authority or private employer, with or without quantities. These standard form contracts are long and detailed and have been drawn up by bodies representing wide-ranging experience in the building industry. The parties using such a contract may choose to adapt or vary the specific clauses of the contract and they are entirely free to do so, but with the warning that in such a complex document, many of the provisions may be interdependent and both parties should be clear what are the repercussions of the changes they have made.

The employer who needs to undertake building work may be totally inexperienced in this field. He may feel that he should have professional advice in making and carrying through the contract. It is a traditional feature of large-scale building work that the employer engages advisers such as architects and quantity surveyors, who will supervise and control the work in progress. The standard forms are drawn up in recognition of this state of affairs and many tasks within the contract are specifically allocated to the architect. It is important to remember that the building contract is made between the contractor and the employer. The architect is *not* a party to that contract. He has a quite separate contract with the employer and he will often have been employed under the Conditions of Engagement published by the RIBA. These are standard terms providing for the amount to be paid to the architect, and setting out his duties and responsibilities in detail. Where several professional advisers such as structural engineers and quantity surveyors are engaged by the employer, it is usual for them to be under the overall control of the architect. The employer may employ all of these professional advisers under contracts of service but it is more likely that they will be in independent practice, and the employer then engages them under a contract for services.

The professional advisers will undertake significant amounts of work before the shape of the final project is determined. They must establish what the employer wants, the means by which it can practicably be achieved, and whether necessary legal hurdles can be surmounted. Then, the drawings and specifications are prepared and tenders sought. After the contract is placed the architect may be required to supervise the work and certify the payments. Inevitably, the supervision and certification will bring the architect into regular contact with the contractor, but the architect is only the agent of the employer, restricted in his authority by the terms of his engagement with the employer.

Where large-scale building work is undertaken it is common for much of the day-to-day administration to be carried out by a clerk of the works, who is employed by the employer, and not the contractor. His role is significant as he is the person who is regularly on site. He works under the direction of the architect and reports to him anything which is out of line with the terms of the contract.

The architect owes duties to the employer arising expressly and impliedly from his contract, and also by virtue of duties established by the law of tort. He is in the same position as other professional or skilled people in that he will be liable if he does not do his work with reasonable care and skill. A failure to show such care could involve design faults, inadequate site examination, delays in furnishing drawings, disregard of building regulations, an injudicious choice of builder, or lack of adequate supervision of the work. Some of an architect's duties are described as quasi-judicial because certain terms of the standard form contract require him to settle matters between the employer and the contractor. This aspect will be considered later in relation to certification.

Tendering for building contracts

Once it is clear what the scope of the building work is to be, the employer or the architect acting on his behalf will invite contractors to tender. The costs of preparing the tender (the offer) are significant, and this cost has to be borne by the contractor. The tender reflects how keen the contractor is to get the job. Even if he puts in the lowest tender, he is not automatically assured of success. His tender is only an offer and the employer may accept or reject it. Indeed, the employer may reject all the tenders. Price may not be the only factor by which he is judging the offers.

Once a tender is accepted, a valid contract can come into existence immediately, but it sometimes happens that the employer makes acceptance subject to conditions such as the execution of a formal contract or the provision of a surety bond. This can be relevant if the contractor starts work at once, and subsequently it is held that a contract never existed. In each case it will be a question of interpreting the precise words used. It must be remembered that the negotiations for building work may be prolonged and it may be essential, for reasons of time constraints, to get the work under way. This sometimes occurs on the basis of a letter of intent, in which one party indicates to the other that it is likely that a contract will be placed, and inviting the other to commence preliminary work. These letters of intent are also found between contractors and sub-contractors, where the sub-contractor has tendered for specialized work, (e.g. heating and ventilation) and the contractor has used that tender as a basis for his own tender. He may then indicate in a letter of intent to the selected sub-contractor that he will place a contract with the sub-contractor if he is successful with his own tender.

Where preliminary work is done on the basis of a letter of intent such work may have to be paid for on a *quantum meruit* basis (as much as it is worth) if no contract is eventually placed. Where a subsequent contract is made the effect of it may operate retrospectively to cover work done under the letter of intent. *Trollope and Colls Ltd v. Atomic Power Constructions Ltd* is a case where this occurred. The letter of intent was sent in June 1959 and the contract was not finally made until April 1960, by which time the contractors had undertaken work worth a considerable amount. They contended that they should be paid for this early work on a *quantum meruit* basis, rather than on the contract price basis. The court did not agree. It was held that when the contract was finally concluded in April 1960 it contained an implied term that the contract would govern what had already taken place.

The JCT standard form contracts

Although the standard form JCT contract runs to some forty clauses it is not in itself the entire contract. It makes provision in clause 2 for the incorporation of 'the contract documents', which include the contract drawings and bills. The parties will also sign 'Articles of Agreement' which is the actual contract between them, and may also use a deed. The advantage of making the contract in this way is to extend the limitation period for breach of contract claims from six years in ordinary simple contracts to twelve years for contracts made by deed. The Articles are dated, they name the parties, define and locate the building work, and are signed and witnessed. They impose an obligation on the contractor to complete the work under the Conditions of Contract for the sum agreed (the consideration); specify the architect and the quantity surveyor; and impose an arbitration agreement.

The Conditions of Contract, referred to in the Articles and therefore incorporated into the contract, contain the essence of the agreement. Clause 2 imposes on the contractor the obligation to execute and complete the works in accordance with the contract documents. 'Completing' for this purpose means practical completion, which is the point at which the defects liability period commences. This is a significant moment because the contractor then ceases to owe certain obligations e.g. insurance obligations. As the contractor has to operate in accordance with the contract documents he does not himself undertake any design responsibility. He is not, therefore, undertaking that the building will be suitable for its intended purpose.

The Conditions of Contract are not laid out in a particularly helpful sequence, and the following appraisal of the conditions attempts to impose a more logical order. It is not intended to deal exhaustively with all of the conditions.

Possession of the site

Clause 23.1 requires the contractor to be given possession of the site on the agreed date of possession. He must then regularly and diligently proceed with the work, bearing in mind that the architect can issue instructions to postpone work, which could in turn provide the grounds for the contractor to claim an extension of time, and for loss and/or expense.

On-site provisions

There must be access for the architect and his representatives at reasonable times (clause 11). The employer may appoint a clerk of works to act as an inspector on his behalf. The clerk of works will be under the supervision of the architect but must be facilitated in the performance of his duties by the contractor. Directions given to the contractor by a clerk of works are of no effect unless confirmed by an instruction from the architect within two days (clause 12). The contractor is obliged by clause 10 to keep a competent person in charge on the site during working hours. Such a 'site agent' is entitled to receive architect's instructions on behalf of the contractor.

The architect must provide the contractor with all the requisite levels and measurements so that the work can be carried out (clause 7). Materials, goods and workmanship must be as described in the Contract Bills and if requested by the architect, the contractor must produce vouchers to prove that materials and goods correspond with the specification. As a further check the architect may issue an instruction for work to be opened up for inspection or to test goods and materials. The cost of this exercise can be added to the Contract Sum, except that if it proves a contractor to be at fault in some way, he must bear the cost.

Clause 16 deals with materials and goods on the site. Such unfixed goods become the property of the employer only when their value has been included in an interim certificate which has been paid by the employer. Before that time their ownership is determined by the usual rules of sale of goods (see page 144). Once ownership passes to the employer the contractor is under a duty to take reasonable care of the goods in order to protect them against damage or theft.

Under clause 29 the contractor must allow entry to the site to persons who are directly engaged by the employer and whose work does not form part of the contract work but where provision was made in the Contract Bills for the execution of their work. The employer is responsible for such persons. For instance an archaeologist may be called to the site to investigate a discovery of fossils or antiquities. Clause 34 makes provision for the action to be taken by the contractor if he makes such a discovery.

Progress of the work

It has already been noted that clause 2 requires the contractor to execute and complete the work in accordance with the contract documents. As the work progresses unexpected technical problems may occur, or it may not be entirely clear what work is intended to be done under the contract. In these cases the ability to vary the contract is important. Under the JCT standard form, authority to order variations is vested in the architect who acts by issuing architect's instructions. Needless to say, instructions can be issued for a range of reasons far wider than just ordering variations but it is convenient to consider these items together.

Clause 13 deals with variations. These give valuable flexibility to the contract. It is basically sound to include such a clause as it permits the employer to require a variation as of right, instead of having to rely on the goodwill of the contractor or having to create a quite separate contract furnished with its own consideration to render the variation valid and enforceable. The variations clause clearly demonstrates that the employer is vesting the architect with authority to order variations by which he, the employer, will regard himself as bound. When the term 'variation' is used in clause 13, it means the alteration or modification of the design, quality or quantity of the works as shown in the Contract Drawings and described in the Contract Bills. It also covers the alteration, addition to, or omission of, any obligations imposed by the employer with regard to limitations on working hours or space, or the execution or completion of the work in any particular order. Clause 13 makes clear that variations ordered or sanctioned by the architect will not vitiate the contract, but this does not mean that the architect can order any change he likes. An excessive change would not come within the definition of a variation. The clause also contains the rules which govern the adjustment of the contract sums following the variation of the work.

Architect's instructions

Architect's instructions generally, and the power to authorize variations are dealt with by clause 4. The contractor is obliged to comply forthwith with instructions issued to him by the architect, but they must be on a matter where the architect is empowered by the contract conditions to issue an instruction. His instructions can cover a wide range of situations such as variations, antiquities, defects and making good defects, nomination of sub-contractors and removal of items not in accordance with the contract. If a contractor does not comply with an instruction within seven days the employer may engage someone else to do that work, pay him and deduct that sum from money due to the contractor. Where a contractor is not clear under what authority an instruction is given, he can ask the architect to specify his authority and, if dissatisfied, the contractor can seek immediate arbitration.

Under clause 4 an architect's instruction is only valid if it is in writing, but the rules do provide for verbal instructions to be confirmed in writing within seven days by the contractor to the architect. If the architect does not dissent from that written confirmation within a further seven days, this becomes a valid architect's instruction. Alternatively, the architect himself may confirm in writing within seven days or, in cases where a contractor has complied with an oral instruction, the architect may confirm it in writing prior to the issue of the Final Certificate.

Extensions of time

Variations which have been ordered by the architect may be one of the reasons why a contractor needs an extension of time. The power for the architect to grant such an extension is contained in clause 25. This could be vital to the contractor because if he does not complete on time, he is likely to become liable under a liquidated damages clause. Extensions of time may not be exclusively for the contractor's benefit. If the employer has failed in some of his obligations, e.g. failure to allow access to the site at the proper time, he may be keen to see the time for completion extended as he would be unable to rely on a liquidated damages clause if he had been the cause of the delay.

Extensions are sought by the contractor making a written application to the architect, which must set out the relevant events which he considers justify an extension. These 'relevant events' are set out in detail in clause 25 and include exceptionally adverse weather conditions, acts of God (i.e. *force majeure*), civil commotion, strike or lockout, compliance with architect's instructions e.g. on variations, non-receipt by the contractor in due time of the necessary drawings or instructions, exercise by the government of statutory powers which affect supplies of labour, goods or fuel essential to the proper execution of the work, or failure or delay on the part of a statutory undertaker in pursuance of a statutory obligation.

Not only must the contractor set out the appropriate relevant event when he seeks an extension of time, but also he must specify the expected effects and the period of expected delay in completing. If the architect decides to grant an extension he does so in writing, specifying such later completion date as he considers fair and reasonable. In his extension, the architect must state which relevant events he has taken into account and the extent to which he has also taken into account variations previously ordered which have cut down on the contractor's task by deleting work originally specified. If no extension or an inadequate extension is granted, the contractor may go to arbitration.

Insurance

Clause 21 provides that the contractor must maintain insurance in respect of liabilities imposed by clause 20, that is liability for third party claims in respect of personal injury and damage to property where the contractor is at fault. Clause 22 determines who should take out an insurance against 'all risks' where the contract relates to a new building. The employer should insure against the specified perils when the work involves alterations or extensions to an existing building. Clause 22 requires the insurances to be effected by joint names policies.

It is eminently sensible to include provisions about insurance in the contract as building work necessarily involves risk. Harm could be caused to people coming onto the site, passers-by, neighbouring landowners or workmen employed on the site. Some insurance is compulsory by law, e.g. employers' liability insurance. Liability for damage or injury may lie on the contractor, or on the employer, sometimes directly and sometimes vicariously, and on some occasions, the contractor and employer may be jointly liable.

Payment for the work

One particular feature of building contracts is that they commonly provide for payment for the work to be made in stages. This is vital when the contractor's expenses and outgoings will be considerable and the overall time for completion may be lengthy. To provide for payments at intervals a system of certification is used. Clause 30 provides for the architect to issue interim certificates, usually at monthly intervals, and the employer must pay the sum due within fourteen days. The architect can seek interim valuation from the quantity surveyor whenever he considers it necessary to determine the amount due on an interim certificate. Following the decision in *Sutcliffe v. Thackrah*, he would be wise to adopt this course. There, the architect issued interim certificates on which the employer paid the contractor. The contractor was subsequently dismissed by the employer and became insolvent. The employer discovered that the architect had certified defective work and work that had not been done at all. The consequence of this certification was that overpayments had been made. The House of Lords held that the architect could be liable for damages for his negligence.

When issuing an interim certificate the architect establishes the amount due by calculating the gross valuation, less any retention and the amount of any previous payments. Gross valuation covers the total value of the work properly done by the contractor, together with the value of materials and goods on site. The retention will usually be a maximum of 5 per cent (though occasionally a higher figure is stipulated), and is held 'on trust' by the employer until its release. Half the retention is released after practical completion and the rest after the issue of the certificate of making good of defects at the end of the defects liability period. Not all items are subject to retention e.g. direct loss due to discovery of antiquities, or losses falling under clause 26.

Clause 26 deals with the contractor's claims for loss and expense where the progress of the work is likely to be materially affected by factors such as not receiving properly requested instructions or drawings in time, or any other of the 'matters' listed in clause 26. If the contractor is aware of such losses or expenses he must make written application to the architect who may determine the amount of such loss and add to it the contract sum. A claim by a contractor under clause 26 is without prejudice to other rights. For example, if the contractor is basing his claim on the architect's failure to provide him with the necessary drawings that would be a breach of contract entitling the contractor to sue the employer for damages.

If sums are added to the contract sum by virtue of clause 26, then clause 3 provides that the adjustment can be included in the calculation of the next interim certificate.

Practical completion and final payment

Once the architect considers that the work is at the stage of practical completion, where the building can function properly and any outstanding work is only minor, he can then issue a certificate of practical completion. This heralds the commencement of the defects liability period. The JCT standard form offers six or twelve months and employers nowadays more regularly choose twelve. The architect will deliver a schedule of defects to the contractor no later than fourteen days after the end of the defects liability period and the contractor must then put the defects right at his own cost, unless the architect is prepared to instruct otherwise. (Of course, these instructions can only relate to defects which manifest themselves within this period. Any which become apparent later must be the subject of a claim for breach of contract under all the normal rules of contract, including limitation periods.) Once defects have been made good, the architect issues a certificate of completion of making good defects. The obligation to make good defects under clause 17 is confined to defects caused by materials or workmanship not in accordance with the contract, or frost damage occurring before practical completion, this being deemed to be the contractor's fault. These certificates of practical completion and on making good defects are significant because they operate to release the retention money, half at each stage.

Non-completion and breach

By the terms of clause 23 the contractor must complete the works on, or before, the completion date. By clause 24 the architect must issue a certificate of non-completion if the contractor does not complete in time, although the time may have been extended by architect's instructions. Once the architect has issued the certificate of non-completion, the employer may, after written notification to the contractor, deduct or recover liquidated damages calculated at the rate stated in the appendix to the standard form.

The provisions of clauses 23 and 24 relate only to delays. It is more likely that other breaches of contract will occur, some quite trivial and others causing significant loss. As an ultimate sanction, clauses 27 and 28 provide for determination by the employer or contractor respectively. But it must be borne in mind that there are arbitration arrangements under Article 5 and clause 41. Many disputes between the parties during the currency of the agreement could be resolved through those procedures. Other provisions may assist in remedying minor breaches. For instance, under clause 8 the architect can order the opening up and testing and, if necessary, the removal of any work consequently found to be defective.

Clause 27 is principally relevant when the contractor cannot go on with performance or where he has committed major breaches. It will operate automatically to determine the contract if the contractor goes into liquidation. In other situations, such as failure to proceed regularly and diligently, or persistent failure to remove defective work, the architect may serve a notice specifying the fault. If the contractor continues the default for fourteen days the employer may send notice by registered post within the next ten days to terminate the contractor's employment. If he takes such a step the employer would then have to engage other contractors to complete the work. Any extra cost incurred in their engagement can be recovered from the original contractor. Clause 27 specifically provides that notice by the employer must not be given unreasonably or vexatiously.

Clause 28 deals with the contractor terminating his employment with the employer. Nothing in this clause affects other rights and remedies of the contractor. The contractor effects the determination by serving notice on the employer or the architect. He may do so if the employer fails to honour a certificate, or if the work is suspended for various reasons, and for longer than a certain period, stated in the appendix. The contractor is then entitled to be paid for work done and materials supplied, and for any loss caused to him.

Clauses 27 and 28 merely have the effect of determining the employment of the contractor. They do not terminate the contract as a whole, and many of the other terms governing the rights of the parties will still be relevant. The ordinary rules of breach of contract will govern all other situations where either party alleges that the other has not complied with his obligations under the contract.

Sub-contracts

In a large building contract the main contractor is likely to organize that some aspects of the work will be performed by sub-contractors. He undertakes the main contract with the employer and then in turn enters into separate contracts with all the individual sub-contractors. For the employer the advantage of this system is that he gets his work done by making only one contract himself, and obligations in respect of the whole job are owed to him by the main contractor, who remains responsible for the work which he lets out on sub-contracts.

Generally, the main contractor may only sub-contract to the extent permitted by his contract with the employer. Sub-contractors will be paid by the main contractor and have no rights against the employer because they do not enjoy privity of contract with him. The employer can enhance his position by insisting on choosing the sub-contractors, in which case they will be known as nominated sub-contractors. When a sub-contractor is nominated in this way, there is the advantageous provision that, should the main contractor fail to pay him within a stated time of certification (seventeen days), then the employer may pay him directly.

Of course, while there may be advantages for the employer in having no direct contract with sub-contractors, this has inevitable drawbacks, principally the difficulty of incorporating the terms of the main contract into the sub-contract. It may be possible to to this, in so far as it is appropriate, by obligating the sub-contactor not to perform in such a way that he causes the main contractor to be in breach. If a subcontractor does default on his obligations the main contractor will be liable. The choice of sub-contractors is thus seen to be of vital importance as a contractor will be anxious in case the sub-contractor causes delay or becomes unable to complete his contract. If anything goes wrong the contractor may be involved in two legal actions: being sued by the employer under the main contract, and then himself suing the sub-contractor under the separate sub-contract.

One difficult problem involving renomination of sub-contractors used to arise where the original nomination failed for some reason e.g. liquidation of the nominee. The employer might then be anxious that if he renominated he would obligate himself to pay any higher charges of the new nominee. It was thought that he could avoid this burden by not renominating, but the House of Lords held, in *Bickerton v. North West Regional Hospital Board*, that the employer was under an implied obligation to renominate. The JCT standard form was altered to take account of this decision. The Joint Contracts Tribunal has also produced standard forms of collateral agreement, to be entered into by nominated sub-contractors and employers, thus creating a direct contract between the two where no privity otherwise existed. This is incredibly cumbersome, as there will be three contracts – employer and main contractor; main contractor and sub-contractor; and sub-contractor and employer – with all the attendant problems of knowing who to sue in respect of breaches. The consideration for the collateral agreement between the employer and the sub-contractor is that the employer promises the nomination and direct payment, should the main contractor default in this respect. The sub-contractor promises to warrant his work in relation to design and selection and specification of materials.

8
Employment law

Employment and self-employment

With a workforce in this country of about 28 million people, the rules governing employment are of considerable importance. The past thirty years have seen a massive increase in new employment rights created by Acts of Parliament. Previously, the law of employment was developed largely through case law, and it reflected the attitude that employers could 'hire and fire' at will, with no recognition of the inequality of bargaining power between employers and employees. To some extent that inequality was remedied by the growth of the power of the trade unions. However, the trade unions have themselves been subject to major changes within the last decade, which have curtailed their activities on behalf of their members, and further changes continue apace.

Most of the important protections of employment law are enjoyed by persons employed under a contract of service. This type of contract can be contrasted with a contract for services which, simply stated, is the distinction between the status of an employed person and that of a self-employed person. It is necessary to make this distinction because:

1. Statutory rights (e.g. the right not to be unfairly dismissed and the right to be compensated for redundancy) only apply to persons employed under a contract of service.
2. Employees under a contract of service are Class 1 National Insurance contributors, and therefore eligible for a wider range of social security benefits than self-employed Class 2 contributors. Class 2 contributors are not eligible for unemployment benefit or disablement benefit.

3 The tax arrangements for the two groups of workers are different. Employees pay tax by means of the PAYE system, whereas the self-employed pay tax by half-yearly instalments in arrears under a different tax schedule.
4 The duties implied by law between the parties to the contract are different, tending to be more onerous between employer and employee in a contract of service.

It will be seen that many of the reasons for distinguishing between contracts of service and those for services have financial implications for the employer as well as the employee. It may seem cheaper for the employer to encourage his workforce to become self-employed, and by that means avoid the increasingly strict statutory control over contracts of service. However, it is not enough for the parties simply to change the name of their relationship. It is a question of fact whether the contract is one of service or for services. In those cases where the courts think a change has been made simply to take advantage of favourable rules they may find no alteration in the true nature of the relationship.

When the courts have to determine which relationship exists between the parties there has been considerable difficulty in establishing a simple test by which the decision can be made. Traditionally the courts used a simple test of control. Could the employer tell the employee what to do *and* how to do it? If so, there was a sufficient measure of control to give rise to the closer contract of service. The drawback with such an unsophisticated test is that it is quite inadequate to cope with more complex employment situations. The test is only concerned with the ultimate possibility of exercising control. The fact that an employer chooses to let an experienced employee get on with the job is irrelevant. Even so, the test proved to be over simple and by the late 1940s the courts were searching for another test. It is only fair to say that even today the control test still gives a reliable answer in simple employment relationships.

In more complex situations the judges thought an answer might lie in asking whether a person's work was done as an integral part of the employer's business. This 'integration' test was not entirely satisfactory, and at times highly artificial. More recently the judges have come to the conclusion that there is no simple question which will provide a conclusive answer. Rather, it is a matter of looking at all the surrounding circumstances, and the modern approach is said to be a mixed or multiple test. Using this approach, the court will consider a range of factors, including:

- What are the employer's powers of selection and dismissal?
- What measure of control is exercised by the employer?
- What agreement is there about the method and amount of remuneration?
- What arrangements exist for the payment of tax and National Insurance contributions?
- Who supplies tools and equipment?
- Does the employee bear any economic risk in the enterprise?
- Does the employee hire his own helpers?
- Does the employee have any responsibility for investment and management?
- Can the employee profit from sound management in the performance of his task?
- How do the parties themselves view their relationship?
- How are the employees usually engaged in the trade or industry?

The mixed or multiple test may in reality be a kind of 'reasonable man' approach. The difficulty in practice is that a situation may arise where the questions listed above give equivocal answers. Consider the example of *Ready Mixed Concrete Ltd v. MPNI*. On the one side, the man in question had to wear a uniform, obey the orders of a foreman and obey company rules. All of these features of his contract pointed towards it being a contract of service. By contrast, he was buying the lorry under HP arrangements with the company, and had to bear the cost of servicing it. He also paid his own tax and National Insurance contributions. On these facts he could be seen to be taking some risk. When all the circumstances were taken into account the court found that he was self-employed.

It is matterless what the parties call their arrangement although, when the factors are very finely balanced, some assistance may be given by looking at what the parties have chosen to call it. This must be treated cautiously though, as two contrasting cases will show. In *Massey v. Crown Life Insurance Co.* one of the branch managers approached Crown Life to suggest a change in his status from employment to self-employment. He took advice about the consequences of the change and his new status was confirmed between the parties, and in a letter sent to the tax office. Subsequently the manager had occasion to regret his decision when certain employment rights were no longer available to him. He sought to prove that, in reality, he was still engaged under a contract of service. The Court of Appeal did not agree. Given how carefully the question of change of status had been approached, the statement by the parties that he was self-employed was of great weight and significance. Compare Massey's case with *Ferguson v. Dawson and Partners Ltd*. Ferguson was injured on a building site and wished to sue his employers for breach of statutory duty. To succeed in his claim he needed to show that he was engaged under a contract of service. The evidence showed that Ferguson had been brought along to the site by some of his friends and had been taken on in an extremely informal manner by the foreman who had informed him that he would be working 'on the lump', i.e. on a self-employed basis. No doubt Ferguson was happy with that situation while everything was going well as he had scope to keep ahead of the tax man and avoid paying appropriate National Insurance contributions! Despite the agreement of the parties to the 'lump' arrangement, the Court of Appeal held that the reality of this situation was a contract of service. It had none of the carefully worked out and evidenced formality of Massey's arrangements, so the view of the parties could hardly be decisive.

Although there are numerous Acts of Parliament conferring rights on employees, none of them gives an exhaustive definition of a contract of service. One usual formulation is to define an employee as a person who works under a contract of employment. That in turn is defined as a contract of service whether express or implied. And as that phrase is not further defined, the rules developed by the judges remain significant.

Creation of a contract of employment

Although contracts of employment are governed by the ordinary rules of contract it is important to consider the role and impact of collective bargaining. This is the name given to negotiations between trade unions and employers, usually concerning the establishment of better terms and conditions of employment for union members. Where a union (or group of unions) has established bargaining rights with an employer (or group or federation of employers), their joint negotiations may result in a collective agreement, which in practice fixes the terms of the employment of relevant employees. Such agreements, together with the increased statutory control of employment, now severely limit the true 'freedom of contract' of the actual parties, the employer and the employee. An employee may ask, 'How can an agreement (the collective agreement), to which I was not a party, affect my individual relationship with the employer?' The answer is that the collective agreement itself cannot alter the terms. But such agreement may well become 'incorporated' into the contract, thereby altering it and have direct force and application to the employer and employee as a result. Two simple methods by which incorporation could occur are:

1 Express incorporation where, by the terms of the original contract, the parties agree to be bound by the appropriate collective agreement for the time being in force.
2 Incorporation by usage, where the employer and employees have acted on the strength of the terms of the collective agreement, e.g. the employer has paid the increased hourly rate referred to in the collective agreement and the employee has accepted the increase.

A significant portion of the working population still belongs to the various trade unions, but the impact of collective bargaining in the 1990s is less extensive, with a move towards individual bargaining. Where collective bargaining is still prevalent, it can have an effect on the terms and conditions of everyone in a firm or industry, whether members of the union or not.

Inevitably, a collective agreement will not be an exhaustive statement of all the terms of the contract. Indeed, it has already been indicated that collective negotiation has no place in some contracts. In the higher echelons of employment, where staff have desirable skills and talents to offer, the terms of the contract may be fixed by much more individual negotiation. In either case, many of the terms will be implied into the contract by statute or by common law. There is even scope for terms to be implied by custom. These will be considered in detail later.

Form of the contract

Apart from exceptional situations, such as that of apprenticeship, there are no rules requiring a contact of employment to be made in writing. However, since 1963, many employees have enjoyed the right to receive a written statement containing the main terms of their contract. The Employment Protection (Consolidation) Act 1978 (EPCA), as amended by the Trade Union Reform and Employment Rights Bill 1993 (TURERB), requires an employer to issue this statement within eight weeks of the start of the employment and it must contain:

1 Names of the parties and date of commencement of employment.
2 The job title and the place of work.
3 Details about pay, including the amount or method of calculating the amount and the intervals at which it is to be paid. (A separate right for employees to receive an itemized pay statement, setting out the purpose and amount of deductions, is established by EPCA 1978.)
4 Details about hours of work, and overtime arrangements.

5 Details about holidays and holiday pay.
6 Details of rules covering sickness and sick pay.
7 Details about pensions.
8 Details about disciplinary rules and grievance procedure.
9 Details about the period of notice to be given by employer or employee to terminate the contract.

The EPCA 1978 specifically states that if there are no agreed particulars under any heading, that fact must be stated. This rather emphasizes a feature of employment which is not always appreciated. Holidays are usually a question for agreement between the parties, and if no holidays have been agreed, then none are due. Holidays with pay in the construction industry might be a problem because of the fact that employees tend to move from employer to employer. This problem is resolved for those who participate in the special Holiday Stamp Card Scheme, negotiated by collective agreement and administered through a separate independent company.

The value of the written statement is that it is evidence of some of the terms of the contract, and may allow disputes between the parties to be quickly and simply resolved. But not all employees are entitled to receive a statement and in particular the following are not eligible:

1 Part-time employees. For this purpose part time means working less than eight hours per week.
2 Crown servants, employees working under certain fixed-term contracts, employees who ordinarily work abroad and employees who are husbands or wives of the employer.
3 Employees who have a written contract.

The EPCA 1978 requires the information to the employee to be kept up to date. If an employer fails to give a statement as required by these rules, the sanction is rather ineffective. The aggrieved employee may complain to an industrial tribunal, which has power to determine what the written statement should contain. Unfortunately, the tribunal has no enforcement powers and if the employer continues to default by refusing to comply with the statement issued by the tribunal, the employee will have to enforce his terms by a separate action in the County Court.

Duties of the parties

Once a contract of employment has come into existence, the very nature of that relationship imposes certain duties on both parties, whether they expressly negotiated on these points or not. These duties are summarized below.

Employer's duties

1 *A duty to pay the agreed wages* – There is no general minimum wage legislation in this country. The amount of the wages is fixed by the parties, although in many cases this will be significantly influenced by collective bargains negotiated within that type of employment. A statutory system of Wages Councils, fixing rates of wages and holiday entitlements for particular groups of employees, enforced by a Wages Inspectorate, has been abolished by TURERB 1993. The amount may also need to reflect the policy of the Equal Pay Act 1970. The Act implies an 'equality clause' into all contracts of employment, whereby men and women who are engaged on 'like work', or 'work rated as equivalent' or 'work of equal value', should receive equal pay and equal treatment with regard to terms and conditions. Employees prejudiced by breach of these rules can bring a claim in the industrial tribunal. English law has long concerned itself with *how* wages are paid, rather than how much is paid. The old Truck Acts 1831–96, regulating deductions from wages and methods of payment to persons employed in manual labour, have now been repealed but in their place the Wages Act 1986 covers all employees. The Act is largely concerned with deductions from wages for reasons like bad workmanship. Such deductions are permitted so long as they are made under the terms of the contract, and those terms must have been notified to the employee in writing. Of course, an employer may always deduct those amounts which are permitted or authorized by statute, e.g. tax and social security contributions. There are also court orders which permit deductions from earnings, called attachment of earnings orders. The actual method of paying the wages, e.g. cash, cheque or directly into a bank account, is a matter for agreement between the parties. Once the method is fixed one party may not change it unilaterally without the consent of the other.

2 *A duty to indemnify the employee for expenses reasonably incurred during the employment* – this point will often be covered by express terms about expenses. In other cases, if an employee is out of pocket as a result of his work, much may depend on how the expenses arose. Where the employee is doing his work in a lawful way and acting under his employer's authority, he should be reimbursed. If, however, the employee chooses an unlawful method of perfor-

mance e.g. employed to drive but driving in excess of the speed limit, then the employee must bear the cost. A more specialized aspect of the duty to indemnify is the rule that if an employee commits a tort in the course of the employment, the employer is usually vicariously liable. The full extent of this rule is considered in Chapter 9.
3 *A duty to take reasonable care for the safety of the employee* – this duty is linked with the whole question of health and safety and is considered in detail in Chapter 10.
4 *A duty to show mutual respect to the employee* – this is an example of a duty which has emerged more recently in employment law, largely from suggestions made in cases being heard before industrial tribunals. It can be said to mean that the employer must treat his employees with appropriate respect and consideration. It is an example of how far the law of employment has moved in recent years from the old philosophy of the employer being able to 'hire and fire' at will.

It is usually argued that in most cases an employer is under no legal duty to provide work. An old case contains the comment: 'If I pay my cook she cannot complain if I choose to eat all my meals out'. In a number of important ways this rule is of practical significance:

1 It allows an employer to pay wages in lieu of notice. He does not usually have to provide work.
2 When there is no work an employer may lay off his employee, although where this occurs over a sufficiently long period the employee may be eligible to claim a redundancy payment.
3 If no work and no pay are available this may amount to the employee being suspended. The usual rule is that an employee may only be suspended without pay if his contract expressly or impliedly provides for this. Suspension with pay normally causes no problems because the employer is under no duty to provide work.
4 If an employee has no work and no pay the employer may have a statutory duty to pay him guarantee payments of fixed amounts for a specific number of days (EPCA 1978).

An employer is under no duty to provide a reference or testimonial, even though an employee may suffer in consequence of this decision. This can be particularly harsh when an employee secures a new job which is dependent on satisfactory references. The employer may feel justified in refusing to give a reference because anything he writes could open up possibilities of legal liability. For example, the employee may sue if the reference is seen to be defamatory. Employers should have little to fear on this point, as an action for defamation can be defeated by the defences of truth or qualified privilege, i.e. the employer can show that without malice or bad motive he was simply communicating information to a person who had a genuine interest in receiving it. Nowadays, an employer is more likely to fear that statements in a reference could expose him to liability for negligent mis-statement. This issue is considered in more detail in Chapter 9. Moreover, an employer may be estopped from denying the truth of statements made in a reference. This can be relevant in cases in the industrial tribunal for unfair dismissal. If the employer has written a glowing reference in the hope that a mediocre employee will move on, he may subsequently wish to dismiss the employee. In any later proceedings it will be difficult for the employer to deny the truth of that glowing testimonial!

Employee's duties

1   A *duty to render personal service.*
2   A *duty to render faithful service* – this means that if the employee in any way damages his employer's interests he is effectively in breach of his duty. Where such a breach occurs there may be justification for the employer to dismiss the employee. Sometimes a breach may cause an employer serious on-going harm, and he may wish to seek an injunction to restrain the particular activity of the employee. In other cases it may also be appropriate for the employer to seek damages from the employee. Some areas of particular importance emerge from this general duty of faithful service:

   (a) The employee should not put himself in a position where his duty to his employer conflicts with his own self-interest. In *Boston Deep Sea Fishing Co. v. Ansell* an employee placed orders for supplies on behalf of his employers. He chose to give the orders to firms which rewarded him for putting business their way. Those firms may well have had the competitive edge anyway, but the 'rewards' may have clouded Ansell's judgment. He was dismissed for this breach of duty.

   (b) The employee should not make a secret profit from his employment. This will often be closely linked with the duty in (a). In Ansell's case he had failed to disclose to his employers what he was making 'on the side'. That was quite a crude example of this problem which crops up regularly in business and commercial life. An employee who is in a position to place large orders for supplies may be courted by hopeful companies. He should take care in accepting any gifts or favours, for he lays himself open to being in breach of his duty, even if his judgment was not affected. Many large organizations tackle the problem by imposing express rules in the contracts of employment, forbidding acceptance of gifts.

(c) The employee must not compete against his employer. An example of such competition might be the employee working as a boiler service engineer who suggests to customers that they approach him personally in his off-duty hours, and he will then undertake their work on his own behalf. In a case like this the employee could be restrained by an injunction and his behaviour would almost certainly warrant dismissal. The reason here is obvious. His spare-time activities are seriously detrimental to his employer. There is nothing to prevent an employee engaging in a second job as spare-time work so long as it does not interfere with the duties he owes to his employers. This rule, that an employee must not compete against the employer, is also relevant when the employment terminates. In appropriate cases, the duty of faithful service can continue to be owed *after* the employment relationship has ended. In such cases it is common to include a 'restraint of trade' clause in the contract, which will be enforceable against the employer so long as it is reasonable (see page 111).

Patents, designs and copyrights

The issue of competing against the employer can also arise where an employee invents something in the course of his employment. It could be very unfair if he then leaves his job, and seeks to patent the invention and profit from it. The rules covering such a situation are now contained in the Patents Act 1977. Before the Act was passed, an employer who thought it likely that an employee might invent something would put an express term into the contract, requiring the employee to assign (transfer) the rights in the invention to the employer. This often produced a very unfair result, with the employee personally getting no reward even if the employer made handsome profits out of the invention. The 1977 Act aims to deal with questions of ownership of the invention and the right of the employee to be rewarded. The invention belongs to the employer if it was made by an employee in the course of his normal duties, or in the course of duties specifically assigned to him (s.39). Any other inventions belong to the employee. If an employer then exploits the invention of an employee, the employee can receive compensation from the court if he can show that a patent has been granted which is of 'outstanding benefit' to the employer (s.40). The Act states that the compensation must secure for the employee a 'fair share' of the benefit derived by the employer, bearing in mind the nature of the employee's duties, the effort, skill and advice of others, and the contribution of the employer in terms of facilities and other assistance. Any attempt to reduce the rights of the employee by agreeing otherwise in his contract is absolutely unenforceable.

Similar problems can also arise in relation to copyright. The Copyright Act 1956 has just been repealed and replaced by the Copyright Designs and Patents Act 1988 which developed out of a government White Paper on intellectual property and innovation. Not surprisingly, the period since the 1956 Act has seen much change and progress, especially in the computer field, and new rules were needed to take account of these changes. Where a literary, dramatic, musical or artistic work is made by an employee in the course of his employment, his employer will be the first owner of any copyright in the work. The work in question could be sales and promotional literature, or a users' manual, or an installers' guide. It is important to bear in mind that such apparently mundane publications are governed by these rules in the same way as novels and pop songs. So complex is this area that the general public is only likely to be aware of the controversy about whether the Act should have included a levy on blank recording tapes.

Design rights are dealt with slightly differently by the new Act. Designs are protected on principles similar to copyrights but lasting only for fifteen years as against the fifty-year period for copyrights. The designer is the first owner of any design right in a design, but if the design has been commissioned for a fee, the person who commissioned the design is entitled to any design right in it. Generally, the rules are the same as those for copyright – when the design was created in the course of the employment the employer will be the first owner of any design right in the design.

3 *A duty to obey the lawful and reasonable orders of the employer* – for an order to be lawful it must not contravene the general law nor must it be contrary to what was agreed by the parties. An employee could therefore refuse to obey an order which required him to contravene safety regulations, or one requiring him to work in another part of the country when his contract plainly does not anticipate that he should be mobile. It may be more difficult to determine whether an order is reasonable. Each case will turn on

its own individual facts. Clearly an order is not reasonable if it exposes the employee to grave risk. Where the employee refuses to obey a lawful and reasonable order he is in breach of his contract, but that may not necessarily be a 'fair reason' for dismissing him. In *Wilson v. IDR Construction Ltd* the employee refused, in breach of his contract, to move to another construction site. Given his past work record, the genuine reason the employee gave for not wishing to move, and the employer's failure to allow the employee to explain, it was held to be unfair to dismiss him for that reason. Cases like this highlight the significant difference between a common law breach of contract and the statutory protections for unfair dismissal. 'Technical' breaches have to be viewed in the overall context of the contract.

4 *A duty to carry out his duties with reasonable care and to take reasonable care of the employer's property* – again, any behaviour short of reasonable care is technically a breach of the contract but whether it would justify dismissing the employee depends on all the circumstances. In any well-structured employment situation the disciplinary procedure drawn up by the employers should provide for sanctions such as written warnings, so that every act of misbehaviour does not necessarily warrant dismissal.

Statutory rights for employees

The duties previously considered arise by virtue of the employer/employee relationship. The past thirty years have seen an explosion of new rights and duties created by statute, but it should be remembered that Acts of Parliament had played an important role in some spheres of employment for much longer, e.g. health, safety and welfare, and methods of paying wages.

The new statutory rights are mainly granted to employees, thereby redressing to some extent the inequality of bargaining power between a single employee and a powerful employer. Some important examples include:

- The right to a written statement of terms.
- The right to receive an itemized pay statement.
- The right, in appropriate circumstances, to receive a guarantee payment or a medical suspension payment.
- The right to a minimum period of notice.
- The right not to be unfairly dismissed, or made redundant, without compensation.
- The right to request written reasons for dismissal.

- The right to maternity leave or maternity pay.
- The right not to be discriminated against.
- Rights in connection with trade union membership.

Most of these statutory protections are confined to persons who work under a contract of service, and each right has a number of qualifying conditions attached to it.

Generally, each right will lay down a 'working-in' period, i.e. a period of continuous service which must have been served before the right accrues. These working-in periods vary from four weeks in the case of guarantee payments to two years for redundancy payments. The concept of 'continuous service' is, therefore, vital to establishing eligibility. In any normal working life there will be times when an employee is not at work, e.g. through holidays or leave of absence or illness. If continuity of service was broken on each occasion, none of these statutory rights would ever be earned. The law provides that where a week is covered by the contract of employment, it will not break the continuity of service. As a general rule, the week in question must always be a week where the employee was contracted to work for eight or more hours.

Not only must the rules establish when there is continuous employment, they must also determine whether the weeks in question count in, or not. Take the example of someone employed for just over two years who now needs to consider eligibility for a redundancy payment. During that two years the employee in question has six weeks of paid holiday, two separate weeks of sickness, one week's absence with the permission of the employer when the employee's mother died, and five weeks spent on strike. It could be critical to the 104 week working-in rule to establish which of these periods count in and which do not. In fact, all of them except for the period on strike would be added in. By law, strikes do not break continuity but the time spent on strike (or absent from work through a lock out) cannot be counted in.

Rules exist to govern the continuity of employment where a business is sold. Where the business is sold as a going concern, employment with the old and the new owner can count as one continuous period. Inevitably the rules about continuity are complex but the employee gets the benefit of a presumption that his service is continuous; it is for the employer to prove a break in continuity. The rules are contained in the Employment Protection (Consolidation) Act 1978, and the Transfer of Undertakings (Protection of Employment) Regulations 1981 as amended by TURERB 1993.

Protection against discrimination

The principal protections against discrimination in employment are contained in the Sex Discrimination Acts 1975 and 1986, the Equal Pay Act 1970 and the Race Relations Act 1976. The Sex Discrimination Act 1975 (SDA) works on a model of defining discrimination, stating where it will be unlawful, determining what defences might be available and establishing procedures for enforcement.

1 Section 1 of the Act states that a person discriminates against a woman on the grounds of sex if he treats her less favourably than a man of the same marital status. This is *direct discrimination*. The Act also recognizes that discrimination can be much less obvious, and has identified *indirect discrimination*. This occurs where a requirement or condition is applied equally to men and women, *but* the proportion of women who *can* comply with the requirement is considerably smaller than the proportion of men, the condition is not justifiable *and* the condition is to the detriment of the person complaining because she cannot comply with it. (Although these examples are couched in terms of discrimination against women, it must be remembered that the Act is drafted to protect 'persons' from discrimination. In the right set of circumstances a man could allege discrimination.) An example of indirect discrimination occurred in the case of *Price v. Civil Service Commission*, where applicants for the particular branch of the Civil Service had to be under twenty-eight years old. Mrs Price complained that the proportion of women who could satisfy this requirement was smaller than the proportion of men, because a significant number of women in their twenties would be out of the job market bearing and rearing children. This argument was answered by the Civil Service seeking to show that, statistically, there were roughly equivalent numbers of men and women who could come forward. On appeal to the Employment Appeal Tribunal, it was determined that the phrase 'can comply' in s.1. of the Act must be interpreted in practice and not in theory for indirect discrimination.

2 Once discrimination is established, the next stage is to establish which acts of discrimination are rendered unlawful by the SDA. So far as employment is concerned, it is unlawful to discriminate against an applicant on grounds of sex in selection arrangements and job offers; and once in employment it is unlawful to discriminate in access to promotion, training or other benefits or facilities (s.6). It is also true that an employee dismissed on discriminatory

grounds could complain under SDA but for technical reasons connected with the burden of proof, a dismissed employee may be better off using the ordinary rules of unfair dismissal (see page 189).
3 Exceptionally, where an unlawful act of discrimination has occurred, the employer may be able to excuse himself by pleading that being a man or being a woman was a genuine occupational qualification in the particular case (s.7). This defence is limited to the situations specified in the Act, e.g. a man or a woman is needed to preserve decency or privacy, or where the work is to be performed outside the UK in a country whose laws or customs preclude one sex from performing the duties.
4 Where a person wishes to pursue a claim for sex discrimination it is possible to complain to the Equal Opportunities Commission, who may carry out an investigation into an employer's policies, and who may give advice and assistance to individual claimants. In exceptional cases, where a major point of principle is at stake, the EOC may finance the bringing of a case. This could be especially useful when it is borne in mind that individual claimants are not eligible for legal aid for industrial tribunal cases.

The more usual enforcement procedure is for an aggrieved person to commence proceedings in the industrial tribunal, usually within three months of the date of the act of discrimination. In appropriate cases, a successful claimant can be awarded compensation up to a current maximum of £10,000. A serious problem facing the claimant is that he or she will have to bear the burden of proving discrimination. Many cases collapse before they are due for hearing because of the difficulties of obtaining evidence, and the pressure a claimant feels under, particularly where he or she is still in the employment of the employer being complained about.

The SDA is drafted in such a way that it appears to be designed to protect women. In fact, it is designed to protect people who are discriminated against on grounds of sex or marriage. Although the rules contained in the Acts are the first line of attack for an aggrieved person, there are also European Community rights, which have played a significant part in the development of this area of law. Where discrimination relates to the amount to be paid for the work the issue will be resolved under the Equal Pay Act 1970. That Act provides for equal treatment regarding terms and conditions whenever men and women are doing like work, work rated as equivalent or work of equal value. An employer can pay at different rates, only if he can show that he is doing so because of a 'genuine material difference' which is not just the difference in sex. The employer may point to factors like length of service, previous experience or additional qualifications. Equal pay disputes are also resolved in the industrial tribunal.

The scheme of the Race Relations Act 1976 is very similar to that of the Sex Discrimination Act 1975

Termination of contract of employment

The parties to a contract of employment do not usually fix a date for its termination, unless they have negotiated a fixed-term contract. If the duration of the contract is not fixed, the law provides for either party to end the contract by giving notice to the other. Many cases are now covered by rules providing for fixed minimum periods of notice, although it is still open to the parties to agree longer periods if they wish. The EPCA 1978 provides that if an employer wishes to terminate, he must give a minimum of one week's notice for each year of continuous employment, up to a maximum of twelve weeks' notice. If an employee wishes to terminate he must give a minimum of one week's notice, although the contract may specify a longer period. If an employee fails to give the proper period of notice he is in breach of his contract. Theoretically he could be sued for damages by the employer. This may seem a little unlikely, but it should be remembered that the employer may be holding moneys due to an employee, e.g. accrued holiday pay, bonuses or expenses, or a week's pay lying on. The employer could certainly delay making those payments to the employee. What is more, the employer could also refuse to give a reference.

There are no particular legal rules about how notice must be given to be effective. This depends on the terms of the contract. It is common to require notice to be in writing, and often the contract will specify that the period of notice must be calculated by reference to a pay day. So a weekly-paid employee who normally gets paid on a Thursday must give notice to expire on a Thursday.

Nothing in these rules prevents either party from terminating with no notice at all, if the behaviour of the other party justifies such action. When an employee receives no notice, this is referred to as summary or instant dismissal. The question will often arise whether such behaviour on the part of the employer is appropriate. An employee who is the victim of such action may be able to pursue a claim for unfair dismissal.

Unfair dismissal

A statutory right to claim for unfair dismissal was introduced in 1971. Before then, an employee who had been dismissed was limited to a claim against his employer in a civil action in the County Court for wrongful dismissal. In many cases, such an action was worthless because the County Court could only award damages, and these were often limited to an amount equal to the ex-employee's wages for the period of notice he should have received. If proper notice had been given, there was little point in an employee bringing an action for wrongful dismissal. However, this common law remedy is still available and may sometimes be useful. This is particularly likely to be the case where the employee is highly paid, as there is no maximum limit on damages for wrongful dismissal; or where the employee is outside the three-month time limit for pursuing a claim of unfair dismissal; or where the employee does not yet have a sufficient period of continuous employment to be within the statutory unfair dismissal rules.

Prior to 1971, even where the proper notice had been given, an employee might still feel a sense of grievance about the dismissal, especially if there had been no good reason for it. The common law attitude at that time still reflected attitudes that the employer could 'hire and fire' at will. A long-serving and loyal employee could be dismissed for no reason at all, and would have no claim, so long as the employer had given the appropriate period of notice. Unfair dismissal protection was created to change this situation. An employee now has a statutory right not to be unfairly dismissed. An aggrieved employee can seek a remedy in the industrial tribunal. The remedy emphasizes an idea which has developed in employment law, that an employee has a 'property right' in his job. If this is lost when the employee is not at fault (e.g. unfair dismissal or redundancy), then he should be protected or compensated.

The current rules on unfair dismissal are now to be found in EPCA 1978 as amended by the Employment Acts 1980, 1982 and 1990. Employees who may claim their protection are those working under contracts of service, except:

- Part-time employees.
- Persons employed outside the UK.
- Persons over retiring age (either the statutory retiring age or that fixed by the employer).
- Persons engaged on certain types of fixed-term contract.
- Persons who have not yet served two years continuously with their employers.

Fair reasons for dismissal

The Act has given employees who are within its scope the right not to be unfairly dismissed, and lists the reasons why a dismissal may be fair (s.57):

1 *Reasons connected with the employee's capability or qualifications* – capability is broadly defined to include skill, aptitude and health, amongst other things. Qualifications includes degrees and diplomas as well as technical or professional qualifications relevant to the particular job.
2 *Reasons connected with the conduct of the employee* – misconduct may range from very trivial acts to acts which can be classed as gross. Inevitably, these will vary in different employment situations. Swearing may be regarded as fairly trivial on a shopfloor of a factory or construction site but may be regarded as extremely serious on a hotel reception desk or in the offices of a professional person. Generally, the more senior an employee, in terms of age and position in the firm, the more the employer can expect from him. The misbehaviour may take place at work, but in appropriate cases, an employer may dismiss because of conduct occurring outside work, where the conduct could have an adverse effect on the employer's business.
3 *Redundancy, where the employee was fairly selected for redundancy* – the ideal situation for an employer faced with inevitable redundancies is to deal with the problem by reference to an established procedure, agreed at some previous time with the appropriate unions. Otherwise, he may be able to show he acted fairly by applying the principle of 'last in, first out'. This approach may still leave the employer having to choose some from a group who all commenced work together. In that case, he could take account of experience and ability, and the hardship likely to be caused by the redundancy. An employee who thinks he was unfairly selected can claim unfair dismissal.
4 *Where the continued employment of the employee would cause the employer to be in breach of a statutory duty.*
5 *Some other substantial reason justifying dismissal* – it is difficult to spell out in detail in the Act all the circumstances where dismissal can be fair, so some residual provision is inevitable. Under this heading, a number of cases have involved employees unreasonably holding out against variations to their contracts of employment. The employer cannot unilaterally change the terms of employment. Indeed, to attempt to do so would be a breach of contract. But where the commercial needs of his business dictate that he must

run it in some different way, he may need to approach employees to seek their approval for altered working hours, or new shift patterns. In such cases, if the employer has set about negotiating the changes in a reasonable way, and one employee holds out against the change, the employer may have little alternative but to dismiss that employee.

Establishing a reason capable of being fair is not sufficient in itself. The employer must also prove that it is fair in all the circumstances of the case to use that particular reason for justifying dismissal. For example, swearing at the employer is a reason capable of being fair, but whether it is fair to dismiss for swearing in any particular case would depend on the words used, the circumstances surrounding the offence, any provocation, previous treatment of employees for swearing, any works rules, any warnings received by the offending employee, and any other relevant factors. The key question facing the industrial tribunal is to determine whether the employer behaved as a reasonable employer would have done. One factor which may be relevant is the extent to which the employer followed his own disciplinary practices and procedures. These should form part of the information given to employees in their written statement. A sensible employer will have disciplinary procedures suitable for his working situation which are in general conformity with the guidelines issued by ACAS (Advisory, Conciliation and Arbitration Service). The House of Lords has confirmed that, in unfair dismissal cases, if the procedure is unfair then the dismissal itself will also be unfair – *Polkey v. A.E. Dayton Services*.

Definition of dismissal

To be able to proceed with a claim for unfair dismissal, an employee must show that he has been dismissed. There can sometimes be doubt whether a dismissal has occurred. Take the case where an employee simply walks out, or an employer asks the employee for his resignation and this is reluctantly given. Section 55 defines dismissal as:

1 Termination of the contract by the employer with or without notice, or
2 termination by the employee in circumstances justified by the employer's conduct. This situation is called constructive dismissal. The employer's behaviour must amount to a breach which is sufficiently grave so that it entitles the employee to treat himself as discharged from further performance of the contract. Not every breach by an employer can be regarded as so significant. Much will

depend on the factual circumstances, but a simple example is the case where the employer proposes to reduce an employee's wages without negotiation. It will always be a question of fact whether a constructive dismissal has occurred.

These cases pose a real dilemma for an employee. He is faced with some kind of breach by the employer, (e.g. the unilateral imposition of a new shift system) and if he leaves, there is the complication of needing to prove constructive dismissal. If the employee delays taking any action the employer could argue that he had impliedly consented to the change by staying on and working under the new arrangements.

Disciplinary rules

The importance of the rules about unfair dismissal can best be appreciated when it is realized that well over 30,000 applications are filed annually by employees who wish to claim against their employers. The very existence of a right to claim, coupled with the possible cost of the claim to the employer if his employee succeeds, has forced employers to give more detailed consideration to their internal disciplinary and grievance procedures. These procedures must be communicated to employees who are entitled to receive a written statement.

When drafting disciplinary rules, employers can seek guidance from ACAS. The rules should aim to encourage better standards in employees, and should not be concerned exclusively with imposing sanctions. Where sanctions are necessary it is better to provide a range appropriate to the different forms of misbehaviour. Not all misconduct necessarily warrants instant dismissal. It may be more appropriate for an employee to receive oral or written warnings, or some other form of 'punishment', such as demotion.

The ACAS Code of Practice on Disciplinary Procedures stresses that disciplinary rules and procedures are necessary for promoting fairness and order in the treatment of individuals. They also help the organization to operate effectively. Employees must know what standard of conduct is expected from them. The responsibility for formulating a policy for maintaining discipline lies with the management, who should aim to secure involvement at all levels when formulating new rules. Rules should not be so general as to be meaningless, but rather they should specify clearly and precisely what is necessary for the efficient and safe performance of work, and the maintenance of satisfactory working relations. Employees should know the likely consequences of breaking the rules, particularly where such a breach is likely to lead to summary dismissal. Disciplinary procedures should

Employment law 193

be viewed as a means of encouraging improved conduct in individuals. Great emphasis is laid on fairness in operating the procedures. This means telling an individual what the complaint is, and allowing him or her the opportunity to state their case before a decision is reached; allowing the employee the right to be accompanied by a trade union official or a fellow employee; ensuring that no decision is reached until the case has been fairly investigated; and ensuring that the employee is given an explanation for any penalty imposed. An employer should keep proper records of all actions taken under the disciplinary rules. Good employers will normally provide for breaches of the disciplinary rules to be struck from an employee's record after completing a further specified satisfactory period of service.

Remedies for unfair dismissal

Claims by an employee under the unfair dismissal rules are made to an industrial tribunal. An action should usually be commenced within three months of the effective date of termination of the employment. Claims are initially referred to a Conciliation Officer of ACAS, whose duty it is to mediate between the parties to see if a settlement can be reached. At this stage the employer may agree to take the employee back, or may agree to pay a sum by way of compensation which is acceptable to the employee. A large proportion of claims are resolved at this stage.

Before a full tribunal hearing takes place, it may be appropriate to hold a pre-trial assessment to consider whether the case has any real substance. The case is not decided at this stage but if it is clear that one party has no real chance of success, the tribunal may warn that party that costs could be awarded against them. Not surprisingly, a number of applications to the tribunal are withdrawn at this stage. If no withdrawal occurs, the case proceeds to a full hearing. An employee found to have been unfairly dismissed may be awarded:

1 *Reinstatement* – the employee's job is restored and he is treated in all respects as though he had never been dismissed.
2 *Re-engagement* – this gives the employee a job comparable to his former job.
3 *Compensation* – all awards consist of two elements, the *basic* award and the *compensatory* award. The calculation of the basic award depends on a combination of factors, namely age, length of continuous service and the amount of a week's pay. An employee's working life is divided into three periods: years worked between eighteen and twenty-one, years worked between twenty-two and

forty, and years worked between forty-one and sixty-four (sixty in the case of women). A maximum of twenty years of employment is taken into account and a financial cut-off point on what constitutes a week's pay is fixed. This is increased from time to time to take account of rising wage levels. Once the relevant facts are known, the basic award is calculated by awarding half a week's pay for each year of service between eighteen and twenty-one, one week's pay for each year of service between twenty-two and forty and one and a half weeks' pay for each year of service between forty-one and sixty-five. The basic award may be reduced if the employee has caused or contributed to his dismissal, or if the employee has been guilty of bad conduct before dismissal which is only revealed after his dismissal takes effect. The compensatory award is an amount which the tribunal considers just and equitable bearing in mind a claimant's losses and expenses, including loss of future earnings. The calculation is often speculative, since it is impossible to know how long a claimant may remain unemployed. This part of an employee's compensation may be reduced where his conduct has contributed to his dismissal, or, where he has not taken reasonable steps to minimize his losses, e.g. if he has failed to seek new employment. The maximum figure for the compensatory award may be increased exceptionally where additional awards can be made. An additional award is relevant if an employer refuses to comply with an order to reinstate or re-engage the employee. The amount could be between thirteen and fifty-two weeks' pay, over and above the basic and compensatory awards.

Note that both reinstatement and re-engagement are discretionary remedies. The tribunal must consider the wishes of the parties and the practicability of compliance before making such an order. Where either order is made, if the employer refuses to comply with it he will be obliged to pay an additional amount of compensation.

Tribunal procedure

Procedure at the tribunal hearing may be relatively formal, perhaps as a result of the 'adversarial' nature of these proceedings. The tribunal consists of a legally qualified chairman, who sits with two lay members who are intended to be representative of employers and employees, respectively. All three play an equal part in the decision-making process. Evidence is given on oath. Although the tribunal may take a more informal approach than a court, it is usually better to bring witnesses who can speak directly about the facts than to rely on hearsay evidence. Applicants should bring along relevant documentary evidence, such as their written statement of terms, or their written reasons for dismissal, or documents relating to their financial position which may be relevant when calculating the compensatory award. It may be difficult to persuade people to come to give evidence, or to get hold of documents considered vital to the case. In these circumstances an applicant can ask the tribunal to grant a witness summons order, or a discovery of documents order.

The parties will each be invited to present their case and to call any necessary witnesses, who may be cross-examined and who may also be questioned by members of the tribunal. Where an applicant is unrepresented, the members can do much to assist by the type of questions they put. The length of each case will depend on factors such as the number of witnesses and the complexity of the evidence, but it will not normally last more than two days. The tribunal retires to consider its decision, which is given to the parties, with full reasons.

Although large sums of money are at stake, and the law to be applied is complex and constantly changing, no legal aid is available to pay for representation in the tribunal. A claimant who is dissatisfied with the outcome of the hearing may have a right to appeal. Appeals on a point of law lie to the Employment Appeal Tribunal, then in turn to the Court of Appeal and the House of Lords.

Redundancy payments

A dismissal is due to redundancy if:

1 The employer has ceased to carry on business for the purposes of which the employee was employed, or
2 the employer has ceased to carry on business in the place where the employee was employed, or
3 the requirements of the business for employees to carry out work of a particular kind have ceased or diminished.

Where a redundancy situation exists, there is no entitlement to a payment if the employee is dismissed for misconduct, or if he unreasonably refuses suitable alternative employment which has been offered to him in accordance with the Act. Employees who have been given notice because of redundancy are entitled to reasonable time off with pay during working hours to look for a new job or to make arrangements for retraining. What is a reasonable amount of time off will vary according to the differing circumstances of employers and employees. During such absences employees should be paid at their normal rate.

Payments are calculated by reference to rules similar to those used to calculate a basic award for unfair dismissal. Many employees are disappointed at the size of the payment they receive. This may be because they have read about especially generous redundancy schemes operating for particular industries, or because they have failed to understand the basis on which the calculation is made. Vital to the calculation of both redundancy payment and unfair dismissal compensation is the concept of 'a week's pay'. Where an employee has normal working hours and the pay does not vary from week to week, a week's pay is simply the basic weekly wage. Overtime earnings are not included unless overtime is obligatory under the contract of employment. If the employee has variable earnings because of a shift pattern, or piece rates, or productivity bonuses, these are usually averaged over a twelve-week period.

An employer is under a duty to consult with trade unions if sizable numbers are to be made redundant over a short period. This duty is quite separate from his obligation to notify the Department of Employment. If the employer fails to consult, a special protective award may be made against him, which will safeguard the remuneration of employees for the period for which it is made.

Hearing redundancy and unfair dismissal claims forms the largest part of the work of industrial tribunals. However, the trend has been to increase this workload in employment-related spheres. Since 1974, the tribunals have heard appeals against Prohibition and Improvement Notices issued under the Health and Safety at Work etc. Act 1974. The equal pay and sex discrimination rules can also be enforced in the industrial tribunal. Extensions to the work of the tribunals were made by the Employment Act 1980 in relation to unreasonable exclusion or expulsion from a trade union. The Trade Union Reform and Employment Rights Bill 1993 makes provision for the jurisdiction of tribunals to be extended to breach of contract claims with appeal to the Employment Appeal Tribunal.

This brief account of employment law is sufficient to show it to be a complex and constantly changing area. Many firms do not have a legal or personnel department which can guide them through the maze of rules. However, the Department of Employment publishes a large range of explanatory booklets, and the officers of ACAS are always ready to give help and advice on questions which are within their particular sphere.

Employees and trade unions

Employees governed by a contract of service may enjoy a further contractual relationship which is of major significance in employment, i.e. membership of a trade union. Several million people belong to trade unions in this country, ranging from the very large general unions like the Transport and General Workers Union to small specialist unions with quite small numbers of members.

The position of the trade unions under the law is unusual, some times accounted for because of the history of their development. Trade unions could only begin to exist and operate as they do today once they were freed from the problems of laws like the Combination Acts and the civil and criminal rules about conspiracy. These shackles were not removed until 1871, when the growth of the movement into what it is today could get under way. After conceding the right of trade unions to exist and organize workers, the law originally did very little to interfere with, or control, their internal activities. The result of their early policy of non-interference is that trade unions have kept control of their internal affairs, to a considerable extent, until the passing of the Trade Union Act 1984. Under provisions introduced by that Act, unions are required to hold secret ballots for trade union elections, secret ballots before industrial action, and secret ballots about political funding, if any. Of the statutory rules relating to trade unions, most have been framed to regulate their external activities, e.g. political activities and industrial activities. The main reason why a union engages in industrial action is to support the claims it is making for its members in collective negotiations with employers. The process of voluntary collective bargaining is recognized in this country as the principal means of fixing terms of employment. If a union is to be in a position to bargain effectively with an employer, it must gain recognition for that purpose. Between 1975 and 1980 the rules of the Employment Protection Act 1975 permitted trade unions to use the services of the Advisory, Conciliation and Arbitration Service to gain recognition from employers who were reluctant to grant bargaining rights. These procedures proved ineffective, and were repealed by the Employment Act 1980. Now, a union wishing to be recognized for bargaining purposes must establish that right for itself, by taking industrial action if necessary. Moreover, under the provisions of TURERB 1993, ACAS will no longer be under a duty to encourage the extension and reform of collective bargaining.

198 Building and Land Management Law for Students

Consolidation of legislation

Throughout the remainder of this chapter, it should be remembered that the Trade Union and Labour Relations (Consolidation) Act 1992 consolidates the Employment Acts of 1980, 1982, 1988 and 1989, the Trade Union and Labour Relations Act 1974 and the Trade Union Act 1984.

Definition of a trade union

The modern definition of a trade union was found in the Trade Union and Labour Relations Act 1974, now consolidated into the Trade Union and Labour Relations (Consolidation) Act 1992 (TULRCA). 'An organisation consisting wholly or mainly of workers... whose principal purposes include the regulation of relations between workers and employers'. It will be realized that trade unions are defined both by reference to their membership and their purposes. No formal act of creation is necessary. Although an organization complying with this definition qualifies for the name 'trade union', the best legal rights and protections are afforded only to 'independent trade unions' (ITUs). The status of being independent is achieved by a trade union applying to an official called the Certification Officer (CO) for a Certificate of Independence.

As a first stage towards achieving certification, a union applies to the CO to be listed. Listing is conclusive proof of status of a trade union. The next is to apply for a Certificate of Independence. The list of advantages conferred by independence is impressive, particularly where the union is also recognized by the employer for bargaining purposes. It includes:

- An independent trade union has the statutory right to receive certain information for bargaining purposes.
- ITUs have the right to be consulted about proposed redundancies.
- ITUs can appoint safety representatives under the Health and Safety at Work etc. Act 1974.
- Members of ITUs may have time off work to take part in union activities.
- Members of ITUs have the right not to be dismissed because of taking part in union activities at an appropriate time.
- ITUs may seek state subsidies for postal ballots.

When the CO considers a union's application for a Certificate of Independence, he must take into account the criteria laid down by

TULRA 1974. These provide that the union must not be under the domination or control of the employer, nor liable to interference from the employer. In other words, the advantages of independence should only be conferred on genuine unions, which can exercise a proper, vigorous and independent role on behalf of their members. Unions least likely to gain certificates are 'house unions' or staff associations which have often emerged from a firm's social club.

When the CO assesses the independence of a union he considers:

1 How is it funded? A subsidy from the employer would tend to show control and domination by him.
2 Does the employer provide facilities, e.g. office space, telephones, typing services?
3 What is the history of the union? Did it spring from an organization created by the employer?
4 Is the union based only in one company?
5 How 'robust' is the union in negotiations with the employer?

The CO is also permitted to hear the views of other unions about granting the certificate. Another union may have a vested interest in the outcome, as its own position may be jeopardized by the emergence of a new independent union. This could particularly affect the recruitment of new members.

As the Certificate of Independence is the key to claiming all the statutory rights and protections, a union which is refused this status may wish to appeal against the CO's decision. A right of appeal against the refusal of a certificate lies to the Employment Appeal Tribunal. If the application is still unsuccessful the union may apply again, after it has put right those matters specified as reasons for refusal of the certificate.

The legal status of a trade union

Strictly speaking, a trade union is an unincorporated association but so many special rules have been created in relation to unions, that their position more closely resembles that of a corporation. The trade union is an association of individuals, bound together by their rule book, which is effectively the terms of a contract between all the members. The greatest difficulties caused by lack of corporate status would usually be connected with holding property or making contracts. Special rules in TULRCA cover these points. A trade union may make contracts in its own name, and its property will be vested in trustees.

The rule book

Generally a union is free to draft its own rules subject to the following limitations:

1 The rules must not offend the discrimination laws or EEC regulations.
2 The rules may not exclude the principles of natural justice, nor may they seek to oust the jurisdiction of the courts.
3 Every member has a right under TULRCA not to be subjected to unjustifiable discipline.
4 The rules of a union which has a political fund must provide for members to be able to opt out of contributing to the political fund.
5 The rules may not exclude the duty of every trade union to secure that its principal executive committee has been directly elected by the membership.

As the rule book constitutes the terms of a contract between the members, the courts have power to enforce the rules if an aggrieved member brings an action. It must, however, be borne in mind that not all the rules may be contained in the rule book; some may exist because of custom or practice.

Membership of a trade union

The right to become and remain a trade union member has been very significant especially when a 'closed shop' was in operation at a place of work. Basically that meant either that the worker must be, or become, a trade union member. The closed shop (which the law calls a union membership agreement) has been gradually eroded by legislation, most recently by the Employment Act 1990. All statutory support for closed shops has now been removed, although they can still exist or be created by agreement between employers and employees.

When an applicant applies to a union for membership, the situation is controlled by the union's rule book (which also controls questions of the discipline of members, including possible expulsion from the union). It has already been seen that trade unions have great control over the contents of their rule book. If the rules are clear, unambiguous and properly applied, there should be no need for recourse to the courts for their interpretation by an applicant refused admission.

As the consequences of an exclusion can be so grave, the excluded worker may refer his case to the industrial tribunal, under rules introduced by the Employment Act 1980. These rules provide that (where a union membership agreement is in force) persons have the right not to be unreasonably excluded from membership of the appropriate union. The tribunal must look at the 'substantial merits of the case'. The mere fact that a union has slavishly followed all its own rules is not in itself evidence of reasonableness. If the tribunal finds the claim well founded, it can make a declaration to that effect, although this will not suffice to get the worker into the union. The advantage of getting such an order from the tribunal is that the worker may then apply for an award of compensation against the union. The amount is fixed to try to compensate him for the loss he will sustain as a result of the union's refusal to admit him. The worker must have taken reasonable steps to keep his losses as small as possible – this is known as the duty to mitigate loss. He may have the compensation reduced if his own behaviour in any way caused the refusal to admit him.

Freedom to join a trade union

The rules which have just been considered operate only when a closed shop exists. Not all workers are faced with such a situation, and normally they are free to join a union if they choose to do so. (The exceptions to this rule are members of the police and armed forces, who cannot form themselves into unions.) What is more, workers are often free to choose between several appropriate unions. This apparent choice may be restricted where the Bridlington Agreement operates. This agreement, made in 1939 between unions affiliated to the TUC, regulates inter-union competition for members, and the transfer of members between affiliated unions. The main object is to prevent unions from 'poaching' members from each other. Where a union wishes to complain about such activities, the dispute will be referred to a Disputes Committee of the TUC. Again, this is a body with no legal sanctions to impose. If it finds the claim of poaching is well founded, it will usually tell the offending union to expel the member in question, in the hope that he will then rejoin his original union. The Bridlington Agreement may contravene Article 11 of the European Convention on Human Rights which guarantees freedom of association. If this is so, then the Bridlington Agreement cannot affect the rights of individuals to join the trade union of their choice. Nowadays, changes in patterns of employment have resulted in many employers wanting to deal with only one union. In such a case it may be a question of competing unions holding a 'beauty contest' to determine which union should be recognized by the employer. Examples of this have occured in the print and motor industries.

'Union only' contracts

It used to be quite common for organizations like local authorities to demand of all their suppliers that they should allow their workers to be union members. Indeed, some authorities would go so far as to refuse to invite tenders from suppliers who did not comply with this rule. Since the Employment Act 1982, any term in a contract for the supply of goods or services is void if it purports to require that work done under the contract shall only be done by unionized or non-unionized labour.

A new tort is also created by the Act. It is a breach of statutory duty if, on grounds of union membership or non-membership, a person:

1 Fails to include a particular person's name on an approved list of suppliers; or
2 excludes a person from tendering; or
3 terminates a contract for the supply of goods or services.

Expulsion from membership

A union's rule book will normally provide a range of disciplinary measures, to cover various acts of misbehaviour by its members. Misbehaviour can range from failure to pay union dues, to breach of vague rules like 'behaviour prejudicial to the union'. The discipline imposed can range from a fine, or temporary withdrawal of benefits, to ultimate expulsion from the union. This would be an exceptionally severe form of punishment where a closed shop operates. The remedies for a worker who wishes to contest his expulsion are:

1 *To bring an action against the union in the ordinary courts* – this can succeed if the union has acted in breach of its rules, or contrary to the principles of natural justice (e.g. by failing to give the expelled worker the opportunity to put his side of the case).
2 *To bring an action against the union in the industrial tribunal* – under ss.3-5 of the Employment Act 1988, where the member has been subjected to unjustifiable discipline. This was a new feature introduced in 1988, and could include imposing a fine, suspension or expulsion on a member for refusing to take part in industrial action.

Industrial activities of trade unions

Industrial action represents the means by which employees can collectively bring pressure to bear upon an employer. The theory of this concerted action is that it produces an equality of bargaining power between workers and employers which could never exist between one individual employee and the employer. Common reasons for using industrial action are support of claims for higher wages or demonstrating protest at unsafe working conditions.

Although the strike is the ultimate form of industrial action, a union may achieve results by using a work-to-rule, a go-slow, a boycott of customers or suppliers, or a refusal to work contractual overtime. If a strike is considered to be the most appropriate action, it may not be necessary for the union to call out all its members. Results can often be achieved quickly with selective stoppages, especially if these are by key workers, e.g. those operating computers, or involved in the collection of money for the employer.

All forms of industrial action can give rise to the question: 'Are the workers concerned in breach of their contract of employment?' The answer to this is not easy. For example, an employee working to rule could give little cause for complaint if he was meticulously observing the rules laid down in his contract; but if he seeks to give those rules an unrealistic meaning, so that he is wilfully obstructing his employer's business, then he may be in breach of his contract. If there was a breach, one of the remedies would be for the employer to sue the offending employees for damages for breach of contract. However, any employer conscious of the need to preserve good industrial relations is unlikely to resort to such action. It is far more likely that an employer who believes his business interests are being wrongly interfered with will want to take action in tort against officials and members. A number of specific torts (e.g. interference with contract, and conspiracy, often known as the economic torts), relate almost exclusively to trade union activities. Other torts like trespass and public nuisance can easily be committed while participating in industrial action.

An example of a typical situation will show how these possibilities could occur. A union in discussion with an employer about new wage rates feels that negotiations are making no progress and that the employer could offer more. The employees who are union members are called out on strike after a ballot. The union members account for about 60 per cent of the workforce. While some employees continue to work, the effect of the strike is not fully felt, so the union officials organize a picket line and attempt to persuade those still working not to cross the picket line (i.e. persuade them to break their contracts of employment). What is more, officials realize that the impact of their action can be increased by persuading lorries bringing supplies to turn back without making delivery (i.e. causing a breach of commercial contracts). The number of pickets may well get out of control, spilling on to the road and causing it to be blocked, or into the employer's premises, constituting trespass.

The possibility of the employer suing successfully in tort, either for an injunction to stop the commission of the torts, or for damages, is restricted because of the immunities enjoyed by trade union officials and members. Since 1906, these immunities have existed if the person committing the tort was acting in contemplation or furtherance of a trade dispute. Because of the protection it affords against tort actions, this is frequently referred to as the golden formula, although it is subject to recent limitations in its scope. Two elements of the formula require consideration:

- When is there a trade dispute?
- What constitutes acting in contemplation or furtherance of it?

Trade disputes

A definition of a trade dispute is contained in TULRCA 1992. It must be a dispute between workers and their employers, which relates wholly or mainly to one or more of the following matters:

1 The terms and conditions of employment.
2 The physical conditions in which people work.
3 The engagement or non-engagement of workers.
4 The termination of employment or suspension of workers.
5 The allocation of work between workers.
6 The question of discipline.
7 The membership or non-membership of a trade union by any worker.
8 The facilities for officials of trade unions.
9 The machinery for negotiation or consultation relating to any of these matters.
10 The question of recognition of a trade union.

The trade dispute may also relate to matters outside Great Britain. This is of considerable significance where companies may be multinational, or working on contracts abroad, engaging cheap local labour.

There are many complex cases on the full meaning of the definition, because an employer who sees his business severely affected by industrial action would naturally be keen to prove that the golden formula should not apply on the ground that there was no trade dispute. An interesting example where this argument succeeded involved the BBC. The BBC planned to televise the 1978 Cup Final at

Wembley, and to 'beam' their transmission to many foreign countries. An anti-apartheid movement approached the union organizing the TV engineers and asked for support in not beaming the programme to South Africa. If the engineers were to agree to this, and refuse to carry out the orders of the BBC, they would be in breach of their contracts of employment. The BBC in turn would inevitably be in breach of a number of its commercial contracts with countries overseas who would not then receive the programme. Was the proposed action of the TV engineers in contemplation or furtherance of a trade dispute? The Court of Appeal decided that no trade dispute existed here, although the engineers could easily have brought themselves within the definition. If they had asked the BBC to insert a new term in their contracts which provided that they would not be required to take part in broadcasts to South Africa, the BBC would almost certainly have refused. Then the engineers could have said that their dispute was related to 'terms and conditions of employment' (*BBC v. Hearn*).

Contemplation or furtherance of a dispute

Although the judges in the cases have often emphasized that acts can only 'further a trade dispute' if they are not too remote from the dispute, or if they are reasonably capable of furthering the dispute when viewed objectively, these approaches inevitably limit the scope of the golden formula. The matter was considered by the House of Lords twice, in 1979 and 1980, and on both occasions the court confirmed that the proper approach to interpreting the word 'furtherance' is to apply a subjective test. Did the person in question honestly believe that his acts would further the dispute? If so, he must have the benefit of the golden formula, however unwise or damaging his actions may be.

Limitations of the immunity

The whole question of the scope of trade union immunity from actions in tort is politically very contentious, and developments in this field have continued apace. The present position can be summarized as follows:

1 *The position of the trade union itself* – the union can be liable in tort if the act in question is authorized or endorsed by the president or general secretary, or the principal executive committee or any other committee or official whether employed by the union or not. Where the trade union is successfully sued for damages, the

damages awarded are subject to upper limits dependent on the number of members. A union with less than 5000 members may have to pay damages up to £10,000, whereas a union with more than 100,000 members may have to face a damages bill for up to £250,000.

2   *The need for a ballot of the members* – under the Trade Union Act 1984, immunities will only be extended to industrial action which has been preceded by a secret ballot. The ballot must have taken place within four weeks of the action being taken and a majority of those voting must support the proposed action. Public funds are available to conduct the ballot. Immunities will be lost if no ballot is held or if it does not conform to the statutory rules.

3   *Secondary action* – when an official or member commits a tort, he will be protected if it is one of the torts protected by the golden formula. Which torts are so protected has varied from time to time. Initially the immunity only covered inducing a breach of contract of employment, but was subsequently extended to cover inducing breaches of commercial contracts. The changes introduced between 1980 and 1990 seek to limit the golden formula protections. Where a breach of a commercial contract has been induced, the immunity can only operate if the breach is caused by primary, or in very limited cases, secondary action. This limitation is designed to protect employers far removed from the actual dispute, who may find themselves being used as unwilling pawns by a trade union experiencing difficulty in achieving its objectives by more direct means.

4   Further restrictions on the immunity from liability in tort are contained in TULRCA 1992. Thus, there is no immunity for tortious actions where the reason for the action is that the employer is employing a person who is not a trade union member or who is not a member of a particular union. Anyone who commits an industrial tort while seeking to impose or enforce a closed shop will not have the protection of the golden formula.

5   Where the member or official concerned is picketing, limited immunity is granted as long as:

(a) he pickets in contemplation or furtherance of a trade dispute.
(b) He pickets only in the right places – this is now usually limited to the member's own place of work.
(c) He pickets for the right purposes – these are limited to peacefully obtaining or communicating information or peacefully persuading anyone to work or not to work.

One difficulty about the immunity in relation to picketing is that it is extremely restricted, and bears no relation to the realities of the picket line. Although some changes in the law were made in 1980, they did not put right the worst practical problems, namely that there is no right to stop vehicles, and there is no control over the number of pickets. The 1980 changes were principally aimed at putting a stop to secondary picketing, i.e. picketing on premises other than those of the employer in dispute. It was hoped that the other problems could be resolved by the issue of a Code of Practice. The most recent code was published by the Secretary of State for Employment in 1992.

By its very nature picketing is a public activity which tends to involve the commission of criminal offences, e.g. obstruction of the highway, causing a breach of the peace. Although the immunity is so worded as to cover criminal behaviour ('it shall be lawful'), the scope of the protection is extremely narrow, as proved by the cases on the point. Thus there was no protection for a picket charged with obstruction because he had tried to detain a lorry driver for long enough to speak to him (*DPP v. Broome*) nor for the pickets who blocked a road by circling round continuously in it (*Tynan v. Balmer*).

Picketing must also be conducted in compliance with the provisions of the Public Order Act 1986. Section 14 of the Act empowers the police to impose conditions on 'public assemblies' i.e. assemblies of more than twenty people in a public place in the open air. The police can issue directions as to the place, the duration and numbers involved. The power of the police exists where it appears that the assembly may result in serious public disorder, serious damage to property, or serious disruption to the life of the community. The police also have power to act where they believe the purpose of the assembly is to intimidate others.

9
The law of tort

Introduction

The basis of liability in tort is probably one of the most difficult legal ideas for the layman to grasp. The word itself is of French origin and means a wrong. Basically, the law of tort is concerned with situations where the behaviour of one party causes, or threatens to cause, harm to the interests of another party. The rules of the law of tort determine when one party can be compensated for the behaviour of another. The law of tort is limited in its scope and although new torts do evolve from time to time, there is not necessarily a legal remedy for every wrong suffered. A plaintiff will only succeed in an action if he can show that the defendant's behaviour falls into a specified situation covered by the law of tort. Those specified situations are then given identifying names such as the tort of defamation, the tort of nuisance, the tort of negligence and the tort of trespass.

Much of the law of tort has developed through the cases. Some of the rules have now been given statutory force, for example the Law Reform (Contributory Negligence) Act 1945, the Occupiers' Liability Act 1984, and the Torts (Interference with Goods) Act 1977. But case law continues to play a very significant role in the development of tort principles. Case law allows the rules to operate very flexibly and to be applied to widely differing situations. The cost of that flexibility will sometimes be a lack of certainty. While the rules are continuing to evolve, it may be difficult to say with any precision whether liability will arise.

Compensating plaintiffs under the law of tort is a long established part of our legal system, with cases dating back to the thirteenth and fourteenth centuries. The earliest forms of tort usually gave protection to the person or to property. These were the various forms of trespass, where the injury which had been inflicted was direct. As society became more sophisticated, the law of tort developed to give protection against less direct forms of harm. Torts such as negligence and the torts designed to protect economic interests then evolved.

Tort distinguished from crime and contract

There are circumstances where exactly the same act or behaviour may give rise to liability in both the law of tort and criminal law. Take, for example, the case of the motorist who causes death by dangerous driving. He may find himself charged with offences in the criminal courts, and at the same time sued for damages on behalf of the deceased victim. Criminal law is concerned with the preservation of order in society, whereas the law of tort is concerned to compensate victims who have suffered harm. Although it is easy to state the differing functions of criminal law and the law of tort, nowadays these differences are blurred in practice. For example, the criminal courts have power in certain circumstances to make compensation orders to victims as well as meting out punishment to the offender.

The same set of circumstances can also give rise to liability in both the law of contract and the law of tort. For example, a person who buys a defective hedgecutter which causes him to cut off his finger may sue the supplier for breach of contract, or the manufacturer for the tort of negligence. The most significant difference between contract and tort is that contractual liability is based on consent between the parties, a form of arranged liability, whereas tort liability is imposed by law. Although this distinction may be true in essence, again the boundaries between the two areas are not so clear. In the law of contract many of the obligations are in fact imposed by law. For example, in a sale of goods situation conditions are imposed by the Sale of Goods Act 1979, and the parties are not free to exclude them. Where tort and contractual liability overlap, a plaintiff can take advantage of whichever claim is more favourable to him. That may depend upon the limitation rules governing time limits for bringing actions and the rules for the calculation of damages. His choice of action may also be governed by any possible defences available to the other party. If he brings an action in tort, the defence of contributory negligence may be pleaded and may largely defeat his claim. That defence is not available in contract.

Although there is considerable emphasis on the compensatory function of the law of tort, an award of damages is not the only remedy when a tort has been committed. A plaintiff may wish to put a stop to some form of behaviour being carried on by the defendant, for example repeated excessive noise. He may ask the court to grant an injunction in his favour. An injunction is an order of the court telling the defendant to do, or refrain from doing, certain things. A defendant who fails to obey an injunction is in contempt of court.

When suing in tort a plaintiff does not have to specify the particular tort on which he is relying. In practice, the conduct he is complaining about may amount to more than one tort. For example, the same behaviour may amount to both the tort of nuisance and the tort of negligence. If more than one tort has been committed the plaintiff will not recover double damages; he will simply have the protection of alternative reasons why he should succeed in his claim.

Tort compared with other systems of compensation

In order to succeed in a tort action the plaintiff must usually show fault on the part of the defendant. To do so, it is likely that he will need to bring complex, lengthy and costly court proceedings. If a plaintiff is successful in those proceedings there is no certainty that the defendant will have the means to satisfy the judgment made against him, i.e. he may be unable to pay the damages. As a general rule in English law, there is no obligation to insure against the possibility of tort claims. (There is an exception to this rule in the provisions of the Road Traffic Act 1972, which compel the owner of a motor vehicle to insure against tortious liability owed to third parties.) In view of the problems of proving fault, it has sometimes been argued that victims of accidents should receive compensation from some kind of central insurance fund on a 'no fault' basis i.e. without having to prove liability against a defendant. A Royal Commission considered this idea and its Report was published in 1978. It recommended that the no fault principle should be extended to apply to motor vehicle accidents, but it did not recommend abolition of tort actions in other cases of personal injury. There is great pressure at the present time to compensate victims of medical negligence on a 'no fault' basis.

There is some experience of operating a no fault insurance system in this country. The Industrial Injuries Scheme, which forms part of our social security system, allows a workman who is injured in an accident at work to receive disablement benefits if he suffered personal injury caused by accident arising out of, and in the course of, his employment, whether or not his employer was at fault. Such an injured workman may also successfully sue his employers for damages in the tort of negligence if he can prove that the employer was at fault. In assessing his damages within the law of tort some account will be taken of social security payments received. The same workman may also have private insurance cover where the payments made to him are unaffected by benefits received under the social security system or through tort damages.

The failure to implement any kind of no fault compensation scheme in this country can be contrasted with the experience of New Zealand. The scheme introduced there in 1972 covers all cases of accidental injury and the common law action for damages in the law of tort has been abolished.

Trespass

Trespass is commonly thought to be a tort relating to land, but the law also recognizes trespass to the person and trespass to goods. Trespass to the person occurs if force is directly and intentionally inflicted on another person. For this purpose force means any physical contact. It would equally be a trespass to punch someone on the nose or to give them a kiss. It is a special feature of all forms of trespass that an action can be brought even if no damage has been suffered. This tort is said to be actionable *'per se'*, of itself, without proof of damage. Of course, a person who has suffered no real harm will usually receive nominal damages. People often consent to actions which would otherwise constitute a trespass to the person, e.g. where one willingly holds out a hand for another to shake it, where a patient willingly opens his mouth to allow a dentist to extract his teeth. Where consent has been expressly or impliedly given it is a complete defence to an action. The defence of consent is usually known by the Latin maxim *'volenti non fit injuria'*, he who consents cannot complain of the injury.

Where there is a trespass to the person, the same behaviour could also be a crime. For example, a punch on the nose could also result in a prosecution for assault.

Trespass to goods

Originally, there were a number of forms of tortious action which could be brought to protect goods. The law in this area was reformed by the Torts (Interference with Goods) Act 1977. Wrongful interference with goods can take the form of trespass or conversion. Trespass to goods is an intentional and direct act of interference by a defendant with goods in the plaintiff's possession. The emphasis here is to protect possession. Frequently, the person in possession will be the owner of the goods in question. The defendant's act must be direct and intentional, so tearing a page from your book, or throwing a stone at your car, or breaking a window are all examples of trespass to goods. In these cases the person inflicting the damage had no intention to deny the other person's right to possess the items. If the defendant does commit some act which denies a plaintiff's right to possess the goods this may amount to conversion. So if the defendant took the book from the plaintiff, rather than simply tearing some pages out of it, he has converted the book to his own use. The principal remedy for any wrongful interference with goods is damages, but where a defendant is still in possession or control of the goods, section 3 of the 1977 Act may permit the plaintiff to obtain an order for delivery of the goods and payment of any consequential damages. Where a plaintiff seeks only damages these are based on the value of the chattel together with any consequential loss suffered.

Trespass to land

This tort consists of an unjustifiable interference with the possession of land. The interference may consist of walking over another's land, or throwing things onto the land, or placing a ladder against the surrounding wall or even swinging a crane jib over the land. For the purposes of this tort, land means the surface of the earth itself, anything which is a fixture on it (for example a building), the air space above it to the extent necessary for the use and enjoyment of the land, and the sub-soil below the land. Aeroplanes flying over land would constitute a trespass if the activity were not permitted by Act of Parliament.

Trespass protects possession of land. Where a person is rightfully in possession he may bring an action against even the owner of the land. An example could occur if the landlord rented premises to a tenant without reserving the right of entry to those premises. If he ever entered without the tenant's permission he would commit trespass. The tort is actionable without proof of damage. In those cases where no harm has occurred the plaintiff may be seeking an injunction, for example to restrain a neighbour who regularly takes a short cut through the plaintiff's garden. Cases such as *Anchor Brewhouse Developments Ltd v. Berkley House Ltd* (1987) show that the courts are prepared to grant injunctions to prevent oversailing by tower cranes, even where no damage is caused.

An apparent trespass may be justified because the person is on the land with express or implied permission or under statutory authority or in exercise of a right of way. An example of an implied permission to enter premises is the open door of the supermarket. Having entered the shop, however, if a shopper creates a disturbance he may be asked to leave. He would then become a trespasser if he refused to do so. There are numerous examples of persons having the right to enter premises without the occupier's permission. Acts of Parliament confer powers on officials such as factory inspectors, gas and electricity board officials and public health officials, who have the right to enter premises if particular circumstances exist. It is not a defence to trespass to plead that the defendant did not know he was trespassing. Even if he has lost his way, he could be liable.

There is a general misconception that trespass is a crime. This is not usually the case but in a limited number of cases Acts of Parliament can make trespass into a criminal offence. Typical examples would be trespass on property belonging to British Rail or British Coal. The notice 'Trespassers will be prosecuted' is, therefore, usually meaningless except in these limited cases. Changes have occurred however, in the law of trespass to deal with the position of squatters and, more recently, itinerant groups of hippies travelling to the Summer Solstice festivals.

It is obvious that construction work could involve constant danger of trespass. A contractor would be wise to seek permission for acts which would otherwise cause a breach of these rules. If court action becomes necessary, a plaintiff who has been dispossessed of his land by a trespasser must bring an action for ejectment and he may combine it with an action for damages. If the plaintiff remains in possession, he may claim damages and, if necessary, an injunction to restrain further acts of trespass. The help of the court is not always needed in cases of trespass. The owner of the land can ask the trespasser to leave. If he does not do so, the owner may then use reasonable amounts of force to eject him. What is reasonable will vary with the circumstances of each case.

Nuisance

For the purposes of the law, nuisances fall into three categories: public nuisance, private nuisance and statutory nuisance.

Public nuisance

Public nuisance is primarily a crime, and consists of generating harm to members of the public. Examples of public nuisance include obstructing the public highway, allowing smoke from a burning field to blow across the highway, polluting a public water supply or creating some projection over the highway. An unruly march or demonstration which causes obstruction of the highway could also constitute a public nuisance. Normally such behaviour is controlled by criminal prosecution, but it may be possible for a person affected by the public nuisance to sue in tort. A plaintiff would need to show that he suffered special and particular harm as a result of the public nuisance, greater than that suffered by ordinary members of the public. Where he can prove this, he may succeed either in recovering damages, or in obtaining an injunction, or both. Sometimes, the behaviour of the defendant will constitute both a public and a private nuisance and in such a case, the individual affected will be in an even stronger position to sue. In one case, quarrying operations had caused all the houses in the neighbourhood to be affected by dust and there was held to be both public and private nuisance. A right to sue for public nuisance may be important in those cases where a plaintiff would have no right to sue in private nuisance, for example in those cases where he has no interest in land to protect.

Private nuisance

The tort of private nuisance is committed when one person unlawfully interferes with another's use or enjoyment of land. The range of activities which can amount to private nuisance is very wide and could include harm caused by smells, fumes, vibrations, noise, dust, encroachment by tree roots, escape of sewage and the keeping of animals. Essentially, the law of private nuisance attempts to reconcile conflicting interests. An Englishman's home is his castle, and he is free to do within it as he pleases, so long as he does not spoil his neighbour's enjoyment of his own land.

In seeking to apply this 'live and let live' approach, and when assessing the reasonableness of a defendant's behaviour, the court takes a number of factors into account. The duration of the nuisance is relevant, not only in establishing the existence of a nuisance but also in determining which remedy is more appropriate – damages or an injunction. Where the occurrence is an isolated event or of a temporary nature, that in itself may be evidence that there is not a sufficiently substantial interference to constitute nuisance. As the court takes all the circumstances into account, duration alone is not conclusive. Consider the case of a firm demolishing a building. If the work will be over in a matter of weeks, it may be reasonable to expect the owners of adjoining properties to put up with the noise and dirt over that time. But if the contractor works twenty-four hours each day, seven days each week, with the constant noise of machinery operating, vehicles coming and going and the glare of arc lamps at night, his activities may well constitute a nuisance. Certainly in *Andreae v. Selfridge and Co. Ltd* a plaintiff hotel owner was able to recover damages from a demolition contractor who had created excessive amounts of noise and dust.

The nature of the locality where the alleged nuisance is occurring will be relevant. One judge summed up this aspect of the tort by saying: 'What would be a nuisance in Belgrave Square would not necessarily be so in Bermondsey'. A person keeping pigs on his land might do so without complaint in the country but might be liable for nuisance if he kept them in a town garden.

Although locality is an important factor, the courts will be more inclined to find a nuisance if the offending behaviour has caused actual harm to property as opposed to mere interference with the occupier's comfort and enjoyment of his land. A plaintiff is far more likely to succeed where he can show actual physical damage. If he is simply complaining about behaviour which causes him personal discomfort in the use of his property, then he must show that his complaint is substantial, and one which ordinary men of ordinary sensitivity would complain about. The essence of nuisance is to achieve a balance. If the defendant has behaved reasonably, the plaintiff cannot complain if he has suffered damage only because his own property is extraordinarily sensitive, or the use he wishes to make of it is in some way unusual or special. If a defendant's behaviour would cause no harm to ordinary plants and vegetation, then there is no cause of action for a plaintiff who has suffered harm only to his extrasensitive orchids. But if the defendant's behaviour would have damaged even the daisies on the lawn, then the plaintiff can recover for the damage to his sensitive property as well as to any ordinary property. It is said that you cannot increase your neighbour's liability by using your own property in some special or extraordinary way.

When the court is assessing the reasonableness of the defendant's behaviour it will take into account that he may be doing something which is of value to the general public, for example making early morning deliveries of milk. A rather greater degree of noise and disturbance may then be acceptable before the behaviour amounts to nuisance. This does not mean that such behaviour can never amount to nuisance. Where it does, the public utility or public interest question may be relevant in determining whether an injunction should be granted. That point was in issue in two instructive cases, *Miller v. Jackson* and *Kennaway v. Thompson*.

In *Miller v. Jackson*, the case turned on the playing of cricket on a village pitch which had been in use since 1905. Adjacent land was developed for private housing and Mr and Mrs Miller purchased a house directly in line with the wicket. They then discovered that they had to live under very trying conditions during the cricket season, when it was impossible to use their garden if play or practice was taking place. Although balls had come into the garden and bounced off the house, no one had yet been struck. On these facts there was indeed a nuisance but, partly because of the public interest, an injunction was refused.

In *Kennaway v. Thompson*, the nuisance arose out of water sporting activities on a lake adjacent to the plaintiff's house. The noise from powerful motor boats was excessive and the court agreed that the activities did indeed constitute a nuisance. Taking into account the public interest in taking part in such recreations, the court decided that some balance could be achieved between the conflicting interests of the plaintiff and the club users. An injunction was granted, not in terms of a total restraint on water sports but limiting the size of boats which could be used and the number of times each year when competitions could be staged.

If the defendant can be shown to have acted out of malice, it will usually be difficult for him to prove that he was acting reasonably. In one case where the defendant lived next door to a music teacher, whenever a music lesson took place he would bang tin trays, whistle and shout. His acts were shown to have been done only for the purpose of annoying and interrupting the plaintiff, and were held to amount to a nuisance.

Liability for nuisance

Where a potential nuisance situation exists, problems arise as to who can bring an action, and against whom that action must be brought. Private nuisance is designed to protect a person's use or enjoyment of land. A plaintiff must, therefore, be able to show that he has some legal interest in the land, whether as owner or tenant. No action in private nuisance is available to a mere licensee (a person who only has permission to be on the land but no legal rights to the land itself). A mere lodger or guest cannot sue, although if either has suffered personal injury he may have a right of action in another tort, for example negligence. Actions in public nuisance are not restricted by this 'interest in land' requirement.

Who can be sued

The occupier of the land from which the nuisance arose is the obvious person on whom to impose liability. But there may be a difficulty if the occupier did not actually cause the nuisance. It seems that any of the following persons may be liable:

1 The actual creator of the nuisance. What is more, he remains liable even if he has disposed of the land and is no longer in a position to put the matter right. The reasoning behind this is that he committed the wrong in the first place and should remain liable.
2 The occupier of the land, if the nuisance was created by himself or his employees, in which case he is vicariously liable for their actions. It is often said that an employer is not vicariously liable for the acts of independent contractors, but this is not always the case. An occupier who engages a contractor to do construction work on his land may well be liable for any nuisance that the contractor creates, if the occupier should have realized that the work was likely to cause a nuisance.
3 A trespasser who creates the nuisance is liable but he may not be easily found or identified. The occupier will be liable for a nuisance created by a trespasser, or resulting from an act of nature if, once he knows about it, he adopts it or continues it. In effect this means that if the occupier fails to take reasonably prompt and sensible steps to stop the nuisance, he will be liable. An example which shows how the occupier can become liable for an act of nature is a case involving a tree which was struck by lightning and which then caught fire. The occupier called someone in to chop it down, but took no steps to put the fire out, leaving it to burn itself out. The fire spread to the plaintiff's property. The defendant was liable in nuisance for the damage caused.
4 The tenant occupier. Where premises are leased by a landlord to a tenant, the tenant as the occupier seems the obvious person to be liable for a nuisance arising from the premises. However, the landlord is also made liable in a number of circumstances, particularly where the nuisance existed when he let the premises or where it is caused by disrepair of the premises for which he is responsible.

Unlike trespass, where a plaintiff does not need to prove damage, an action in nuisance is only possible where damage can be proved. The damage may be to the property itself, (e.g. some nuisance which causes windows to break, or masonry to crack) or may consist of interference with the plaintiff's enjoyment of his property. Neither of these heads of damage includes mere depreciation of the land.

Even where all the various elements of the tort of private nuisance can be proved by the plaintiff, a defendant may escape liability if he has a valid defence. For example, he may be able to plead that the conduct complained of is authorized by an Act of Parliament. This defence can often be used by local authorities, or public authorities who have been authorized by statute to conduct certain operations. Whether the defence of statutory authority will succeed depends partly on whether the statute lays down a duty or merely a power. Much will depend upon the precise wording of the particular piece of legislation. However, in 1981 the House of Lords showed themselves prepared to give a wide interpretation to the words of the statute in the case of *Allen v. Gulf Oil Refining Ltd*. In that case, the Act granted the authority to acquire land and build an oil refinery there. It was held that the defendant's powers must be taken to include the power to operate the refinery on the land and such operations would not, therefore, establish any liability in nuisance unless negligence could be proved.

A defendant may also be able to show that he had a prescriptive right to do the act complained of. Rights to commit a nuisance may be acquired under the Prescription Act 1832 if the defendant can show that he has committed the nuisance openly and continuously for more than twenty years. This is limited to those cases where the right is capable of existing as an easement, which is a question to be determined in accordance with land law principles.

A plaintiff may also have consented to the state of affairs (*volenti non fit injuria*), but this defence would not operate where a plaintiff acquired the land with knowledge of an existing nuisance. It is no defence to argue that the plaintiff came to the nuisance – in other words that he must accept the state of affairs prevailing when he arrived. A defendant's behaviour will still be tested by the stated principles and could be found to amount to nuisance.

Remedies

It is true that the law allows a person affected by a nuisance to take appropriate steps to put a stop to it, but this form of self-help should not be lightly undertaken. The judges have said that this is not a remedy which the law favours and is usually inadvisable. The sort of problem a plaintiff could face is shown in one case where branches from the defendant's apple trees encroached over the plaintiff's land. The plaintiff cut off the branches, stripped off the apples and then sold the fruit. Although he was within his rights to lop the branches, he had no right to keep them, or the fruit. He was successfully sued by his neighbour for having misappropriated the fruit!

It will usually be much safer for a plaintiff to seek an injunction where he wants to put a stop to a nuisance. The granting of an injunction is entirely at the discretion of the court and may be refused.

That may happen where the nuisance occurs only very infrequently or in circumstances where the injunction would cause undue hardship to the defendant. Damages are always available as a remedy to compensate the plaintiff for harm suffered as a result of nuisance.

This account of the rules of private nuisance will be sufficient to convince many people that they would be foolhardy to embark on private litigation. Bringing an action may well prove costly in terms of time and money. If the action is against a large company or organization, a private individual may well feel that he is at a disadvantage because of the resources and expertise available to the other side. There can be significant problems of proof, especially in cases where it is hard to show a link between 'cause' and 'effect'. If these arguments are valid in small-scale instances of nuisance, they will be all the more so when the nuisance complained of is creating extensive environmental damage. In such a case it may be especially unrealistic to expect one private individual to sue, and the rules of statutory nuisance may then be more than ever relevant.

Statutory nuisance

Most plaintiffs simply want to put an end to the state of affairs causing the nuisance, but may be put off by the thought of bringing a costly and time consuming individual court action, which may not improve relations with their neighbours. In many cases, the state of affairs about which they wish to complain may also constitute a statutory nuisance under the Environmental Protection Act 1990, which came into force on 1 January 1991. The act replaces many of the rules previously found in the Public Health Act 1936 and the Control of Pollution Act 1984. Examples of nuisance covered by the new act include: premises in such a state as to be prejudicial to health or a nuisance; accumulations or deposits or animals which are prejudicial to health or a nuisance; emissions of smoke, gas, fumes, dust, steam or other effluvia; noise or vibration. The key phrase from the legislation is 'prejudicial to health or a nuisance'. A state of affairs is prejudicial to health if it is likely to cause injury to health. The nuisance element of the phrase can only be satisfied by showing what amounts to a nuisance at common law (see page 214). A statutory nuisance cannot exist where there is merely an interference with the personal comfort of the occupiers.

A local authority is under a positive duty to make inspections within its area to establish whether any statutory nuisance exists. Additionally, the local authority must take such steps as are reasonably practicable to investigate complaints of statutory nuisance made by persons living within the local authority's area. The inspections are made by Environmental Health Officers, who have the right of entry to property for detection purposes.

Where a statutory nuisance exists, or is likely to occur or recur, the local authority may serve an abatement notice on the person creating the nuisance or, in some cases, on the owner of the property from which the nuisance emanates. The notice indicates what must be done and a time limit for action. Non-compliance with such an abatement notice is a criminal offence, for which the offender can be fined on a continuing daily basis until compliance. Alternatively, the local authority may undertake the work necessary and recover its expenses.

The 1990 Act has introduced a new right of appeal against an abatement notice. The appeal must be lodged within twenty-one days and will be heard by the Magistrates Court.

Where there is evidence of a statutory nuisance and the local authority declines to act, or is itself responsible for the nuisance, any person aggrieved may bring proceedings in the Magistrates Court. In such instances, the aggrieved person must give at least twenty-one days' notice in writing to the defendant, stating his intention to bring proceedings, and outlining the matters complained of. When the case comes before the Magistrates Court, the court has power to make a nuisance abatement order, as well as imposing a fine and ordering reimbursement of necessary expenses. It is also possible for the court to make a criminal compensation order in favour of the aggrieved person, up to a maximum of £2,000. Failure to comply with a nuisance abatement order is in itself a criminal offence. The court may order the local authority to perform the necessary work under the order where a defendant has defaulted.

The rules under the Environmental Protection Act do not give rise to any liability in tort. A person suffering harm from a nuisance would still need to pursue a claim in private nuisance if a large claim in damages was in issue. However, in smaller problems, the possibility of an award of criminal compensation under the 1990 Act may suffice. The advantage of the statutory rules is that, where the local authority is prepared to take action, the state of affairs constituting the nuisance can be put right without any need for individual action.

The rule in *Rylands v. Fletcher*

The type of liability created by this tort is strict liability. This means that if a set of given facts fits the requirements of the rules, then the defendant will be liable, whether he took care or not. This tort is always known by the name of the case which gave rise to the particular form of liability. An occupier who engages in hazardous activities on his property will be liable where those activities injure or cause harm to someone. The same set of facts may also give rise to an action in negligence. The importance of the *Rylands v. Fletcher* rule was that it pre-dated the general development of negligence principles. Nowadays, it is often suggested that the rule is of decreased significance, and certainly, its technical requirements mean that it is difficult to prove.

The rule is usually stated as follows:

> Where a person, for his own purposes, brings on to his land, and collects and keeps there anything likely to do mischief if it escapes, he must keep it at his peril, and if he does not do so, he is liable for all the damage which is a natural consequence of the escape.

The use if the word 'escape' is naturally suggestive of animals, which could come within the rule. But liability in respect of animals may also exist under the Animals Act 1971 and the torts of negligence and nuisance.

The rule involves a number of points:

1 The dangerous 'thing' must be brought by the defendant onto his land. It follows that no liability can arise from an escape of something which is naturally on the land, e.g. weeds or rocks.
2 The 'thing' must be likely to cause harm if it 'escapes'. The rule has been applied to a wide range of substances and items from oil, gas and explosives to flagpoles and fairground equipment. There is no requirement that the 'thing' must be intrinsically dangerous.
3 There must be an escape, i.e. the person injured must show that he was not on the defendant's land at the time when the harm occured.
4 The escape must cause damage. This may be injury to the person or damage to property.
5 The defendant must bring the 'thing' onto his land for his own purposes, and in doing so must be making a non-natural use of his land. This latter point is probably the most difficult and contentious. It appears that the defendant must be using his land in some extraordinary and unusual way. The idea of non-natural use is closely bound up with public benefit (or lack of it) and will tend to change with changing social needs. In the actual case *Rylands v. Fletcher*, the non-natural use was bulk storage of water in a reservoir. That was in 1868. Would the court still decide the case in the same way today? In *Rickards v. Lothian*, a non-natural use was defined as: 'Some special use bringing with it increased danger to others, and not merely the ordinary use of land, or such a use as is proper for general benefit of the community.' It has been said that

where planning permission has been granted for a particular use of land, it would be difficult to argue that such a use was non-natural.

Although liability is said to be strict, i.e. not dependent on whether the defendant took reasonable care, a number of defences may be available to the defendant.
These include:

1 *Fault of the plaintiff* – if the damage is caused by the plaintiff's own wrongful act, he cannot recover.
2 *Consent by the plaintiff* – presumably tenants on different floors of a building impliedly consent to the presence of a water supply running through the building. If flooding occurs on an upper floor causing damage to a lower floor there will be no liability.
3 *Statutory authority* – a defendant may be able to point to an Act of Parliament which excuses his behaviour which would otherwise be tortious.

In some ways this tort is wider than nuisance, in that it protects persons even though they cannot show an interest in land, and even if they suffer only personal injuries; but in other ways, it is narrower in that it requires something to be brought onto the land which is a non-natural use. Finally, it must be remembered that the same set of facts could give rise to liability in both nuisance and *Rylands v. Fletcher*.

Liability for the spread of fire

Fire may be one of man's greatest discoveries but its potential for causing harm is obvious. A person may start a fire which he fails to control and which then spreads, or a fire may begin accidentally on his land, which he then fails to check. Such circumstances can give rise to liability in nuisance, *Rylands v. Fletcher* or negligence, according to the facts of the case.

Not surprisingly, some rules were developed at an early date to deal with harm caused by fire. The Fires Prevention (Metropolis) Act 1774 is still in force, and provides that no action can be brought against a person on whose land a fire accidentally begins. The word 'accidentally' is very restrictively interpreted. If the fire is caused by nuisance or negligence, it cannot then be said to be 'accidental'. Moreover, the defence afforded by the Act is lost if an accidental fire is negligently allowed to spread.

The hazards of the spread of fire have been recognized for so long that a large number of cases exist on the point. Some newer hazards, which are potentially far more harmful, have been dealt with by specific Acts of Parliament. Particular examples are:

- Nuclear incidents.
- Oil pollution.
- Poisonous waste.

The specific acts sometimes create civil liability and allow injured persons to claim damages.

These torts so far considered all relate to land, in that they either protect an interest in land or concern liability arising out of the use of land. Negligence as a tort is not so confined, and can compensate for harm caused in a wide range of circumstances.

Negligence

Negligence as a tort consists of the breach of a duty owed to the plaintiff to take reasonable care, which results in damage of the right type which is not too remote. Many more people are injured by careless acts than by acts which are intentional. As the law of tort developed, a number of specific situations in which liability would be imposed on a person for his negligent behaviour were recognized by the law, (e.g. where an employer fails to take reasonable care to ensure the safety of his employees; or where an occupier fails to take reasonable care to ensure the safety of his visitors), but there was no general principle established which could be applied to any and every set of circumstances.

Such a general principle began to emerge in 1932 from the famous 'snail in the bottle' case, *Donoghue v. Stevenson*. In that case, a manufacturer of ginger beer distributed his product in opaque bottles. One bottle was sold by a café owner to a customer who shared it with his friend. A decomposed snail slithered out from the bottle, causing the friend to become seriously ill. As she had not bought the ginger beer, the friend had no claim in contract. She successfully brought a claim against the manufacturer for negligence. The House of Lords decided that a manufacturer owes a duty to take reasonable care to see that his products are not contaminated. The prime importance of the case is that it tried to formulate some principles which would link together all the previously separate instances of negligence and by which new sets of circumstances could be tested.

Over the past fifty years, the separate tort of negligence has overtaken all other torts in importance; first, in the number of actions which are brought under this head, and second, in the way that the tort of negligence has begun to take over some of the liability which would previously have been covered by torts like trespass or nuisance.

The tort of negligence is usually analysed under three main headings:

- A duty of care owed to the plaintiff.
- Breach of that duty.
- Damage resulting of the right type which is not too remote.

The duty of care

Liability in negligence can only arise if a duty to take care is owed. When a court is faced with a situation which has been considered previously, it may be able to follow a precedent from an earlier case in determining whether a duty exists. Alternatively, an Act of Parliament may impose a duty, e.g. the Occupiers' Liability Act 1957. In either case, establishing the existence of the duty is relatively straightforward.

Where the court has to consider an entirely novel situation, how does it decide whether or not a duty to take care exists? One of the judges in *Donoghue v. Stevenson* suggested the 'neighbour test' for answering this question. He said that it was impossible in law to have a rule which states that you must love your neighbour. But there could be a rule that you must not injure your neighbour – you must take reasonable care to avoid acts or omissions which you can reasonably foresee would be likely to injure your neighbour. Your 'neighbour' for this purpose is any person who is so closely and directly affected by your act or omission that you ought to have him in mind when directing your mind to the act or omission in question. This 'neighbour principle', as it is usually described, has often been thought to rest exclusively on reasonable foresight of harm. In a simple example, if you run a bath for a baby, you can reasonably foresee that the baby will be injured if you do not take reasonable care to test the temperature of the water before putting the baby into the bath. It is clear from the cases that mere foresight of harm is not now to be regarded as establishing a duty of care, without more. Emphasis nowadays is more on a test of proximity. Is there a sufficient degree of proximity between the parties so that the plaintiff should reasonably have contemplated injury to the defendant? The advantage of using a test of proximity is said to be that it can be more stringent where the circumstances require it. Certainly, the very wording of the 'neighbour principle' contains the test of proximity. A neighbour is a person 'closely and directly affected by the act or omission.'

The House of Lords subsequently refined the neighbour principle in *Anns v. Merton London Borough Council* in 1977, by the addition of a 'public policy' element. This consisted of the court weighing factors such as social conditions, a balance of interests and fear of inflicting excessive liability. This produced a result that in establishing whether a duty was owed, the test became a two-stage process. First, apply the neighbour principle, and if a duty appears to be owed, then ask whether there are any policy considerations which may justify the exclusion of, or restriction of the duty.

By the mid 1980s, it had become clear from the cases that the judges were no longer happy with this two-stage formulation of the test for establishing duty. It left the introduction of policy considerations until too late a stage in the assessment of whether a duty existed. It would be more appropriate to take account of such factors when deciding whether the relationship between the parties was sufficiently proximate. A key factor to be taken into account by the court is whether it is just and reasonable to impose a duty. The use of this new approach is made much easier by the fact that the Anns case has now been overruled by *Murphy v. Brentwood District Council* in 1990.

In the light of recent cases, it is therefore better to state that before a duty of care can exist, it must be shown that there is foresight of harm, that there is a proximate relationship between the parties, and that it is just and reasonable to impose a duty. Although it is convenient to list the elements in this way, it should be borne in mind that they are inevitably interrelated. Moreover, in determining what is just and reasonable, account may be taken of factors which previously were considered under the 'policy' head.

In essence, the purpose of requiring a plaintiff to establish a duty of care is simply to determine whether the defendant will have to pay for the harm he has caused. In cases where it is not considered appropriate for a defendant to have to pay, he can be insulated from the need to do so by manipulation of the duty concept, which is inherently flexible when expressed in the three-fold way seen above.

There are certain well-recognized situations where the courts have faced particular problems in establishing whether a duty should exist because of policy considerations.

These include:

- Cases involving pure economic loss.
- Cases involving negligent mis-statement.
- Judicial immunity situations.
- Nervous shock situations.
- Negligent exercise of statutory powers.
- Pure omissions to act.

It should be remembered that these areas constitute 'problem' areas of the law of negligence, and may well prove to be the fields where new developments occur.

1 Cases involving pure economic loss

As a general rule, no duty of care is owed where a plaintiff suffers only pure economic loss. An example can be found in the case where a research institute negligently allowed a virus to escape which caused foot and mouth disease, and as a result cattle needed to be slaughtered over a wide area. Markets were closed to stop the spread of the infection. The plaintiff was a cattle auctioneer who lost profits as a result i.e. his only loss was pure economic loss. He could not recover damages for negligence. No doubt the court feared that if his claim was allowed, every cattle transporter, supplier of cattle food and dairyman would come forward to claim. Liability would have been excessive in terms of the numbers of claims and the possible amounts involved (*Weller v. Foot and Mouth Disease Research Institute*).

The judges are particularly anxious not to create liability 'in an indeterminate amount for an indefinite time and to an indeterminate class'. They also keep in mind that a person who suffers mere economic loss can frequently correct the situation, for example by accepting a smaller level of profit or by increasing charges. This may justify treating economic loss differently from personal injury which is usually irreparable. They may also be taking a realistic account of which party could more easily have insured against the loss.

The rule can undoubtedly be very harsh, and sometimes almost illogical in application. Take, for example, the case of *Spartan Steel and Alloys Ltd v. Martin and Co. (Contractors) Ltd*. The defendants had negligently cut the electric power cable under a road. This resulted in power being cut off in the plaintiff's metal foundry. Work was in progress (a 'melt') which had to be aborted, resulting in damage to the metal. That was physical damage and the plaintiffs could recover for that harm plus the resulting loss of profit on that 'melt'. As there was no power, they lost the possibility to carry out other 'melts' on which they lost profits. That was classed as pure economic loss for which the plaintiff could not recover damages.

Pure economic loss therefore means loss which is not injury to the person or damage to the plaintiff's property. However, there have inevitably been cases where the courts have sought to 'stretch' the definition of the phrase, sometimes with disastrous results. One such infamous case is *Dutton v. Bognor Regis UDC*. In that case, the defendant's negligence had caused a house to be built with defective foundations. The foundations were defective from the very moment of building. When a subsequent purchaser bought the house, could it be argued that the purchaser was suffering from physical damage to his property as a result of the negligence, or alternatively was he merely suffering loss which was purely economic. The house was damaged from the outset. The subsequent purchaser did not therefore 'suffer' damage to the property itself, but only to its value. Yet in the Dutton case, the damage was accepted as damage to property. No doubt the court in that case was anxious to allow the house owner to recover, but in achieving that result, the case created many problems and was clearly wrong. Fortunately, it too has been overruled by the House of Lords decision in Murphy.

During the period when Dutton was regarded as good law, it had to be seen as an exception to the usual rules about pure economic loss. The courts appeared to have created another exception in the very difficult case of *Junior Books Ltd v. Veitchi Co Ltd* which came before the House of Lords in 1982. The plaintiffs had nominated the defendants as sub-contractors to lay a new factory floor. The floor laid by the defendants was defective and the plaintiffs claimed to have suffered losses totalling in excess of £200,000. The figure consisted largely of loss of profits while the factory had to be closed for the floor to be relaid. No one had been injured by the floor being defective, and no harm had been caused to the property, i.e. the losses were pure economic loss. Nevertheless, the court thought that the plaintiffs should recover damages. This decision seems to have been based largely on the 'close relationship' between the plaintiff and the defendant. There was no contract between them but the defendant was a nominated sub-contractor. The court also took account of the fact that making the defendant liable would not open up liability to 'an indeterminate class'. It is certainly true that, in nominating the defendants, the plaintiffs knew of them and relied on them, and although they had no direct contract with the defendants, the defendants were doing their work under the main contract so that was the next best thing. But it could be argued that by allowing the plaintiffs to sue in tort in this case, when they could not sue in contract because of the privity of contract rules, the courts may be undermining the very rules of the law of contract – and that in itself may be a policy consideration for not allowing recovery for pure economic loss in a situation such as the Junior Books case!

Any idea that the Junior Books decision might see the beginning of the end of the economic loss rules has not been borne out by subsequent court decisions. In *Leigh and Sillivan v. Aliakmon Shipping* in 1986, where a shipper damaged goods through his negligence, the buyer of the goods (to whom ownership of the goods had not yet passed) sued in respect of the damage. As the goods did not yet belong to the buyer, the harm was not damage to his property, but mere economic loss. The buyer's problem was that under his contract to buy the goods, he had to undertake to bear the risk of their loss or damage. The House of Lords would not allow the buyer to succeed against the shipper. Although the shipper had been negligent, he had caused the buyer only pure economic loss. Choosing not to try to bring this case within any exceptional principle, the court found that as an aspect of policy, it would be a bad thing to allow the buyer to recover, as that would effectively render useless the protection which the shipper enjoyed under his contract with the seller. In other words, allowing the buyer to succeed would undermine the rules of the law of contract.

In a trio of cases concerning building and building work, (*Simaan General Contracting Co. v. Pilkington Glass Ltd* (No.2), *Greater Nottingham Co-op Society Ltd v. Cementation Piling and Foundations Ltd* and *D & F Estates Ltd v. Church Commissioners for England*), all decided by the Court of Appeal and the House of Lords, it seems clear that the possible trends suggested by Junior Books have been halted. Great emphasis in these cases is laid on the role of the contracts, and claims should be pursued down that route where possible. Within contracts, the parties have the opportunity to establish their responsibilities. Once they have done that, then it would be wrong to impose any further liability.

The difficulty created by Dutton's case, and perpetuated by the decision in Anns, can best be illustrated by reference back to the facts of *Donaghue v. Stevenson*. In that case, the negligent act (putting the snail into the ginger beer) caused harm to the plaintiff. What if, instead of complaining about harm she had suffered, the plaintiff were merely complaining about some harm to the ginger beer itself, e.g. that the presence of the snail had turned it sour? That would have been a complaint about the quality of the product, in which case the plaintiff would not have succeeded in her action in tort. Now translate that example into the building situation. A builder builds a house with defective foundations. When this is discovered by the owner, it has not caused him any physical injury, nor has the house damaged other property. The house merely suffers from a defect (defective foundations – snail in bottle) which damages the property itself (damaged house – damaged ginger beer). Yet in the Dutton and Anns cases, the courts chose to draw a distinction, and to allow the plaintiffs to recover.

The halt to this line of cases drawn by *Murphy v. Brentwood District Council* is greatly to be welcomed. In that case, Lord Keith said of the Anns case: 'It has engendered a vast spate of litigation, and each of the cases in the field which have reached this House [the House of Lords] has been distinguished. Others have been distinguished in the Court of Appeal. The result has been to keep the effect of the decision within reasonable bounds, but that has been achieved only by applying strictly the words of Lord Wilberforce and by refusing to accept the logical implications of the decision itself. These logical implications show that the case properly considered has potentiality for collision with long-established principles regarding liability in the tort of negligence for economic loss. There can be no doubt that to depart from the decision would re-establish a degree of certainty in this field of law which it has done a remarkable amount to upset.' The facts of the Murphy case are considered on page 236.

2 Negligent mis-statements

The policy of excluding pure economic loss claims caused particular problems in those cases where the negligence consisted of a negligent statement rather than a negligent act. In such cases economic loss is the most likely form of harm to result. For example, if a firm wishes to do business with a new and unknown customer, it may have reservations about extending credit to that customer. To protect itself, the firm may ask permission to approach the customer's bank for a credit reference. If the reference is negligently given by the bank and the firm relies on it, extends credit, and sustains heavy economic loss as a result, can the firm sue the bank for negligence when the loss it has suffered is purely financial?

In a case in 1964, *Hedley Byrne v. Heller*, where the facts were similar to the example given above, the courts drew a distinction between negligent acts and negligent statements. If the negligent statement results in physical damage, the normal rules of negligence operate. So, in one case where an architect negligently stated that a wall was safe to be left standing, he was successfully sued in negligence by a workman who was injured when the wall collapsed (*Clay v. Crump*). Where the negligent mis-statement causes economic loss, liability may now arise either under the principles set out in the Hedley Byrne case or by virtue of the Misrepresentation Act 1967. (Liability for false statements in the law of tort had previously been covered only by the tort of deceit. In order to succeed in an action for deceit, it is necessary to prove that the false statement has been made knowingly, without belief in its truth, or recklessly, careless whether it be true or false. There has to be dishonesty, mere carelessness was not enough. Some relief can also be obtained from the law of contract in cases where the false statement does not amount to a term of the contract, but induces the plaintiff to enter the contract. In those circumstances, the plaintiff may be permitted to rescind the contract. It would be useful at this point to refer to the relevant material on misrepresentation in the law of contract.)

The general effect of the Hedley Byrne case is that certain negligent mis-statements are now actionable, even if the only loss is pure economic loss. The judges in the case were no doubt cautious in view of the sizable step they were taking to change the law. In consequence, they wanted to impose restrictions on those negligent statements causing economic loss which should be actionable. Not only should the maker of the statement reasonably foresee that it was likely to cause harm; there must also be a special relationship between the plaintiff and the defendant. The judges in the Hedley Byrne case were not at all clear exactly what would amount to a special relationship. It seemed it would exist if the party making the statement knew, or ought to have known, that the other was trusting him and was going to rely on what he said. Much of the need to establish a special relationship under the Hedley Byrne case is caused by the judge's desire not to open up too wide a potential liability.

The possible scope of the Hedley Byrne principle is very wide. It could apply, for example, to professional advice given by architects, surveyors, engineers, accountants, lawyers or doctors, amongst others. It is clear from the cases that some considerable overlap can occur between the rules of the law of tort and the law of contract. Most professional people are likely to have a contract with their client, so it is more likely that they will be sued for breach of contract in respect of any negligent service. However, two recent cases serve to illustrate how reliance on a negligent mis-statement made outside the context of any contract can cause extensive loss. In *Smith v. Eric Bush*, a prospective purchaser of a house sought a mortgage. The building society instructed a surveyor to carry out a house valuation for them, and on the strength of that survey, the building society loaned money to the purchaser. Relying on the survey, the purchaser decided to go ahead with the purchase. There was no contract between the purchaser and the surveyor, who had been engaged by the building society. However, the court held that the surveyor was liable to the purchaser, as the situation was akin to contract, in that the surveyor knew that the consideration he received derived from the purchaser. He also knew that the purchaser would rely on his report in determining whether to buy the house. The valuation report was negligently prepared, and in consequence, the purchaser was successful in a claim for damages against the surveyor. The surveyor sought to rely on a disclaimer of liability, but the court further held that it would not be fair and reasonable to allow reliance on the disclaimer, as the surveyor must have known that the purchaser would be supplied with a copy of the valuation report and would be likely to rely on it.

In *Caparo Industries plc v. Dickman*, the auditors of a company issued inaccurate and misleading accounts for a company, Fidelity, in which Caparo held shares. Caparo bought further shares on the strength of the audited accounts, which were circulated to all shareholders. Later the same year, he mounted a takeover bid for Fidelity, but once he acquired the company, Caparo discovered the true state of its finances. He alleged that the auditors owed him a duty of care when making statements, both as a shareholder and as a potential bidder to take over the company. In the House of Lords, it was held that no duty was owed to him in either capacity. As a member of the general public, looking at the audited accounts with a view to takeover, there was no sufficiently proximate relationship between the auditors and Caparo. As a shareholder wishing to buy more shares in the company, no duty was owed as an individual shareholder should be in a position no different from that of a member of the general public. The accounts were prepared and circulated to members of the company to enable them to better manage the company, not to permit them to make a personal profit. To find otherwise would be to impose a virtually unlimited and unrestricted duty on the auditors, and would suggest that foresight of harm, without more, was sufficient. Such a finding would run contrary to all the other recent decisions.

Looking back to the Hedley Byrne case itself, the actual claim of the plaintiff failed because the defendants had expressly disclaimed responsibility for their reference. The possibility of using such a disclaimer successfully is limited now by the provisions of the Unfair Contract Terms Act 1977. Under s. 2, where a person seeks by a notice to exclude or restrict his liability in negligence which results in loss or damage other than personal injury, he may only rely on that disclaimer if it satisfies the requirements of reasonableness. It should be noted, however, that the Unfair Contract Terms Acts 1977 applies only to a business liability situation.

The law relating to negligent statements is complicated because of the overlap of the rules of the law of tort and the law of contract. Certainly the Hedley Byrne principles can be important where a false statement has been made which does not induce a contract. Where a contract has resulted it seems likely that a plaintiff would prefer to use the rules of the Misrepresentation Act 1967, to take advantage of a more favourable burden of proof which would then lie with the defendant. The law is probably much more confused than it need be, due in part to the fact that the Misrepresentation Act 1967 was passed on the basis of earlier Law Revision Committee reports and suggestions, made at a time before Hedley Byrne was decided in the courts. The legislators went ahead with the Act, making no attempt to blend in the developments which had occurred subsequently in the Hedley Byrne case.

3 Judicial immunity situations

As a result of Hedley Byrne, the courts have recognized that there are times when there will be liability in tort for negligent mis-statements. Nevertheless, no liability attaches if the statement is made by a judge, or barrister, or arbitrator in the course of legal proceedings. Quite simply, the reason for this is public policy. It is regarded as being against the public interest to have the constant possibility of litigation being reopened. Take the case of a person accused of criminal offences who is represented by a barrister. The accused is found guilty and sent to prison. He believes his barrister handled his case negligently. If the accused could now sue in tort for negligence, this would in effect reopen many of the issues considered in his criminal trial, and would run counter to the principle that there must be an end to litigation.

The immunity of arbitrators was considered in the case of *Sutcliffe v. Thackrah*, where an architect who had negligently certified work claimed to be acting as an arbitrator in the certification process. That claim was not accepted by the court but, had it been correct, no action in negligence would have lain against the architect.

4 Nervous shock situations

Nervous shock in this context means actual illness like psychiatric disorder or depression. Nervous shock as a consequence of negligent behaviour causes no problems where the victim has also sustained physical injuries. So, a victim of a road accident, run over by a negligent motorist, may be claiming in respect of broken limbs, fractured skull, punctured lungs *and* nervous shock. The cases which cause problems are those where the only harm is the nervous shock. The problem is more acute in those cases where the victim of the nervous shock was never in any danger himself of suffering physical harm.

Take the example of the mother who looks out of her bedroom window and sees her small son run over by a reversing lorry. No doubt it is likely she will suffer nervous shock but she was in no personal physical danger. Or, take the example of the person who sees nothing of an accident to a loved one, but reads of it later, or hears of it from a third party. Are such people owed a duty of care by the defendant? Inevitably, these cases raise the usual policy issue, that liability may come to be owed to an 'indeterminate class'. Moreover, the courts may be anxious that these cases could give rise to false claims, as medical knowledge about shock and mental illness is not a particularly exact science.

Many of the issues in this area were clarified by the House of Lords in 1983, in the case of *McLoughlin v. O'Brien*. A horrendous road accident had occurred, in which the plaintiff's husband and children were badly injured, and one child was killed. A friend came to tell the plaintiff the news of the accident and took her to the hospital. There, the plaintiff learnt the full extent of the tragedy and saw her family in harrowing and distressing circumstances. It was held that her nervous shock was reasonably foreseeable, and she should be permitted to recover damages.

A significant number of the anxieties about the development of this area of negligence were aired in the McLoughlin case. Should claims be limited to those people who saw an accident or its immediate consequences? Or should claims be limited to those people who had some reasonably close relationship with the victims of the accident? (If this point is regarded as significant, that could eliminate claims from persons who suffer nervous shock as a result of assisting in an accident or disaster, where they might see and experience dreadful injuries and suffering). Should claims be limited to those people who are reasonably close in time or place to the accident?

If damages for nervous shock are dependent on reasonable foresight of harm to the plaintiff in question, that in itself will act as a significant filter to ensure that only a proportion of cases are likely to succeed. In a recent action brought by relatives of spectators killed in the Hillsborough football stadium disaster, in respect of nervous shock suffered as a consequence of watching the events unfold on television, the House of Lords rejected their claim for damages. It seemed that watching TV miles away from the catastrophe could not satisfy the test of proximity.

5 Negligent exercise of statutory powers

Dutton v. Bognor Regis UDC opened up a significant development in the law of negligence, when an owner of a building sued the local authority, alleging negligence with regard to the inspection of the foundations of the building. In many of these cases, the builder himself will not be worth suing if he has gone into liquidation or where there are doubts about his solvency. Local authorities look like richer pickings in these circumstances!

Where a local authority acts under statutory powers, as for example in the building inspection processes, there may be liability for negligence if, once having decided to inspect, the inspector fails to carry out the inspection properly. Of course, where there is merely a statutory *power* to act, as opposed to a statutory *duty*, the local authority has some discretion about whether to do anything at all. Its resources may be so scarce that it has to determine priorities about which inspections it will make. In these circumstances, if it decided to make no inspection, the question then arises whether the authority could be negligent for failing to inspect. It seems that there would be no liability in negligence, so long as the local authority has given proper consideration to the question of whether or not to inspect.

These issues were all considered in *Anns v. Merton London Borough Council* in 1977. This was another case involving defective foundations. Although the decision has now been overruled, some consideration of the approach taken by the House of Lords in Anns is vital to an understanding of the later developments. So far as local authorities are concerned, Anns appeared to decide that a local authority may be liable in tort to a building owner in relation to its statutory control over building operations. The extent of the liability would be the cost of remedying a dangerous defect in the building which had resulted from a negligent failure by the local authority to ensure that the building was constructed in conformity with the appropriate standards laid down in building regulations.

The difficulty created by this broad-ranging approach can best be illustrated by comparing a defective chattel with a defective building. If you make a chattel which contains some latent defect which causes damage or injury to persons or property, you may be liable in negligence if you made the chattel negligently – *Donoghue v. Stevenson*. But if the chattel is merely defective in quality, then your liability in that regard lies only in contract. Anyone acquiring the chattel who has no contract with you may suffer economic loss as a consequence of having to have the chattel repaired or having to simply abandon it, but of course, you would owe no liability in tort in respect of mere economic loss.

Now apply these same principles to a defective building. If a builder constructs a building with a latent defect (e.g. defective foundations), the builder may be liable in tort for injury to the person or damage to property arising from the dangerous defect. But if the owner of the house gets to know about the defect before it causes any injury or damage, then the owner would have rights in contract if the house had been built for him, but would have no rights in tort, because his loss would be mere economic loss – i.e. the cost of the repair, or the cost of abandoning the building if it cannot be repaired. To hold otherwise would be to impose on the builder a warranty as to the quality of his work which would be owed not only to the original owner but also to all who subsequently acquired an interest in the property.

The Anns case, following the earlier decision in Dutton, erred in failing to take account of this distinction between defects causing external harm and defects which merely harm the product itself. If, however, the distinction is a proper one, then it raises the question about liability of local authorities. If the builder's duty should be limited as suggested above, then it seems altogether wrong to make the local authority liable to a greater extent than the builder. The function of the local authority in inspecting building work is to ensure that there is compliance with building regulations. If the local authority were to negligently fail to ensure compliance with those regulations, it seems absurd that it would then incur a liability greater than that of the builder himself. In the Murphy case, overruling Anns, the Lord Chancellor, Lord McKay, makes the point that it is not a proper exercise of judicial power for the courts to create large new areas of responsibility on local authorities in respect of defective buildings. That, properly, is the role of Parliament. He was strengthened in his view by the consideration he gave to the Defective Premises Act 1972 (see page 250).

During the period whilst the Anns case was still good law, it was clear in a number of other cases that the judges were anxious not to let the Anns principles in relation to the exercise of statutory powers run too far. In *Curran v. Northern Ireland Co-Ownership Housing Association Ltd* a house extention on the plaintiff's home had been financed by a grant provided by one of the defendant organizations. It was allegedly so badly built that the work had to be completely redone. The plaintiff argued that he had bought the house in reliance on the fact that a grant had been given, assuming that a grant would only have been paid to the previous owner if the work had been done properly. The defendant organization had a statutory power to make grants to improve houses. A grant could only be made if the property was in stable structural condition on completion of the work, and if the work had been done to the satisfaction of the defendant organization. If the work was defective, could a subsequent owner of the property hold the defendant liable for negligence? Their Lordships found no negligence here, and indicated that caution should be exercised about extending duties of care to statutory bodies exercising their statutory powers to control third parties. Such an extention should only occur if:

1   The statutory power was directed to safeguarding the public, as would be the case in situations like Anns where the building regulations could be seen to have such a purpose;
2   it is clear that a proper exercise of the statutory power would have prevented the danger; and

3   if the negligent exercise of the statutory power created a latent defect which could not have been discovered and remedied before harm occurred. The whole object of the power in the Curran case was to protect the public purse, not to protect the health and safety of the general public. In consequence, the case could be easily distinguished from the Anns case.

Thankfully, the Anns case has now been overruled by *Murphy v. Brentwood District Council*. This case again arose out of defective foundations. A pair of semi-detached houses had been built on an infilled site on a concrete raft foundation to prevent damage from settlement. The local authority approved the plans. Some years after the house was completed, the owner noticed serious cracks, and he then discovered that the raft foundation was defective, and had become distorted. Necessary repairs would have cost £45,000, which the plaintiff could not afford. Instead, he sold the house, and realized £35,000 less than the market value, had the house been in sound condition. The plaintiff sued the local authority to recover that sum. The House of Lords held that the plaintiff was unable to recover from the local authority, as his loss was mere economic loss.

Although the Murphy decision has been greeted with relief, in that it seems to take the law back to a point from which it should never have moved, the case does still leave some questions unanswered. For example, would the local authority be held liable in negligence if a defect in a building, which their negligence has failed to prevent, then caused actual physical injury to someone, or physical damage to some other property? This point was not decided in the Murphy case and must await a future decision. Cases are already emerging at first instance which are raising some of these questions.

Inevitably, these complex cases are often the result of an aggrieved person looking for someone who is financially worth suing. It has already been suggested that the erring builder may be insolvent, but the other difficulty which a plaintiff faces is the problem of limitation periods. In all tort actions, a claim must be commenced within the appropriate time limit laid down by the Limitation Act 1980. An action relating to damage to property must be commenced within six years of the cause of action accruing. Usually, this is the moment when damage occurs. That does not necessarily mean when the defective building was constructed. Since the decision of the House of Lords, in *Pirelli General Cable Works Ltd v. Oscar Faber and Partners* in 1983, the cause of action arises when physical damage actually occurs to the building, and it is matterless whether the plaintiff knew or could have known about it. This rule is undeniably harsh, especially as it fixes the date for the cause of action accruing for subsequent as well as present owners.

It runs counter to an earlier test proposed by the courts, that time should only run from a point when the plaintiff could have discovered the defects. Some help with this problem has now been given under the Latent Damage Act 1986. In those instances where damage cannot be discovered immediately, (i.e. defective foundations), the plaintiff must bring his action within three years from the time when he discovers the damage, or when he should have discovered it, if he had been using reasonable care, subject to an overriding rule that the action must be commenced within fifteen years of the building being completed.

6 Omissions to act

As a broad general rule there can be no liability in negligence for a pure omission to act. If a road accident occurs where victims are trapped in a car, a passer-by commits no legal wrong if he makes no effort to save those trapped. This is described as pure omission, pure in the sense that there was no existing duty to act. If there is some existing duty, then an omission can be as wrongful as a negligent act. So, a parent of a small child has a duty to care for the child. If he omits to feed the child, such behaviour could be actionable.

If there is no duty to act, but someone decides to act as may be the case in the example above, where the passer-by decides to try to release the crash victims from the car, then if he behaves negligently, the law generally takes the view that there will be no liability for the negligence unless the passer-by makes matters worse. As there is no duty to act, then if you do act, there is no duty to effect an improvement.

This principle can be regarded as being subject to a most important exception in relation to the exercise of statutory powers. In such a case, although there is no duty to act, the defendant had a duty, if he chooses to exercise the power, to effect an improvement.

Breach of the duty of care

If a duty to take reasonable care is owed to the particular plaintiff, the next stage is for the plaintiff to prove that the defendant was in breach of the duty, i.e. that the defendant failed to take reasonable care. The obvious question must then be asked – how much care should the defendant have taken? What would amount to reasonable care in the particular circumstances? If the defendant falls short of a reasonable standard, then he is in breach of his duty.

Inevitably, there is no absolute precision in measuring what amounts to reasonable care in any given situation. A court will need to weigh up factors like the degree of risk involved in the defendant's behaviour; how serious was the harm which was likely to be caused; was there any particular social value or utility in the defendant's activities; and how costly and inconvenient would it have been to take precautions to eliminate or reduce the risk? The standard of care expected is the care of a reasonable man, although no such person may exist in fact! This means that the defendant's behaviour must be assessed against an objective standard, and it is matterless that the defendant tried hard or did his best. Of course, where a defendant holds himself out as possessing a particular skill or capability, he will be expected to display reasonable amounts of that skill or capability. So a surgeon must operate with the care that a reasonable surgeon would take; a builder must build with the care a reasonable builder would take; and so on. If there is a suggestion that a child has been negligent it may be appropriate to ask whether the defendant child took the care that could be expected of a reasonable child of the same age.

It is important to remember that, in formulating a standard of reasonable care, everyone is assumed to have knowledge about common facts, e.g. fire burns, water can drown, knives cut. In particular fields, a person with expert knowledge will be expected to display a reasonable amount of such knowledge. When judging whether a defendant has taken reasonable care, some guidance may be given by looking at what is usual or common practice in the activity or industry or operation in question. So an employer who fails to provide protective clothing for employees may be doing exactly the same as other employers in the same line of business. Arguably, all the employers may be being negligent, so general practice is not an absolutely reliable guide!

In assessing how much risk is involved in the defendant's activities, one judge has said that 'people must guard against reasonable probabilities, not fantastic possibilities'. If you can only conceive of harm occurring in some obscure or bizarre turn of events, then the risk may be so small that you are justified in ignoring it. After all, the law only requires reasonable care, not an absolute guarantee of safety in all circumstances.

In practice, it may be very difficult to separate questions like degree of risk from seriousness of harm likely to be caused. Obviously, even in cases where a very small risk is involved, it may be necessary to take more elaborate precautions where the harm caused, if any, would be very serious. If an employee has huge tanks full of liquid, he may be able to foresee that employees may fall in, or objects may fall in causing the liquid to splash out. If the liquid is cold water, some barrier to prevent employees falling in may be all that is necessary in order to show reasonable care. If the liquid is some caustic solution, likely to burn on the slightest splash, then the employer may have to consider some protective cover for the tanks. The precautions which are reasonably necessary may also be determined by asking whether employees are meant to be in the vicinity of the tanks or not, and whether employees have been issued with protective clothing.

It will already be apparent that reasonable care will fluctuate according to the exact circumstances of the case. Failing to test the temperature of the bath water may well be negligent if the bath is intended for a baby or someone lacking the mental ability to perceive the dangers of scalding water for themselves; in other circumstances such a failure may not be negligent, because the user of the bath should have checked for himself. A defendant engaging in some activity designed to save life and limb may well escape liability for negligence even though his behaviour may amount to a failure to take reasonable care in less socially useful circumstances. In *Watt v. Hertfordshire County Council*, firemen rushed to the scene of an accident where a jack was needed in order to free a woman who was trapped. Because of the emergency nature of the call, the jack was transported on an unsuitable vehicle, and was not secured during the journey. It shifted on the lorry, and a fireman was injured. There was held to be no negligence here, but it was made clear that if the jack had been transported in a similar way in a commercial enterprise, there would have been a failure to take reasonable care.

Even in those cases where some minute risk is foreseen, it may be possible to eliminate that risk entirely, but only by massive expenditure. The law does not impose such an impossible standard. If the risk could be eliminated or reduced by reasonable expense or practicable precautions, then the defendant should have done what he could. But again, it must be emphasized that he is only required to behave reasonably. In *Latimer v. AEC Ltd*, oil normally ran away in channels in a factory floor. It had been spread over the surface of the floor when the factory was flooded. The factory occupier scattered some sawdust, mopped up as much of the oil as possible, and gave a warning to employees to take care as the floor was slippery. The plaintiff slipped and was injured but the defendants were held not liable as they had taken reasonable care. The risk was not particularly great, any harm they could foresee was not likely to be particularly serious, and short of closing down the factory (which would have been extremely costly), there was nothing more they could have done.

The discussion of reasonable foresight of harm may raise the idea that what the courts are actually concerned with is hindsight! Inevitably, when a court tries to consider what the defendant should have foreseen, the judges will be influenced by policy factors, as they are in so many aspects of the development of the rules of negligence. So, for example, in cases involving motoring accidents, the fact that the defendant is insured could well have a quite distorting influence on the judgement of what amounts to reasonable behaviour. In *Nettleship v. Weston*, it was held that a learner-driver must achieve the standard of a reasonably competent driver. Likewise, in medical negligence cases, policy factors like a fear of encouraging defensive medical practice may play a part in determining what is reasonable.

It has already been indicated that it is up to the plaintiff to show that the defendant was in breach of duty, i.e. that the defendant had failed to take reasonable care. The plaintiff may be assisted in proving negligence by s.11 of the Civil Evidence Act 1968. If the defendant has been found guilty of a criminal offence involving negligence, that fact may be used as evidence of negligence in subsequent civil proceedings. This is obviously a very useful rule where a person has already been convicted of certain driving offences, but it could also be relevant where an employer had been found guilty of offences under the health and safety legislation.

A plaintiff seeking to prove negligence may also rely on the maxim '*res ipsa loquitur*', the facts speak for themselves. This could be pleaded in cases where the facts argue of no explanation other than negligence. In one case, a man was walking along past a warehouse when sacks of sugar fell on his head. What other explanation could there possibly be than that someone was being negligent? It is only appropriate to plead *res ipsa loquitur* if the defendant had management or control of the thing which caused the harm. This tends to depend on whether there was any possibility of some external interference. Where a plaintiff relies on the maxim, it is open for the defendant to rebut the finding of negligence, if he can do so.

Damage resulting of the right type which is not too remote

The type of harm which must be suffered in order to succeed in a claim for negligence has already been considered. The two significant issues which must now be addressed are the question of causation and remoteness. Causation is concerned with the problem of whether the defendant's conduct caused the plaintiff's damage. Remoteness is concerned with the cut-off point at which the law regards the defendant as no longer liable to compensate the plaintiff. If such a cut-off point were not established, a defendant could owe limitless liability.

1 *Causation* – a defendant may behave negligently towards a plaintiff and yet still not be the cause of his injuries. In *Barnett v. Chelsea and Kensington Hospital Management Committee*, a doctor was negligent in failing to examine a patient who had been brought into casualty, complaining of feeling sick after drinking tea. In fact, the patient was dying from poison and nothing could have saved him, so the doctor's negligence did not *cause* his death. It is usual to approach the causation question by using a 'but for' test. But for the defendant's negligence would the plaintiff have died in the Barnett case? Unfortunately, this test is by no means conclusive, and it may be particularly ineffective in giving an answer where the harm caused to the plaintiff was inflicted by two or more people, in quite separate incidents.

2 *Remoteness* – assuming there is no difficulty in proving causation, the plaintiff must then show the harm he suffered was not too remote from the defendant's negligent act. Any act can give rise to endless consequences, and one can often remark with hindsight that if only a particular event in the past had not happened, none of the later events would have resulted.

As a feature of not imposing excessive liability, the law has created the rule that a plaintiff can only recover for harm which is not too remote a consequence of the negligent act. As soon as such a rule comes into existence, there must be principles which will assist in finding the 'dividing line'. The test used by the court is that a defendant is only liable for such consequences of his negligent act as could reasonably be foreseen. An objective test is used. Would a reasonable man reasonably foresee a particular consequence? As an example, if it is intended to light a bonfire, it is reasonably foreseeable that people may fall into it, or that a spark from the fire may set alight nearby fencing, trees or property. But is it reasonably foreseeable that fumes from the bonfire will choke to death battery chickens being reared on adjoining land?

In a case which is accepted as establishing the approach to issues of remoteness, the Wagon Mound, a ship called the *Wagon Mound* negligently discharged oil into Sydney Harbour as it was sailing out. By the action of the wind and tide, the oil eventually floated all around the wharf area where the plaintiffs were carrying out welding operations. The welding work was suspended until the plaintiffs checked whether the oil would ignite. They discovered that this was unlikely as it was furnace oil with a high ignition point. Welding work recommenced and an extensive fire was caused when the oil did in fact ignite. The defendants, the owners of the *Wagon Mound*, were held not liable. Although they owed a duty to take care in relation to the plaintiffs, and were in breach by negligently discharging the oil, the damage they had caused was not a reasonably foreseeable consequence of their negligence.

The damage caused must, therefore, be of a reasonably foreseeable type. In the Wagon Mound, no doubt, the ship owner would have been liable if the plaintiffs had been complaining about oil fouling their slipways. This refinement can only be pursued so far. Take the example of *Bradford v. Robinson Rentals Ltd*, where a van driver was sent on a long journey in the depths of winter in a defective and unheated van. The employers owed him a duty, and were in breach. Could the employee recover for his injuries, which consisted of frostbite? No doubt that is quite an unusual complaint in this country but nevertheless some harm to his health was reasonably foreseeable and frostbite was therefore in the same class of risk. Compare that example with *Doughty v. Turner Manufacturing Co. Ltd* where the employer had a vat of acid maintained at a very high temperature. It was reasonably forseeable that employees might be injured by falling in or by being splashed if any object fell in. The employer would certainly be liable if an accident had occurred in either of those ways. What actually happened was that the lid of the vat, made from asbestos, slid into the acid. A chemical reaction was caused which resulted in an explosion and employees were injured by the acid erupting. The employers were held not liable, the damage was too remote. It had occurred in a manner which was not reasonably foreseeable.

When deciding the remoteness question a court will bear in mind the following points:

1. *The defendant must take the plaintiff as he finds him* – if a defendant driver is unfortunate enough to knock down a plaintiff who suffers from a weak heart and dies as a result, he may argue that, whereas the plaintiff's injury has proved fatal, any normal victim would only have been shocked and bruised. Again, he may crash into the back of another vehicle, which may be old and rusty or an expensive Rolls-Royce. But in each example the defendant will be liable to the full extent of the injury or damage caused, if injury or damage of that type could have been foreseen.
2. *The consequences of the defendant's breach of duty may have been 'overtaken' by some act which intervenes* – the defendant's breach may satisfy the 'but for' test, yet another event is regarded as the sole cause of the plaintiff's damage. For example, an employer's breach of duty towards his employee causes the employee a slight injury, and an ambulance is called to take the injured man to hospital. The

ambulance driver is negligent, causes a crash and the employee is killed. The crash is then the new act intervening.

If the court takes the view that the harm which the employee has suffered is not too remote from the employer's act, it would consider the question of contribution, if more than one person is liable to the plaintiff. The situation is governed by the Civil Liability (Contribution) Act 1978. The Act provides that where two or more persons are liable in respect of the same damage, each may recover from the other, or others, a contribution of an amount regarded as 'just and equitable', taking into account his or her share of responsibility for the damage.

Even if the plaintiff has satisfied all the above aspects of proof, his claim may still fail if the defendant has a complete defence (e.g. volenti). More usually his claim may fail in part, if it could be shown that the plaintiff was partly to blame for his own injuries, i.e. contributorily negligent. The present rules are to be found in the Law Reform (Contributory Negligence) Act 1945. If a plaintiff suffers damage partly through his own negligence and partly through the negligence of the defendant, the compensation will be reduced to the extent which the court thinks just and equitable, taking into account the plaintiff's share of the responsibility.

To use the defence, the defendant must show that the plaintiff failed to take reasonable care for his own safety. As with negligence itself, the standard of care varies according to the circumstances. Thus, an experienced mature workman can be expected to take more care when carrying out his work than a new apprentice. There has been a modern extension of the use of this defence in motoring cases when an injured plaintiff has failed to wear a seatbelt or crash helmet. In such cases, he may well find himself regarded as contributorily negligent (*Froom v. Butcher*).

If the plaintiff has consented to the risk, he should have no claim if injured. The scope of the defence is very limited and a defendant is more likely to plead contributory negligence. The defence can be relevant, however, when a defendant alleges that the plaintiff impliedly consented. Common situations where it may be raised arise in relations to sporting events, both for spectators voluntarily sitting there to watch, and participants in the sport.

Any account of the rules of negligence shows that there are many obstacles in the way of a successful claim. Often, cases turn on their own individual facts and there are numbers of conflicting precedents. The present process of claims through the courts has been called 'a lottery' and the time may be approaching when a different method should be found to compensate injured persons. The question was considered by a Royal Commission on Civil Liability for Death and Injury, which reported in 1978. Changes recommended by the Commission have not been made, but it was suggested in its report that motor vehicle injuries should be dealt with on a 'no fault' basis. This would mean that an injured person could be compensated out of a central fund, regardless of whether he could prove anyone to be at fault. Naturally, such a system removes the vagaries and imponderables always likely to be present in a negligence action. Despite the introduction of such rules in other countries, e.g. New Zealand, there seems to be no current intention to change the law in the UK.

Occupier's liability

Although the law can now be said to have general principles relating to negligence, there are some categories which pose their own special problems and for which particular rules have been created. An important example is the liability of occupiers of premises, whose duties are now laid out in the Occupiers' Liability Acts 1957 and 1984. The importance of the rules is obvious. Where a person enters premises and trips on a worn stair or collides with an obstacle because the lights are not working, liability can occur under the Acts. The main duty under the 1957 Act is a common duty of care owed by occupiers to their visitors. This requires the following analysis:

- Who is an occupier?
- What are premises?
- Who are visitors?
- What is a common duty of care?

The occupier

This word is not defined by the Act, but is generally understood to mean the person who has occupational control of the premises. He is the person in the best position to know who is likely to be using the premises, and to know about, and be able to put right, anything wrong with the premises. The occupier does not need to be exclusively in occupation; it is possible for premises to have more than one occupier. For example, in *AMF International Ltd v. Magnet Bowling Ltd*, a building contractor was held to be a joint occupier.

The premises

This word is wide enough to include not only buildings, but also land itself. What is more, by s.1(3) the Act extends its meaning to cover vessels, vehicles or aircraft.

The visitors

The range of people 'using' premises can be very wide, from a canvasser, a postman, an invited guest, a gasman, a deliveryman, even a burglar. The most important point is to distinguish between people lawfully on premises and trespassers. A person's presence on premises could be lawful because:

1   He has express or implied permission given by the occupier.
2   His entry is authorized by law (e.g. a policeman with a search warrant, an electricity official to turn off the supply, or factory inspector to investigate a work accident).
3   His entry is by virtue of a contract (e.g. going into a cinema after buying a ticket).

Even where entry on to premises is initially lawful, a visitor can subsequently become a trespasser if the permission to be there is withdrawn. For example, a shopper might enter a department store in a drunken state, create a disturbance and then be asked to leave. If he refuses to go, he becomes a trespasser. What is more, a visitor must only use the premises for the purposes for which he is invited or permitted to be there.

It should be noted that persons who enter under a public or private right of way are not visitors for the purposes of the 1957 Act.

The common duty of care

Under s.1 of the 1957 Act the occupier's duty is owed in respect of dangers due to the state of the premises themselves *and* due to things done or omitted to be done on the premises.

By s.2(2), the duty owed under the Act consists of taking such care as is reasonable in all the circumstances of the case to see that the visitor will be reasonably safe in using the premises for the purposes for which he is invited or permitted to be there.

The Act also sets out a number of factors to be taken into account when determining whether or not there has been a breach of duty:

1   When an occupier is considering how careful a visitor can be expected to be, he must be prepared for children to be less careful than adults (s.2(3)).
2   He can, however, expect visitors who use the premises in the

exercise of their calling (e.g. where a window-cleaner comes to the premises to clean the windows) to appreciate and guard against risks incidental to such calling (s.2(3)). For example, if the occupier asks a lift engineer to repair his defective lift, he will not be liable for damages caused to the engineer by that lift, but he will be liable if the engineer cuts his hand on an already broken pane of glass in the front door when entering the premises.

3   If the occupier has given warning of a danger (e.g. by a notice stating 'cliff edges crumbling – danger of collapse') this is not enough, in itself, to protect him from liability (s.2(4)). But it may in all the circumstances be sufficient to enable the visitor to be reasonably safe, and by this means the occupier may therefore have discharged his duty.

4   The occupier may know of a state of affairs on his premises which he cannot personally remedy, which will involve him in asking a contractor to undertake work of construction, maintenance or repair. Section 2(4) provides that a visitor injured as a result of the faulty execution of such work by a contractor will not succeed in a claim against the occupier so long as:

(a) The occupier acted reasonably in entrusting the work to a contractor, and
(b) the occupier had taken reasonable steps to satisfy himself that the contractor was competent, and
(c) the occupier has taken reasonable steps to satisfy himself that the work had been properly done. Note that this could require the occupier to employ someone to supervise or check the contractor's work e.g. where building work is being carried out, by the employment of an architect.

This section was extensively considered by the House of Lords in *Ferguson v. Walsh and others*. The plaintiff had been injured while working for Walsh, the first defendant, on a demolition job. Walsh had been asked to undertake the work by X, who in turn had been given the job by the district council who owned the site in question where the demolition was to be carried out. By the terms of the contract between X and the district council, X was not allowed to sub-contract the work without the council's permission. Nevertheless, that was exactly what X had done by his contract with Walsh. The first issue in the case was to decide if Ferguson was a visitor for the purpose of the Act. It was decided that he was, because he was effectively there by X's authority, and X was a joint occupier. Whether the council as joint occupiers were liable to Ferguson depended on s.2(4). The council had acted reasonably in entrusting the work to X and were held not to be liable. On its wording, s.2(4) appears to be limited to construction, maintenance or repair, but the House of Lords said that those words should be construed broadly and purposively, by which interpretation they were sure that 'construction' would embrace demolition.

5 The liability of the occupier may be limited or excluded, to the extent permitted by law. Section 2 provides that the occupier owes a common duty of care 'except in so far as he is free to do, and does, restrict or exclude his duty by agreement or otherwise'. There are two principal limitations to an occupier's right to exclude liability:
   (a) Even where he can successfully limit his liability he can only cut it down to the level of responsibility he would owe a trespasser. Otherwise a person entering lawfully could have less protection legally than a trespasser.
   (b) More important, any limitation of liability must be read subject to the Unfair Contract Terms Act 1977. In a 'business liability' situation an occupier cannot exclude or limit his liability for death or personal injury caused by negligence. Nor can he limit his liability for other damage caused by negligence unless the restriction satisfies the requirement of reasonableness.
6 When a person enters or uses premises by virtue of a contractual right, then unless the contract states otherwise, it is implied in it that the occupier owes a common duty of care. If the contract does contain express terms about the occupier's liability, these may be subject to the Unfair Contract Terms Act 1977.

A visitor to premises may suffer injury to himself or his property, or both. When he claims against the occupier, the defences which may be pleaded are:

1 Consent – s.2(5) expressly preserves consent as a defence.
2 Contributory negligence.

Occupier's liability to trespassers

As the 1957 Act extends protection only to visitors, it has no relevance when the person injured on premises is a trespasser or non-visitor. The rules which had been developed by the courts, culminating in the House of Lords decision in *British Railways Board v. Herrington*, have now been overtaken by the Occupiers' Liability Act 1984. Cases like Herrington had determined that an occupier owed a lower standard of care to a trespasser, a standard described as a duty of common humanity. The lower standard took account of the fact that a trespasser was uninvited. Nevertheless, common humanity was a variable standard, depending on factors such as whether the occupier knew of the likely presence of the trespasser; the kind of trespassers they were likely to be, e.g. children, burglars, poachers, squatters; the seriousness of the risk or danger; and whether or not the occupier knew of the risk.

The interplay of these factors can be seen in a case like Herrington. A child trespassed on the railway line and was electrocuted. The fence guarding the line was in a state of disrepair. The stationmaster knew that children were trespassing and had alerted the police. The line was something of an enticement, as the broken fence allowed children to take a short cut by that route to the nearby park. It was held that British Railways Board had failed to show common humanity. In view of the high level of danger, the likelihood of trespass, and the relative ease and cheapness of eliminating the danger for an organization with such resources, the child succeeded in an action for damages.

Under the 1984 Act, 'premises' and 'occupiers' are defined in the same way as for the 1957 Act. The duty under the Act is owed to persons 'other than visitors', and the duty consists of taking such care as is reasonable in all the circumstances of the case to see that the non-visitor does not suffer injury on the premises by reason of the danger concerned. For these purposes 'injury' is not to include loss or damage to property; it covers personal injury only. In this respect, the scope of the act in relation to a non-visitor is narrower than the scope of the 1957 Act to a visitor.

The duty under s.1 is only owed by the occupier of premises if:

1 He is aware of danger or has reasonable grounds to believe it exists.
2 He knows, or has reasonable grounds to believe, that the non-visitor is in the vicinity of the danger or may come into that vicinity.
3 The risk is one against which, in all the circumstances of the case, he may reasonably be expected to offer the non-visitor some protection.

As the occupier owes a duty to take such care as is reasonable in all the circumstances, factors which were previously relevant like the age of the trespasser, the nature of the premises, the character of the entry and the extent of the risk, will all still play a part in determining whether there is a breach of duty. One significant difference now is that the subjective element under the old common humanity rules may have gone. It used to be the case that the trespasser had to take the occupier as he found him, so if an occupier had limited financial resources, that might have been relevant in determining whether he had exercised common humanity. However, in determining what care is reasonable in all the circumstances, it is no doubt still open for the court to consider how burdensome it would be for the occupier to eliminate the risk.

Section 1(5) allows the occupier to discharge his obligations in appropriate cases merely by giving warning notices. It is doubtful if a notice saying 'keep out' would suffice, as it would merely alert persons entering the premises to the fact that they were trespassers, and not to the fact that there was some danger. Warning notices may be totally ineffective if the trespassers are children. And of course, as the Act is concerned largely with trespassers, they may choose to enter premises at some unconventional place where no notice is displayed. The Act specifically retains the defence of *volenti* (consent).

One unsatisfactory feature of the Act is that it does not make clear whether it is possible to exclude or limit the liability it creates. If an occupier puts up a notice stating that he is excluding all liability under the 1984 Act, it would be normal to test such a notice against the rules of the Unfair Contract Terms Act 1977. That Act seems not to be relevant here, as it limits its definition of negligence so far as the statutory duty of care is concerned to the Occupiers' Liability Act of 1957. This aspect of the 1984 Act awaits clarification.

It is widely believed that the 1984 Act will make little difference to the sort of trespasser plaintiff who will succeed against an occupier. In cases like *Pannett v. McGuiness & Co.*, decided on the old 'common humanity rules', demolition contractors were occupiers of a site in a busy urban area where it was reasonably to be expected that there would be large numbers of children. It became necessary to burn large quantities of rubbish and, realizing that this would attract children, three workmen were employed to keep them away. Children, including the plaintiff aged five, were chased away from the site on a number of occasions. While the fire was still burning, the three men left the site and the plaintiff re-entered and was injured by the fire. The nature of the activity was hazardous, the likelihood of trespass was great and the attraction to children was obvious. Applying the principles of the 1984 Act, the defendants would be aware of the danger, would know that trespassers were in the vicinity of the danger and in all the circumstances the risk would surely be one against which the contractor could reasonably be expected to offer some protection. The defendants were held liable under the pre-Act rules and the same outcome would be likely to occur today.

Liability of non-occupiers

For complicated reasons connected with the overlap of duties in contract and tort, the law has been slow to develop protection in tort for the person who suffers damage because of a defective building, where the defects are due to a non-occupier. A simple example of this is the case of a builder who built a house in 1970 for X. X sold the house to Y in 1980. The house has defective foundations and Y now discovers cracks appearing and distortion of door and windowframes. Y has no contract with the builder – he bought the house from X. The builder had been negligent. Can Y succeed in a claim against the builder?

In 1972, the Defective Premises Act was passed, imposing a duty on persons who take on work for the provision of dwellings. That duty is owed to the person ordering the work and to any person who subsequently acquires an interest in the dwelling. The extent of the duty is to see that the work is done in a workmanlike and professional manner with proper materials so that the dwelling will be fit for habitation when completed.

Some very significant limits operate, however, to cut down the apparent effectiveness of the protection given by this Act.

- The protection does not apply to dwellings already protected by an approved scheme of the National House Building Council.
- The Act only applies to dwellings, not to all types of buildings.
- Any action based on the Act must be speedily brought as there is a limited period of only six years, which runs from the date of completion of the dwelling.

Although the protection afforded by this Act is useful, the very limited extent of it has encouraged claimants to seek alternative ways of obtaining redress. Earlier cases such as Dutton and Anns demonstrated a willingness of the courts to allow claims against local authorities to succeed, based on negligent inspections of defective foundations, despite the loss in these cases being properly classifiable as pure economic loss. The trend in those cases, already discredited as they were in *D & F Estates Ltd v. Church Commissioners for England*, has now been halted by the decision of the House of Lords in *Murphy v. Brentwood District Council*.

In that case, one of the judges, Lord Mackay, the Lord Chancellor, said: ' . . . I am of the opinion that it is relevant to take into account that Parliament has made provisions in the Defective Premises Act 1972 imposing on builders and others undertaking work in the provision of dwellings obligations relating to the quality of their work and the fitness for habitation of the dwelling. For this House in its judicial capacity to create a large new area of responsibility on local authorities in respect of defective buildings would in my opinion not be a proper exercise of judicial power'. Lord Mackay was supported in this view by Lord Jauncey, who added: 'Parliament imposed a liability on builders by the Defective Premises Act 1972, a liability which falls far short of that which would be imposed on them by Anns. There can therefore be no policy reason for imposing a higher common law duty on builders, from which it follows that there is equally no policy reason for imposing such a high duty on local authorities'.

It seems now that a first purchaser of a defective building is therefore limited to contractual remedies and/or the protection provided by the 1972 Act. A subsequent purchaser of the property will have no contract with the original builder, and will therefore have to rely exclusively on the 1972 Act.

A practical application of the Act can be seen in *Andrews v. Schooling*, which was decided in 1991. An Edwardian house had been converted into flats, where the ground floor flat was bought by the plaintiff. Her flat also incorporated a basement. When the conversion work was undertaken by the builders, no damp proof course was installed. Once the plaintiff took up occupation, she then found that her flat was affected by penetrating damp in the basement. She sued the builders under s.1 of the Act. The builders sought to argue that they would indeed be liable if they had put in a damp proof course and done it badly (misfeasance) but that they could not be liable if they had done nothing (non-feasance). The court rejected that argument, indicating that it made no difference to the plaintiff whether it was misfeasance or non-feasance, as in either case, the builders had not provided a house which was fit for habitation. Of course, in awarding damages to the plaintiff, we have to accept that this was to compensate her in respect of economic loss, and it is arguable whether that was ever the intention of the Act.

To date, there is no sign of Parliament tackling the issues raised by the Murphy case, and the limitations indicated on the 1972 Act.

Breach of statutory duty

Many Acts of Parliament create obligations or duties, where a breach may result in the imposition of a penalty, very often a fine. For example, the Factories Act 1961 imposes a duty to fence certain machinery, and an offending factory occupier can be prosecuted and fined. Sometimes, the courts have allowed the Act of Parliament to be used as the basis for a civil action in tort. Such an action is called breach of statutory duty. The difficulty lies in knowing which Acts of Parliament can be used in this dual way. Most industrial safety legislation e.g. Factories Act 1961, has been so used, yet by contrast the Health and Safety at Work etc. Act 1974 makes it clear that its duties give rise to no civil liability (s.47). To succeed in an action for breach of statutory duty a person must, therefore, show:

1 The act in question was intended to create a civil action. This is the biggest hurdle unless a plaintiff can rely on a precedent showing that the Act has been used before as the basis of a civil claim.
2 The act imposes a duty on the defendant which he has broken.
3 The breach of duty has caused harm which is not too remote.

In general, this action is most used by employees in relation to Acts of Parliament concerning industrial safety, when the result is often to give an injured employee two heads of claim – an action based on negligence, and an action based on breach of statutory duty. The attraction of this tort for the employee is that the standard of behaviour expected from the employer is prescribed by the particular statute.

An example of breach of statutory duty of particular relevance to the construction industry is to be found in the Building Act 1984. This is the Act under which the Building Regulations are made. These regulations relate to the design and construction of buildings and are designed to secure the health, safety and welfare of people in the building. Section 38 specifically provides that where a breach of any duty imposed by the Building Regulations causes damage, including death or personal injury, a civil claim for breach of duty will lie. This is in addition to any action in negligence which may be available in relation to the construction of the building. (Note that this section is not yet operative.)

Defective products

Where a person is injured by using a defective product and is looking for some legal redress, there can be considerable overlap between the law of contract and the law of tort. If the injured person purchased the product himself, then he may sue the seller under the Sale of Goods Act 1979 if there has been a breach of any of the implied terms relating to reasonable fitness for purpose and merchantable quality. Moreover, in a consumer transaction, the seller cannot exclude or limit his liability for breach of these implied terms, because of the provisions of the Unfair Contract Terms Act 1977. In exceptional cases a buyer of defective goods may also have some limited contractual rights against the manufacturer, where there is a guarantee in operation. This 'contract route' to gain legal redress is limited to persons who can prove that they bought the goods, so it cannot assist a person who receives the goods as a present, or who is injured when using borrowed goods. Moreover, it is rather inefficient, in the sense that it may set in motion a chain of legal actions – buyer v. seller, seller v. wholesaler, wholesaler v. importer, importer v. manufacturer. Several costly actions may thus be necessary to lay liability at the door of the manufacturer. Moreover, the 'contract route' relies on suing a retailer who owes a strict liability, who will be liable whether he is at fault or not. This does nothing to ensnare the manufacturer, who is the person best placed to remedy such faults for the future, and who could redistribute his losses much more widely if made liable for the defective product.

The 'tort route' for damages in respect of defective products finds its origins in *Donoghue v. Stevenson*, where the defective product was contaminated ginger beer. It was consumed by a friend of the purchaser. The friend had no contractual rights. The House of Lords recognised in that case that a manufacturer owed a duty to take reasonable care to the ultimate consumer of his product where he could foresee injury and it was unlikely that there would be any intermediate examination of the goods. Of course, under this rule, the range of people who can sue is potentially much wider and the manufacturer's liability is not strict. He owes a duty to take reasonable care. The plaintiff will bear the burden of proving negligence against the manufacturer. That may not prove easy, although the plaintiff may be able to rely on the *res ipsa loquitur* principle. A plaintiff's claim may be defeated or reduced if the manufacturer proves that the plaintiff was contributorily negligent, e.g. by failing to follow the instructions issued with the product. And of course, a plaintiff needs to show that he suffered harm of the right type. In an action for negligence, we have already seen that the harm must normally be physical harm, such as personal injury or damage to property. A plaintiff will not usually succeed if he has suffered mere economic loss. But earlier discussions in the area of negligence have already focused on the fine dividing line between physical injury and economic loss. In some cases, there has been a suggestion that a plaintiff can recover for economic loss e.g. *Junior Books Ltd v. Veitchi Co. Ltd*; but that case turned on a peculiarly proximate relationship between the parties which is not usually likely to exist between a manufacturer and a consumer.

Thus the tort and contract routes can both be seen to be problematical. This area is now also governed by legislative controls. This is particularly important as there are inevitable policy decisions to be taken into account. Moreover, English law has had to come into line with the European Community which issued a directive in 1985 requiring member states to harmonize their rules on product liability. This has been achieved by the Consumer Protection Act 1987 which imposes strict liabilities which cannot be excluded or limited, contractually or otherwise (s.7) The Act provides that a producer of defective products is liable for any damage caused by the defect (s.2). The terms 'producer', 'product' and 'defect' are all defined by the Act.

The producer is defined by s.1 as the person who manufactured the product; or in the case of a substance which has been won or extracted, the producer is the person who won or extracted it; or in the case of a product where its essential characteristics are due to an industrial or other process being carried out, the person who carried out the process.

A product, defined by s.1, means any goods or electricity, and includes a product which is comprised in another product as a component part.

The Act recognizes that it may be easy to identify the person who sold goods but not so easy to establish who is the producer. This problem is met by s.2(3), under which a supplier of a product is liable if the plaintiff asks the supplier within a reasonable time of damage occurring to identify the producer, in circumstances where it is not reasonably practicable for the plaintiff to identify the producer, and the supplier fails to satisfy the request within a reasonable time.

A defect in a product is defined by s.3. There is a defect if the safety of the product is not such as persons generally are entitled to expect. In determining what persons generally are entitled to expect, a court could take account of:

1 The manner in which, and purposes for which, the product has been marketed, and any instructions for use, or warnings in relation to the use of the product.
2 What might reasonably be expected to be done with, or in relation to, the product.
3 The time when the product was supplied by its producer to another.

Nothing in this section is to be taken to infer that a product is defective merely because a later issued product adheres to an even higher safety standard. The factors specifically listed in s.3 are not necessarily exhaustive. In determining whether a product is defective a court is bound to take account of a balance which needs to be struck between manufacturing a product with some deficiencies, and the cost of eliminating those deficiencies. If that cost is so great, it may not be reasonable. The type of product will no doubt also be relevant when weighing up such a balance. Any cost may be considered necessary when manufacturing toys for very young children. Conversely, a drug with known adverse side effects may well still be better than no drug at all.

Where a producer is sued in respect of a defective product, s.4 provides that it is a defence for him to show (amongst other things) that at the relevant time (usually when the producer supplied the product to another) the state of scientific and technical knowledge was not such that producers could have been expected to discover the defect. This is often described as 'the state of the art' defence. It is thought to be vital if technical and scientific innovation is not to be discouraged. It was one aspect of the Directive where member states had some choice about whether to incorporate such a defence.

In order to sustain a claim under the Act, a plaintiff would need to prove damage. This is defined by s.5 as including death or personal injury or any loss or damage to any property including land. It expressly excludes loss or damage to the product itself. The property lost or damaged must be property intended for private use or consumption. Where the harm to property is valued at less than £275, no claim lies, but this restriction does not operate where the damage is personal injury.

The Act provides rights which are in addition to existing common law rights. Section 2 specifically states that it is without prejudice to any liability arising otherwise than by virtue of the Act. This could be relevant in cases involving pure economic loss. The Act places an embargo on recovery for such loss but cases on the common law rules of negligence may develop differently. (See Junior Books *v.* Veitchi.) An action under the 1987 Act is obviously advantageous to a plaintiff because the burden of proof is primarily on the manufacturer.

Vicarious liability

The word 'vicarious' means 'in place of another person', or 'in substitution for the proper person'. A rule has emerged in tort that an employer is vicariously liable for torts of his employees committed in the course of their employment. The rule is one of policy. Its importance is that is ensures that an injured person can sue the employer instead of suing the employee. The employer is likely to have greater financial resources, and may have taken out insurance against such claims. There is no particularly satisfactory justification for the existence of the rule, although it is often said that an employer profits from his employee's work and should, therefore, bear the risk of it. The two main criteria for the operation of the rule are:

1 The person who commits the tort must be an employee. This, however, can be misleading, as there are a number of important situations in which an employer will also be vicariously liable for the torts of an independent contractor.
2 The employee must commit the tort in the course of employment.

The course of the employment

There is no single test to determine when an employee is in the course of his employment. The decisions of the courts tend to turn on the individual facts of the case. Certain situations seem definitely to be covered.

1. Torts committed while the employee is doing what he is employed to do. For example, a man employed to drive a bus does so negligently and injures a pedestrian.
2. Torts committed while the employee is doing what he is employed to do, but doing it in a manner forbidden by his employers. For example, a petrol tanker driver is forbidden by his employers to smoke while making deliveries. He disobeys this order and causes an explosion which injures a passer-by.
3. Torts committed while the employee is doing what he is employed to do, but doing this in a criminal manner. For example, a solicitor's clerk employed to advise clients, fraudulently advises the client to transfer property to him.
4. Torts committed while the employee is doing acts which are reasonably incidental to his work. For example, an employee opens his office window for ventilation and in doing so negligently knocks a potted plant out of the window onto a passer-by below.
5. Torts committed while the employee is acting in an emergency for the protection of his employer's person or property, when his actions would normally be outside the scope of his employment. For example, an employee negligently moves a vehicle which he has no authority to drive, but which he reasonably fears is about to be damaged in a fire.

Where the employee's behaviour has amounted to a 'frolic of his own', the employer will not be vicariously liable. Reported cases cover the bus conductress who decides to drive the bus, the delivery driver who varies his proper route to visit a friend, and the garage attendant who assaults a customer with whom he has had an argument. In all of these cases, there is (at least temporarily) no course of employment.

The rule of vicarious liability is often said to be one of the distinguishing features of an employer/employee relationship. Generally, an employer is not vicariously liable for the torts of an independent contractor. If I enagage a builder to build me a house, he is liable for any acts of trespass or negligence he may commit, but I am not. However, in some cases either the liability imposed is strict, or the duties owed by a person are so onerous that there can be no delegation to someone else. Then the employer is liable in addition to the contractor. Although these are often spoken of as examples of vicarious liability, it is more appropriate to regard them as examples of an employer continuing to be personally liable. This can arise in the following situations:

1 When the employer is negligent in choosing the contractor, he remains personally liable. For example, an occupier might need repair work of a technical nature undertaken on his premises. He knows that the work involves electrical rewiring and yet he appoints an 'odd-job-man' with no training or expertise in this sort of work. The job is not properly done and causes a lawful visitor to the premises to receive a severe electric shock. The occupier would be in breach of his duty under the Occupiers' Liability Act 1957 and could not plead the benefit of s.2(4) of the Act, because he has not taken reasonable steps to satisfy himself that the contractor was competent.
2 When the liability of the employer is strict (e.g. as in the tort of *Rylands v. Fletcher*) and can occur even if he has taken reasonable care, the employer is liable whether he commits the tort personally or through an independent contractor.
3 When the employer undertakes a particularly hazardous activity, especially when it is on or near the highway, he is liable for torts committed personally or by his independent contractor. In one case, a heavy lamp suspended over a footpath fell into disrepair. The employer was held liable when a passer-by was injured by the activities of the contractor called in to repair it. In another case, a contractor was engaged to thaw out frozen pipes and chose to use a blow-lamp. The employer was held vicariously liable when a fire broke out as a result of the contractor's negligence.
4 When the employer authorizes or instructs the contractor to do something involving the commission of a tort, he remains liable. When a gas company engaged a contractor to dig up part of a street over which they had no such authority, the gas company was held liable when a passer-by fell over a heap of earth left by the contractor.

Contractors' acts of collateral negligence

Even in the above cases where both employer and contractor are laible, the former will not be held responsible for the contractor's collateral acts of negligence. In other words, an employer will only be liable if the risk of harm arises from the work itself, rather that the negligent performance of the work. For example, if an employer engages a competent contractor to install replacement frames on windows overlooking the highway, he will be liable if injury to a passer-by is caused by the glass being knocked out of the old windows (injury due to the work itself) but he will not be liable if the injury is caused by a workman negligently dropping his hammer (injury caused by the performance of the work).

10
Health and safety

The background to legislation

The following comment from a report published by the Health and Safety Executive on safety in the construction industry sets the scene for the problems of accident prevention in the industry:

> Construction management has to contend with a number of problems which vary enormously from site to site during the life of the site – climate, regional attitudes, geology, the time of year, the type of contract, the scale of the job, the methods of payment, the type of employment, the rapid turnover of labour. There is also the very complex problem of the relationship between a main contractor and an ever-changing group of sub-contractors. All these factors affect management in all its aspects, including the management of safety.

For a long time the law has played a part in controlling how employers must 'manage' safety. Acts of Parliament laying down duties and standards in specific working places date back to 1802. The earliest rules related to the employment of the very young in the mills, where the particular evil was the importing of large numbers of pauper children and engaging them in appalling working conditions. The procedure of making rules to control specific workplaces was established, and it will be seen that this became one of the worst features of the safety legislation.

With the pattern set, whenever a new loophole, crisis or new hazardous work area was discovered, special rules would be made to cover the situation. The end result has been piecemeal legislation covering mines, factories, offices and shops, agriculture and so on, culminating in a series of specific workplaces each governed by its own Act of Parliament. The Acts were not sufficiently detailed in themselves to cover every conceivable situation, so they all had various sets of regulations (delegated legislation) made under them. The Factories Act 1961 alone has more than 200 sets of regulations made by virtue of its authority.

The main aim of the safety legislation was to establish standards and duties for the workplaces covered. Failure to abide by the prescribed standards and duties would lead to employers being prosecuted and fined. By operating in this way the rules could, therefore, be seen to fulfil an accident prevention role. When an employer's breach of duty led to a workman being injured the various Acts were interpreted by the courts as giving the injured workman the right to sue for compensation.

The defects in the system of piecemeal legislation ultimately led to new thinking on health and safety – implemented by the passing of the Health and Safety at Work etc. Act 1974. Although the intention is for this Act to replace all previous specific legislation, this process will take time. Currently the 1974 Act must, therefore, be read in conjunction with all the earlier Acts and Regulations.

About 1000 people are killed at work annually and about half a million suffer injury as a result of accidents at work. There has been a drop in the number of fatal accidents in the construction industry recently, but this may be attributable to the recession in the industry. However, one significant form of injury, to the head, shows a reduction of 30% since the introduction of the Construction (Head Protection) Regulations in 1990.

The health and safety question can be seen to have three dimensions:

1 *The point of view of an injured employee* – he will want to know the answer to several questions. Should he report the accident? Can he get compensation? Is his employer insured? Will he get state benefits? Does it matter that he was partly to blame?
2 *The point of view of the employer* – he will want to know whether he will be prosecuted? Will his insurance premiums be increased? Will the employee claim damages? Should he alter existing safety arrangements?
3 *The point of view of the state* – here the issues will include the cost of health care for the injured person, lost productivity, the cost of state benefits and whether there are any lessons to be learnt for future safety policy. The answers to some of these questions come from the interaction of:

- The employer's duty in common law to take reasonable care for the safety of his employee. This is concerned exclusively with compensation.
- The employer's duties imposed by various specific Acts and Regulations e.g. Factories Act 1961, Construction (Working Places) Regulations 1966. These Acts and Regulations are concerned with accident prevention and compensation.
- The general duties imposed upon employers, employees, the self-employed and others under the Health and Safety at Work etc. Act 1974, which are concerned with accident prevention.
- The rules governing the Industrial Injuries Scheme, which is part of the system of social security and primarily concerned with compensation.
- The rules introduced to bring the UK into line with the EC Directives. The approach adopted in the Directives is to require the employer to be more proactive, and more aware of his duties to the wider world.

The Robens Report

At present, the 1974 Act has to operate side by side with the old Acts and Regulations but eventually these will be repealed. The Act was based on recommendations contained in a report on safety and health at work (The Robens Report). That report found the following major defects in the pre-1974 situation. There was too much law, written in a style which was unintelligible, relying too heavily on state regulation. Administration of the rules was fragmented, with seven different inspectorates. The complexity of the system led to apathy amongst employees, which needed to be redressed. Emphasis on particular workplaces meant that many did not enjoy the benefit of statutory protection.

The Health and Safety Commission

It is worth remembering that safety is not just an issue affecting those in employment. The general public is awakened to safety issues from time to time when a major disaster occurs such as the Kings Cross tube station fire in 1987. At such a time, there are often demands for safety standards to be improved, and a public inquiry (held by virtue of s.14 of HSWA 1974) may highlight changes that are necessary. The government has overall responsibility for introducing legislation to promote health and safety, and several of its departments, such as Transport and the Environment, will see this as a significant part of their work. Other organizations (e.g. the Royal Society for the Prevention of Accidents) exist to promote health and safety, and although their work may be more general, it can have useful implications for health and safety in employment.

The most significant organization, the Health and Safety Commission, was established by the Act in 1974. It has up to nine members, who are required by the Act to be representative of employers, trade unions, local authorities and other bodies, such as RoSPA, concerned with health and safety. The role of the Commission is to undertake, by whatever means it considers appropriate, to secure the health, safety and welfare of people at work, and to protect other people from risk to their health and safety created by work activities. The Commission may carry out research, provide training, disseminate information and advice, and submit proposals to the Minister for new regulations.

The Commission's accident investigation or inquiry role is significant, as it tends to have a high public profile and creates significant pressure for change or some government action. An investigation is less formal than an inquiry, so whereas the latter is the proper method to adopt in the Kings Cross situation, an investigation may be more appropriate for, say, a serious accident at a fairground. The person appointed to conduct an inquiry is given formal powers to compel the attendance of those he requires to be present. Inquiries are usually held in public and may well involve visits to the site.

One major initiative of the Health and Safety Commission has been to establish advisory committees on issues such as major hazards, nuclear safety and toxic substances, as well as industry-based advisory committees designed to promote the health and safety of workers in a particular industry. One example is the Construction Industry Advisory Committee. This kind of specialist committee can play an important role in assisting the Commission in tasks such as drawing up guidance. For instance, the booklet *Managing Health and Safety in Construction* (HMSO 1987) was drawn up in consultation with the Construction Industry Advisory Committee.

The most significant development which resulted from the establishment of the Health and Safety Commission was the new unified focus for safety issues and policy-making, which had previously been lacking. The Commission operates on a day-to-day basis through the Health and Safety Executive, which is the enforcing authority. Some enforcement duties are allocated to local authorities, and to avoid duplication between Health and Safety Executive inspection and local authority inspection, some types of premises e.g. shops, catering establishments and offices, have been specifically allocated as the responsibility of the local authorities. However, local authorities do *not* inspect their own offices! That task is performed by health and safety inspectors.

Much of the current work of the Health and Safety Commission is dictated by our membership of the European Community. Article 118A of the Treaty provides that member states shall pay particular attention to encouraging improvements, especially in the working environment, as regards the health and safety of workers. To help achieve this objective, directives will set out the minimum requirements for implementation. Such a directive 89/391/EEC was adopted in 1989, and we can now expect to see sets of new regulations introduced here to implement it.

The Health and Safety at Work etc. Act 1974

The 1974 Act adopted the recommendations of the Robens Report, placing emphasis on greater involvement and self-regulation rather than reliance on state regulation. The Act provides the framework of a new system, and under this 'umbrella' new codes of practice will be created. As this happens, the intention is that old Acts and Regulations will gradually be repealed.

The main objectives of the Act are set out in s.1:

1 To secure the health, safety and welfare of persons at work.
2 To protect persons other than persons at work against risks to health and safety arising out of work activities.
3 To control the keeping and use of dangerous substances.
4 To control the emission of noxious and offensive substances.

To achieve these objectives, the Act lays general duties on employers (s.2), employees (s.7), the self-employed (s.3) occupiers of workplaces (s.4) and manufacturers and suppliers (s.6).

The Offshore Safety Act 1992 now extends the 1974 Act to cover the health, safety and welfare of persons on offshore installations.

Duties owed by the employer – sections 2 and 3.

So far as is reasonably practicable, an employer must ensure the health, safety and welfare of his employees at work. This broad generalist approach is certainly a remarkable contrast with the detailed format previously adopted by safety Acts and Regulations. Perhaps unnerved by the 'baldness' of the duty, the Act goes on to spell out its nature more specifically. The employer must (so far as is reasonably practicable):

1. Provide and maintain plant and systems which are safe and without risk to health.
2. Make proper arrangements for the safe handling, storage, use and transport of articles and substances.
3. Provide necessary information, training, instruction and supervision.
4. Maintain the place of work and access to it so that it is safe and without risk to health.
5. Provide and maintain a safe and adequate working environment.

The Act does not define the terms 'health', 'safety' or 'welfare' nor the phrase 'so far as is reasonably practicable'. The latter phrase is familiar to the lawyer, because it is also used to qualify certain duties created by the Factories Act 1961. For example, s.28 of that Act provides that floors and stairs must, so far as is reasonably practicable, be kept free from obstruction. The meaning of this phrase has been judicially considered and has a narrower meaning than 'physically possible'. It implies that: 'A computation must be made in which the quantum of risk is placed on one scale and the sacrifice involved in the measures necessary in averting the risk is placed on the other. If the risk is insignificant in relation to the sacrifice, the defendants discharge the onus on them.' The following case provides an illustration. As a result of an accident at work, the Health and Safety Inspector issued an Improvement Notice, requiring a dairy to provide protective footwear free of charge to all employees involved in operating hydraulic trolley-jacks. The employer successfully appealed against the notice. The expense of providing the footwear free of charge was disproportionate to the risk to the employees. The dairy's existing arrangements, whereby employees could buy the footwear at cost price and pay by instalments, were adequate to satisfy the firm's duty to secure its employees' safety so far as was reasonably practicable (*Associated Dairies v. Hartley*).

Nothing in the Act allows these duties to be used as the basis for a claim for compensation. The sanctions for their enforcement are criminal. However, an injured workman would be able to proceed against the employer under ordinary negligence principles.

Section 2(3) provides that an employer must prepare and, when necessary, revise a written statement of his general policy with regard to health and safety, and he must bring the statement to the notice of his employees. The introduction of such written policy statements is seen as an important application of the ideas about greater involvement and self-regulation expressed in the Robens Report. If employees are to be involved, they must know what the safety policy is. The Act gives no guidance on the form or content of the safety policy. This will naturally vary with the size and type of organization and the specific hazards involved. The lack of a model form is sensible, as it compels employers to think positively about safety problems. Failure to abide by the rule about issuing a safety policy is an offence usually punishable by a fine.

Section 3 imposes on employers and self-employed persons a duty to conduct their undertakings in such a way as to ensure, as far as is reasonably practicable, that persons not in their employment who may be affected are not exposed to risks to their health and safety. This section is a valuable protection for members of the public but an unusual application of it can be seen in R. v. *Swan Hunter Shipbuilders Ltd*. In that case a fire broke out on board a ship being built by Swan Hunter. In one part of the ship the fire became particularly intense because a sub-contractor had left an oxygen hose there. Swan Hunter was aware of the dangers of using oxygen equipment, particularly because oxygen enrichment could occur in confined spaces. In order to alert its own employees to this danger, Swan's had distributed a booklet to them, setting out practical safety rules. These booklets were not given to all sub-contractors as a matter of course. As a result of the fire, eight men died. Swan Hunter was convicted of a breach of its duties to provide a safe system of work and to provide information and instruction (s.2). It was *also* convicted of a breach of its duty under s.3. By not providing information to non-employees, Swan Hunter had failed to ensure that persons not in its employment were not exposed to risk to their health.

The duties under s.2 and s.3 are now considerably reinforced by the introduction of the Management of Health and Safety at Work Regulations 1992, which are the first health and safety regulations made to implement the EC framework and directives. They create specific management responsibility to implement the safety policy, and are backed up by an Approved Code of Practice.

The employer must carry out a 'suitable and sufficient' assessment of all risks to health and safety, both of employees and anyone else who might be affected by the work. 'Suitable and sufficient' is explained in the code of practice.

Under the 1974 Act, once the employer had drawn up his written safety policy, there could be a danger of complacency. Now, Reg. 4 requires the employer to implement good health and safety management. He is required to appoint a person to assist in carrying out the health and safety duties. Such a person could be an employee or a consultant. Other new duties relate to giving information to employees about job risks identified by the assessment, and providing training on health and safety matters.

Duties laid on controllers of premises (s.4)

Section 4 imposes duties on controllers of premises, where the premises are used by persons who are not employees, and where those persons may use plant or substances provided there for their use. A frequently quoted example of such premises is a coin-operated dry cleaners. Here, the controller must take such measures as are reasonable to ensure that the premises, plant and substances are safe and without risk to health so far as is reasonably practicable. (This is the criminal law provision corresponding with an occupier's duties at civil law under the Occupiers' Liability Act 1957.)

Section 4 can also be significant where a site is in the control of the main contractor. That site will be used by sub-contractors and their employees, i.e. people who are not employees of the main contractor. In one case a prosecution was brought in London under s.4. The employee of a sub-contractor was driving a small dumper truck out of a service lift when the lift suddenly rose as he was half way out. The driver was killed when he became trapped between the dumper truck and the lift shaft. The tragedy could have been avoided if the interlock on the lift had not been jammed with pieces of timber, thus tricking the machine into believing that the lift gates were shut. The main contractor was fined £1,000.

Duties laid on manufacturers and suppliers (s.6)

These duties are regarded as among the most important methods of securing an improvement in safety standards at work. Generally, manufacturers or suppliers of any article or substance for use at work must, so far as is reasonably practicable:

1 Ensure that the article or substance is safe and without risk to health when properly used.
2 Carry out such tests and examinations as are necessary to achieve (1) above.
3 Take such steps as are necessary to ensure that adequate information is available about how to use the article or substance, so that when properly used, it is safe and without risk to health.

These duties allow faults and defects to be traced back to their source, and prosecutions can be brought against persons with the ability to remedy them. The effect of s.6 is not to absolve the employer from his duties, but it is recognized that the power of the employer to do much is often limited. These duties do not create any new civil liability. Any civil proceedings must be brought under ordinary negligence rules. However, the possibility to prosecute under s.6 is a significant increase in the scope of the inspectorate's power. Note also that similar duties are imposed on designers, importers, erectors and installers.

Duties laid on employees (s.7)

Under the old Acts and Regulations duties were placed directly on employees, and this section merely extends an old principle. The duties which an employee owes while at work are:

1 To take reasonable care for his own health and safety.
2 To take reasonable care for the health and safety of others affected by his acts or omissions at work.
3 To cooperate with the employer where necessary to enable the employer to fulfil his duties.

As with the other duties under the Act, s.7 cannot give rise to a civil claim for compensation but an offending employee can be prosecuted. Employees are most likely to be prosecuted in those cases where the employer has fully complied with his duties. The most notable feature about prosecutions brought against employees is that fines tend to be smaller than those imposed on employers, although the maximum fine in the Magistrates Court is £1,000. In a case brought against two nurses under s.7, where they had put an elderly patient into a scalding bath without testing the water, and the patient had died as a result of the accident, both were fined £1,000.

Although s.7 relates specifically to employees, they could also, with others, be prosecuted under s.8 which provides that no person shall intentionally or recklessly interfere with, or misuse, anything provided under the Act for health and safety purposes.

Under the new Management of Health and Safety at Work Regulations 1992, employees are under a duty to use machinery and equipment as they have been trained and instructed to do so, and they must inform the employer of any situation which is an imminent danger, or where there is noted to be a gap in the health and safety arrangements.

Safety representatives

In safety matters, the Robens Report considered that management and employees had an 'identity of interest' which should lead to 'participation in working out solutions'. Indeed they found evidence in some working situations of existing participation by employees, e.g. the representatives appointed by coalminers to inspect mines. Many firms and organizations had also established voluntary safety committees before 1974. This voluntary effort needed statutory support, and the 1974 Act permits regulations to be made providing for the appointment of safety representatives and for the establishment of safety committees. Such regulations have been in force since 1978 and provide for recognized trade unions to appoint safety representatives from amongst the employees. Generally, representatives must have been employed for two years by the employer or have two years' experience in similar employment. The main tasks of the representative are:

1. To be consulted on safety matters.
2. To inspect plant and premises, usually at three monthly intervals but more often if there has been a change in work methods, or an accident.
3. To investigate complaints from employees with regard to health, safety and welfare.
4. To investigate potential hazards and dangerous occurrences.

To perform these functions effectively, the representative needs time and training. The Regulations provide for time off with pay for carrying out their responsibilities.

The appointment of safety representatives is a step designed to overcome the apathy on safety matters which had been identified by Robens. This can be regarded as enforcement through cooperation. Two points may be noted, however:

1. The representatives have no obvious 'teeth' for enforcement. Given that they are appointed by recognized trade unions, it could be argued that any necessary 'teeth' would be the usual forms of industrial pressure or action. If this is what is envisaged, it seems to contradict the idea of a 'common interest' in safety matters.
2. By limiting the power to appoint safety representatives to recognized trade unions, no statutory machinery exists to appoint representatives in non-unionized areas of employment. Presumably, employees in these areas are to be left to 'flourish' in the state of apathy which the Robens Report identified. The initiative in these cases may need to come from management.

Safety committees

Regulations made under the Act provide for the establishment of safety committees if the employer receives a request in writing from two safety representatives. Once a committee is established, s.2 lays down that it should keep health and safety measures under review. The recommendation about health and safety committees is that they should not be given any other functions or responsibilities. There are no prescribed rules about membership, but the make-up of the committee, and the seniority of the management representatives chosen to sit on it may well give a good indication of a firm's general attitude to safety questions!

One useful function which a safety committee can fulfil is to monitor the advice available which is disseminated by bodies like the Health and Safety Executive. That advice can then be pursued and applied more specifically to the particular work situation. For example, in demolition work, an area notorious for the number and seriousness of accidents, the committee might consider the Executive's guidance notes, and then recommend to the employer that he should adopt a policy of drawing up a 'written methods statement' for each demolition job. This would focus the employer's attention on the hazards of the particular job to be undertaken (e.g. the presence of asbestos, unusual structures, toxic chemicals, safe access for work to be carried out at height, dust levels, and the need for protective clothing). If this is prepared after a site visit and at tender stage, it will allow the contractor to tender at a price level which takes account of doing the job safely.

By monitoring all the reports and advice available, the safety committee can also play a useful role in alerting employers to wider health and safety issues. Again, taking demolition as the example, it may be thought that the hazards are all too obvious. But this type of work may be even more hazardous than conventional construction work because of hidden dangers. It may seemingly have been fungal spores which caused severe pulmonary illness in Lord Caernarvon and some of his team who had excavated the tomb of Tutankhamun. Indeed, the illness was referred to as the 'curse of the Pharaohs'! Some old buildings could present similar hazards, calling for breathing equipment or protective clothing.

It should be remembered that the inspectors employed by the Health and Safety Executive have always seen advice giving as one of their key functions, and a safety committee can act as an impetus to the employer to take the benefit of their advice.

Enforcement of the act

A system of inspectors for the enforcement of safety legislation has existed for more than 150 years. Indeed it is the oldest established inspectorate in the world. With the establishment of the Health and Safety Executive, all inspectors have been brought together under one umbrella, but there are still different branches of the inspectorate. The factory inspectorate, which has responsibility for construction, is by far the largest branch. Its inspectors are responsible for almost half a million sets of premises, and the focus of their work has been dramatically broadened by the 1974 Act. In one of their publications for young people who may be contemplating a career with the inspectorate, they say:

> As well as gaining an awareness and understanding of the practical problems that industry faces, a Factory Inspector must be able to deal firmly and diplomatically with a wide range of people. Many situations will require delicate handling and the successful inspector is one who is able to adapt his approach and meet the needs of the situation without losing sight of his main objective – an improvement in health and safety conditions. The Factory Inspector has to be conversant with a wide range of technical issues, ranging from the nature of toxic chemicals to the techniques for guarding a whole range of machinery. He must also have a thorough understanding of health and safety law, as well as having the tact and diplomacy to deal with the sensitive issues that arise from time to time.

Training for new inspectors is partly 'on the job' and partly by formal training courses. A warrant is issued immediately to a new inspector. This is his authority to act. It specifies the powers conferred on him and he must produce it on request when seeking to exercise those powers.

Although inspectors have always had power to prosecute, there has been internal conflict over their role, which they have seen to be the improvement of standards of safety, rather than the mere enforcement of the law. Indeed, the time and effort needed to mount a prosecution was often felt to be wasted because of the low level of fines imposed. The inspectorate was fragmented with separate inspectors for factories, mines and quarries etc. Valuable time was often wasted making routine inspections of places where there was no great danger, leaving insufficient time to concentrate on places where there were real hazards. In its analysis of the problem of enforcement and the inspectorate, the Robens Report found that:

1 There should be a unified inspectorate.
2 The principal objective of the inspectorate should be to give impartial advice to industry.

Health and safety   271

3   The services of the inspectorate should be more selectively used foregoing the system of routine visits.
4   Where sanctions needed to be used by inspectors, the penalties imposed should be more severe.
5   A range of alternative procedures should be available to inspectors, allowing them to use sanctions which were immediate and constructive.

These proposals were implemented by the Act, and the scope of an inspector's powers is now set out in s.20. To enable him to perform his duties effectively, he may:

1   Enter premises (if necessary with a constable) taking necessary equipment with him.
2   Make investigations and examinations.
3   If necessary, order the premises to be left undisturbed.
4   Take photographs and measurements or recordings, and if necessary take away samples.
5   Question persons whom he believes can give him information relating to his investigation.
6   Inspect records, accident books, test certificates etc.

When establishing how his time can best be spent, an inspector involved in the construction industry might prioritize as follows:

1   Investigating fatal accidents and large incidents.
2   Visiting demolition sites and sites where steel erection is taking place. This is seen as particularly important, not only because of the high level of accidents, but also because these may involve examples of employers using untrained employees.
3   Investigating any site involving asbestos. This takes a high priority because of public perceptions of likely harm.
4   Investigating non-fatal accidents, where there has been a breach of law.
5   Investigating complaints from the general public. Very few of these come from trade unions. Some come from groups such as parents who are worried about an adjacent building site. Others come from rival contractors, possibly disgruntled that they failed to get the job!

Each year, the inspectorate focuses on one particular type of inspection, for instance roofing or transport. This is seen as an important part of their preventative role. The inspectors also lay considerable stress on their instruction role and point to guidance notes issued e.g. on use of ladders, or demolition practice. These guidance notes are less formal than codes of practice, of which there are very few. Ironically, codes and guidance are not always popular with 'customers' who actually seem to prefer strict regulations!

Most inspectors emphasize that, by and large, they get a good reception from employers, who accept how wide the powers of the inspector are. This may be due in part to their training in the skills necessary to present new ideas and persuade people to make changes in their working practices. The inspectors deal with people from the shop floor and people at director level, with back-street garages and multinational corporations. In all these dealings, they hope to advise and persuade. Often the threat of action by an inspector is the only 'enforcement' necessary.

If that fails, or the situation otherwise merits it, an inspector may use the powers granted by ss.21 and 22 to serve either an improvement notice or a prohibition notice. The advantage of both of these notices is that they can be issued by the inspector immediately.

Improvement notices (s.21)

The inspector may discover a breach of HSWA 1974 or the old Acts and Regulations and decide to serve an improvement notice on the person in breach. This notice will set out the breach of duty which the inspector thinks is occurring, specify the Act or Regulation in question, and give the inspector's reasons for his opinion. The notice will also set out the time limit for correcting the breach, the matters to be put right and the way in which this is to be done.

The earliest date which an inspector can set for compliance is twenty-one days after service of the notice. This is to allow time for an appeal to be lodged but it creates the biggest drawback about improvement notices – it means that it is impossible to get something done immediately. This can sometimes undermine the inspector's authority. Take the example of a site with no welfare hut. This is certainly not likely to be a situation warranting the use of a prohibition notice, but if the inspector serves an improvement notice, the employer can continue to 'cock a snook' for the next twenty-one days.

There is a right of appeal against the notice:

- Against the time limit imposed.
- Against the substance of the notice.

The effect of lodging an appeal is to suspend the operation of the improvement notice. On appeal, which is heard by the Industrial Tribunal, the notice may be affirmed cancelled or modified. The most frequent modification is an extension of the time limit. There must be proper grounds for an appeal. Mere lack of finance to carry out the

Health and safety 273

improvements is doomed to failure as an excuse. Similarly, arguments that there has never been an accident, that the employees have not complained, or that the breach is trivial, have not found favour with tribunals.

Prohibition notices (s.22)

This type of notice can be served when activities being carried on involve a risk of serious personal injury. Unlike an improvement notice, no specific breach of the Acts or Regulations need be specified, but the inspector must give reasons to support his decision. The notice directs that the activities in question must stop until matters have been remedied. One significant difference about a prohibition notice is that lodging an appeal does *not* suspend its operation. The repercussions of a prohibition notice are serious for an employer, as his business may be brought to a standstill. In consequence, these appeals are usually heard as a matter of urgency.

Anyone served with either variety of notice who fails to comply with it commits an offence for which he may be prosecuted. As well as imposing a fine, the court may order him to attend to the matters specified in the notice.

Prosecutions (ss.33-42)

The two types of notices already mentioned may not be considered appropriate by the inspector. His knowledge of an employer's past record may convince him that immediate prosecution is necessary. Prior to 1974, proceedings were always summary in the Magistrates Courts, and the levels of fines tended to be quite low. The Act has now created the possibility of summary proceedings in the Magistrates Court or proceedings on indictment in the Crown Court.

Prosecutions in the Magistrates Courts must be brought within six months of the date of the offence, but there is no time limit if a case is tried on indictment. In the Magistrates Court, the inspector will conduct the prosecution himself. Commonly, the charge may be failing to comply with s.2 of the 1974 Act. Although the maximum fines in the Magistrates Courts have now been increased to £5,000 for many offences, the average penalty imposed in 1991–92 was £1,134. That figure includes some large fines against BR (£250,000) and Shell UK (£100,000).

Where the prosecution is brought on indictment in the Crown Court, the court has the power to impose unlimited fines and in certain circumstances can also impose a term of imprisonment not exceeding two years.

Prosecutions are only brought in extreme cases. In any one year, less than 3000 employers may be prosecuted, whereas in excess of 12,000 prohibition and improvement notices will be issued. In an industry where more people are killed each year than in all the rest of manufacturing industry, weapons such as notices and prosecution form a useful part of the inspector's tools.

Control of substances hazardous to Health Regulations 1988

These rules were enacted in response to an EC Directive on hazardous agents. Substances are defined as 'any natural or artificial substance whether solid, liquid, gas or vapour'. Harm may be caused by inhaling the substance, or ingesting or absorbing it.

The employer is required to assess the risk presented by the substances, both to employees and others. He must control exposure to substances, and monitor it, and carry out health surveillance of employees exposed to risk. The regulations impose obligations on the employer in regard to training and information for employees.

The Factories Act 1961

It has already been seen that the Health and Safety at Work etc. Act 1974 provides a framework which will eventually operate with new Codes of Practice and Regulations. Until the intended changes are completed, the Health and Safety Executive created by the Act is also responsible for enforcement of the old Acts and regulations, which remain in force. The Factories Act 1961 is a useful model serving to show the distinction between the old format and the new approach. It is also the parent Act under which important Regulations for the construction industry derive their authority.

The duties created by the Factories Act 1961 and the Regulations made under it have been interpreted to give rise to both criminal and civil liability, with the result that the provisions have been the subject of much judicial interpretation. The civil action is in respect of breach of statutory duty.

What is a factory?

It is not surprising that the Robens Report did not like the existing system of legislation for specific places when the committee members encountered the definition of a factory! It covers almost two pages and has ten sub-sections. Moreover, it is recognized that the main definition is neither adequate nor exhaustive, because s.175 of the Factories Act 1961 also lists as factories an assortment of places, in case they do not fall within the general definition. Basically to constitute a factory there must be:

1 Premises,
2 in which persons are employed,
3 in manual labour,
4 for a variety of listed purposes (e.g. making articles, altering articles, adapting articles for sale),
5 where the work is carried on by way of trade or for purposes of gain,
6 where the employer has the right of access or control.

The definition can produce some odd results, e.g. a florist's shop, where the back part was curtained off and flowers were made into wreaths and bouquets. The front part of the premises was a shop and the back part was a factory. This breakdown of premises into parts which are covered by different statutory requirements can occur frequently. For example, a factory is also likely to have parts which are offices, and these will be covered by the Offices, Shops and Railway Premises Act 1963.

Once it is established that the place is a factory, the duties owed in those premises fall under three main headings: health, safety and welfare. Generally, the duties are owed by the 'occupier'. This term is not defined, but in most cases the occupier will be the employer.

Health provisions cover cleanliness, overcrowding, temperature, ventilation, lighting, drainage and sanitary conveniences.

Safety provisions cover fencing of machinery, safe places of work and safe means of access to them and fire escapes. The fencing provisions require that prime movers, transmission machinery and other dangerous parts of machinery must be securely fenced while in motion or in use. The problems of interpretation of these rules are obvious. What are dangerous parts of machinery? What does 'securely fenced' mean? When is a machine in motion or in use?

Welfare provisions cover washing and sitting facilities, first aid, accommodation for clothing and provision of drinking water. Many of these duties are duplicated by more detailed provisions in the Construction Regulations.

Construction regulations

The vast range of workplaces and working situations covered by the Factories Act 1961 means that the rules, though detailed, cannot be very specific. Each working situation will tend to have hazards peculiar to it. In the construction industry statistics show that many accidents are caused by falls – from ladders, scaffolds, platforms, or roofs – or by materials falling. Another significant cause of accidents, sometimes fatal, is the use of lifting equipment and machinery. Significant numbers of less serious accidents occur when employees step on, or strike against, objects. There is extra danger for workers in the industry when excavation and tunnelling takes place. To take account of these special risks, there are several sets of Regulations in the construction industry:

- Construction (General Provisions) Regulations 1961.
- Construction (Lifting Operations) Regulations 1961.
- Construction (Working Places) Regulations 1966.
- Construction (Health and Welfare) Regulations 1966.

These Regulations are made by the Secretary of State for Employment under powers granted by the Factories Act 1961. In any situation covered by the Regulations, an employer will owe duties under the Health and Safety at Work Act 1974, the Factories Act 1961 and the Regulations. The same breach of duty could, therefore, constitute an offence under all of these different provisions.

General provisions regulations

These Regulations apply whenever there are building operations or works of engineering construction. The two terms are used in all the sets of Regulations and their definitions are contained in the Factories Act 1961.

'Building operations' means the construction, structural alteration, repair or maintenance of a building (including repainting, redecoration and external cleaning of the structure), the demolition of a building and the preparation for, and laying the foundations of, an intended building.

'Works of engineering construction' means the construction of railway lines or sidings or the construction, structural alteration or repair or demolition of things such as docks, harbours, tunnels, bridges, viaducts, reservoirs, pipelines, sewers or gas holders.

Where more than twenty persons are employed, the Regulations

Health and safety 277

require the appointment of safety supervisors, whose duties include promoting safe conduct of the work generally. These supervisors are a specialized example of the use of safety representatives well before the 1974 Act took effect.

The Regulations principally cover:

1 Excavations, shafts and tunnels.
2 Coffer dams and caissons.
3 Use of explosives.
4 Work in dangerous or unhealthy atmospheres.
5 Dangers connected with transport.
6 Demolition.
7 Miscellaneous protections, e.g. fencing, protection from falling materials, lighting, or lifting excessive weights.

The duties are imposed on contractors and employers of workmen, and their nature varies according to the form of words used. For example, a strict duty is imposed by the regulation that 'Explosives shall not be handled or used except by, or under the immediate control, of a competent person . . . ' Compare 'In any excavation, shaft or tunnel where there is reason to apprehend danger to persons employed therein from rising water, there shall be provided, so far as is reasonably practicable, means to enable such persons to reach positions of safety'. In the latter case the duty is qualified.

Lifting Operations Regulations 1961

These are also made by the Secretary of State to apply in building operations and works of engineering construction. The rules cover:

1 Lifting appliances – these must be of good mechanical construction, sound material, adequate strength and free from patent defect, properly maintained, and inspected at least once a week by a competent person who must make a report in the prescribed form. The rules go on to deal with the support and anchoring of the appliance, with safe working loads, and the testing and certification of appliances.
2 Chains, ropes and lifting gear.
3 Hoists.
4 Carrying people on lifting appliances.
5 Secureness of loads.
6 Keeping of records.

Again, the strictness of the duty depends on the exact words used e.g. 'no crane which has any timber structural member shall be used', imposes a strict duty, whereas 'Where reasonably practicable, the cabin shall when in use during the cold weather be adequately heated by suitable means', imposes a qualified duty.

The Working Places Regulations 1966

Scaffolding is one of the principal matters covered by these Regulations. The accident statistics for the industry show that a high proportion of accidents and deaths involve scaffolding, and the importance of the Regulations will be obvious. Their scope again includes building operations and works of engineering construction. Among the duties imposed are:

1. Providing safe and suitable access to places of work so far as is reasonably practicable.
2. Providing scaffolds or ladders where work cannot be safely done from the ground or from another part of the building.
3. Erecting and dismantling scaffolding under the supervision of a competent person.
4. Providing scaffolds of good construction, suitable and sound material, and adequate strength for their purpose. The timber used for the scaffolds must be of suitable quality in good condition.
5. Properly maintaining the scaffolds and ensuring that they are stable.
6. Special duties providing for sling scaffolds, suspended scaffolds, cantilever scaffolds, bosun's chairs etc.
7. A number of duties relating to working platforms, gangways and runs. These should be closely boarded, planked or plated with material of specified thickness and width resting on sufficient supports and if the boarding projects at an end, precautions should be taken to prevent tipping.
8. Duties with regard to guard rails and toe-boards.
9. Duties with regard to the construction, maintenance and use of ladders.
10. Duties relating to work on sloping roofs.
11. Where there is special risk of workmen falling, duties with regard to safety nets and safety belts.

The Health and Welfare Regulations 1966

The main duties under these Regulations relate to matters such as:

1. First aid equipment and training in first aid treatment.
2. Shelters and accommodation for clothing and taking meals.

3   Washing facilities and sanitary conveniences.
4   Protective clothing for persons required to continue working in bad weather.

It is undoubtedly the case that all of the rules stated above will be of little effect without proper attention by employers and employees to safety questions. The reports of the Health and Safety Executive indicate that accidents are more frequent in smaller firms, which may take a less positive interest in safety. Motivating employees to be safety conscious is an especially difficult problem if the workplace is constantly changing and at some distant location, making constant supervision impossible. One suggestion of the Executive may have some impact. Clients about to engage a contractor are advised not only to satisfy themselves about his technical and financial ability to do the job, but also to enquire into the contractor's health and safety record!

Compensation for injuries at work

Where a workman is injured or is killed in an accident at work, he or his dependants may wish to sue for damages. Damages are only available where an employee can show that the employer was in some way at fault. Generally, this means that the employee must prove the tort of negligence. Although negligence is still developing as a tort, the example of the duty owed by an employer to take reasonable care for the safety of his employees was recognized very early. Not only may the injured employee be able to sue in negligence, he may also have a claim for damages for breach of statutory duty. Negligence and breach of statutory duty are considered in detail in the chapter dealing with tort. Apart from claims for damages, the injured employee may also be entitled to disablement benefits under the Industrial Injuries Scheme. He may be entitled to benefits under his own private insurance arrangements, or from his trade union.

When the court is considering a claim for damages from an employee, it will expect the employer to have taken reasonable care, given the state of knowledge and recognized practice in particular industries and trades. An employer will be expected to keep abreast of developing knowledge. The duties owed are not absolute so the employer must weigh up the risks involved and set these against the effectiveness of precautions he could take, bearing in mind the cost and inconvenience of taking those precautions.

An employer owes the duty to take reasonable care to all of his employees as individuals. Where some of his employees are inexperienced or do not easily understand English, or are newly recruited, he may have to take more precautions in their case than for other workmen. However, the courts have said that they do not see the relationship of employer and employee as that of nurse and imbecile child!

The employer's duty to take reasonable care has been analysed regularly by the courts and they have been inclined to break down the duty into a number of broad heads:

- A duty to provide reasonably safe plant, equipment and premises
- A duty to provide reasonably safe fellow workers.
- A duty to provide a reasonably safe system of work.

In all of these aspects the duty is on-going. A safe system of work, once introduced, must be regularly checked and monitored. It may need to be changed to meet new standards created by new knowledge.

An employer is in breach of his duty if he fails to take reasonable care. The duty is not absolute; he does not guarantee an employee's safety. It will be for the injured employee to prove negligence. In some cases this will be admitted by the employer and the subsequent court action will concern itself only with the measure of damages. The employer may seek to defend himself by pleading that the employee was partly to blame – contributory negligence. If that plea is successful, it will operate to reduce the damages payable by that proportion which the employee is found to be at fault.

Inevitably, employers have insurance cover to provide for the damages in such cases. Large sums may be awarded to the employee, but that is not, in itself, a particular reason for an employer wanting to improve his accident record. The employer may be embarrassed by the publicity given to the case but more often than not, potential civil claims by employees may encourage the retention of bad practices, as the employer may not want to change his practices in case such steps are seen as an admission of negligence.

11
Land law

Classification of property

Land is one of a number of kinds of property which can be owned, so it is useful to establish the meaning of the word 'property' and then to attempt to classify types of property. When considering sale of goods we saw that the word 'property' was used in the special sense of 'ownership'. The more general meaning of the word is, simply, anything capable of being owned.

The various categories of property are so diverse (e.g. land, jewellery, cars, stocks and shares, patents, money) that it is not surprising that the law has needed to develop different rules for each type to cover buying, selling or other transfer. Of course, if you buy a loaf of bread, it can be handed over to you. The same is not true when you buy land. A transaction involving land may be far more complex, especially when several people may have different rights over the same property at the same time. A person may own a house in which a tenant is currently living. The house may be mortgaged to a building society, and the owner may have leased shooting and fishing rights on the land and granted rights of way. This is a simple example but it serves to illustrate why the rules relating to land can be complex.

Obvious ways to divide property of all types are into things which are movable or immovable, or things which are tangible or intangible. Unfortunately, English law has not used either of these classifications, but instead divides all property into that which is real and that which is personal. This might be simple, if real property consisted of land, and personal property was everything else. But real property consists only of freehold interests in land, so (for historical reasons) leasehold interests are classed as personal property. As the terms 'real' and 'personal' do not conjure up as clear a picture of what they cover as the other classifications, it is interesting to note the derivation of the names. Historically, English law has always given preferential treatment to land and interests in land. If a person had his freehold land wrongfully taken from him the law would ensure that he could recover it by means of a 'real' action. If any other type of property was wrongfully taken from him the law merely allowed him to make a claim for damages against the wrongdoer, by means of 'personal' action. So property came to be classified according to the type of court action needed to protect it, which accounts for the anomalous classification of leasehold interests in land.

What is land?

Land is the surface of the earth, the airspace above it and the ground below. An owner of land could expect the law to protect his interests if someone unlawfully tunnelled under the land. However, Acts of Parliament limit a landowner's rights in this respect. Certain mining and mineral rights have been granted by statute to public corporations e.g. coal mining rights to British Coal, formerly the NCB.

As for the airspace above the surface of the earth, it used to be assumed that the landowner owned the space up to the heavens. This rather romantic idea has now been restricted by cases such as *Bernstein v. Skyviews and General Ltd* where it was held that an owner owns only as much of the airspace above his land as is reasonably necessary for his ordinary use or enjoyment of the land. In that case, a firm specializing in aerial photography flew over the plaintiff's house to photograph it. This act was held not to constitute a trespass, given the height at which the aeroplane flew. Most normal aircraft activity is excused under the Civil Aviation Act 1949, which precludes actions for nuisance or trespass if the aircraft flies over at a reasonable height.

It is difficult to think of airspace as being a valuable commodity but it may have significant commercial worth. For example, it may be possible to rent out space for erection of an advertising hoarding. The commercial value of airspace was at stake in a case, *Woollerton and Wilson Ltd v. Richard Costain Ltd* where a builder was using a tower crane. The arm of the crane swung over adjacent land, trespassing to the owner's airspace. The builder had not sought permission for this activity. When it became clear that he was trespassing, he offered to pay a weekly sum to the landowner. The landowner obviously thought he could hold out for more. However, his greed was 'rewarded' when the court granted him an injunction to prevent the builder using the crane, but its operation was postponed for several months, by which time the need for the crane had disappeared! An injunction was granted with immediate effect in similar circumstances in *Anchor Brewhouse Developments Ltd v. Berkley House Ltd*.

Is it clear that the law now sees some limit on the extent to which land extends into the air. The precise limit will depend on the type of land and the sort of use to which it could be reasonably expected to be put. An owner of a grouse moor may need a considerably greater height of air than the owner of a suburban bungalow where the only likely use for airspace will be for the erection of a TV aerial.

On the surface of land there are likely to be structures such as houses or factories. Buildings form part of the land, and so do fixtures. A fixture is something on or attached to the land for its improvement, which is regarded as forming part of it, e.g. a garden gate, rose bushes or a dry-stone wall. A chattel is an item of personal property which does not form part of land, e.g. a watch or a pen. The same items may, in different contexts, be either fixtures or chattels. Thus, a pile of stones in a builder's yard is personal property, but the same stones in the form of a dry-stone wall would be a fixture.

It may be important to determine whether an object is a fixture or a chattel. When land is sold, all fixtures pass with the sale unless otherwise agreed; when a tenant installs fixtures in his landlord's property, the fixtures become the property of the landlord; and, where land is mortgaged, all fixtures form part of the mortgage security, even those installed after the mortgage was created. Further, where a testator in his will leaves all of his real property to one person and all of his personal property to another, fixtures will pass with the real property and chattels with the personal property.

The question of whether an item is a fixture can also be relevant where a landowner becomes insolvent. In the case of *Lyons Co. v. London City Bank*, the owners of chairs had hired them out to the owner of a hall. In order to comply with local bye-laws on safety, the hall owner had fixed the rows of chairs to the floor. He then became insolvent, and the receiver was keen to show that the chairs formed part of the hall owner's land, as fixtures. The chair owner was equally keen to show that they were still his chattels, as he would be limited to a fairly hopeless claim for their value in the insolvency proceedings unless he could take the chairs back. The court found that the chairs were chattels.

In determining whether an item is a fixture or a chattel, the basic question is whether the item is physically attached to the land. If an item is attached to the land, this raises a presumption that it is a fixture. A statue which is cemented in place would thus be presumed to be a fixture. If an item is not attached to the land, the presumption is that it is a chattel. Thus, a free-standing statue, held in place only by its own weight, would be presumed to be a chattel.

It is possible to rebut the presumption that things attached to the land are fixtures, by showing that the item in question was fixed only to allow it to be enjoyed as a chattel and not for the improvement of the land as land. Thus, in *Leigh v. Taylor*, where a valuable tapestry was fixed to battens on a wall, the tapestry was held to be a chattel because, in order to display and enjoy the tapestry, it was necessary to fix it in some reasonably robust way appropriate to its considerable weight.

Equally, it is possible to rebut the presumption that an item which is not attached to the land is a chattel, by showing that the item does form part of the land. Thus, a dry-stone wall would clearly be regarded as a fixture, because it is there to improve the land, although unattached other than by its own weight. Similarly, a free-standing statue which formed part of an overall landscape garden design might be a fixture.

Clearly, it may at times be difficult to determine whether an object is a fixture or a chattel. Thus, in relation to a sale of land, it may be sensible for the parties to reach an express agreement about those items to be included in the purchase price and those which the vendor can remove. If such an agreement were always reached, there would be fewer disputes over greenhouses and sheds, TV aerials, dishwashers, carpets and garden ornaments!

In relation to tenancies, whilst items installed by a tenant in his landlord's property become the property of the landlord, the tenant may be entitled to remove 'tenant's fixtures' before or within a reasonable time of the termination of the tenancy. Tenant's fixtures are fixtures installed by the tenant which are domestic and ornamental (such as stoves and paintings), trade fixtures (such as petrol pumps and industrial machines), or agricultural fixtures. (Special rules under the Agricultural Holdings Act 1986 apply to agricultural fixtures, which are outside the scope of this book.) In removing tenant's fixtures, the tenant must do as little damage as possible, and must repair any damage done.

As ownership of land includes the surface and the ground below, a landowner might imagine the he owes everything in or on the land. The special position of mineral rights has already been mentioned, but it is also worth noting that any 'treasure trove' found on the land belongs to the Crown. Objects will only be held to be treasure trove if they consist in large part of gold or silver which have been deliberately hidden, and the true owner is unknown. If necessary, the coroner must hold an enquiry into findings of treasure trove. Items of value do get found from time to time, particularly where land was previously occupied by the Romans. Discoveries are more likely nowadays because of the use of metal detectors. In 1981 a farmer in Lincolnshire was approached by a man asking if he could go over the land with a metal detector. The farmer agreed but insisted that anything found should be brought to him. The man found several thousand Roman coins and a broken Roman pot but he handed over only the pot to the farmer! He was subsequently found out and convicted of theft. The farmer was allowed to keep the coins because they contained only trace amounts of silver and were held not to be treasure trove. (*Attorney General v. Overton*).

Wild animals living on a person's land are incapable of being owned. Once the animals are dead, they belong to the landowner, so a poacher should hand over the fruits of his labours! Percolating and flowing water cannot be owned, but a landowner is permitted to abstract quantities of it for uses connected with his land. Similarly, if his land is adjacent to a river, he may take water for domestic purposes but if he takes it for other purposes connected with the land, he must be careful to observe the rule that landowners further downstream are also entitled to the flow of the river unaltered in quantity and quality. If a landowner pollutes the water which he returns to the river he can be sued in nuisance.

Freehold and leasehold estates

People are often surprised to learn that in England individuals do not own land. All land belongs to the Crown, and the greatest interest which a landholder can own is the freehold interest or estate. This peculiarity of English law is explained by reference to history. When Willaim conquered England in 1066 he claimed all the land for himself, and then parcelled it out to his barons as a reward for their support. In return, they owed allegiance to him and undertook various services, some merely ceremonial, some valuable. This was the feudal system. It was not until 1925 that our legal system was rid of most of the consequences of feudalism. Modern land law is based on the Law of Property Act 1925.

As every landowner is technically a tenant of the Crown, it is common to refer to his interest in land as an 'estate'. The law recognizes two estates – freehold and leasehold. The major difference between a freehold and leasehold is the length of time for which the land will be held.

Freehold is the nearest equivalent to absolute ownership of land, and is more technically described as a 'fee simple absolute in possession'. Basically, a fee simple continues to exist as long as there are persons entitled to take it on the owner's death either under his will or on intestacy. (Intestacy is the name used to describe a situation where there is no will to govern the disposal of property of a deceased person. Statutory rules then operate to determine who will inherit.)

By contrast, a leasehold estate is described as a 'term of years absolute'. This means that a landlord (who may be the owner of the freehold or may himself be a tenant) grants to a tenant exclusive possession of property which the tenant is to enjoy for a fixed period. If the period is quite short it is usual to call it a tenancy (for example, a monthly tenancy). Longer terms are usually referred to as leases (for example, a 99-year lease), but there is no precise or rigid rule.

Both freehold and leasehold estates are property which can be sold and can be inherited under a will or on intestacy, and are the only two legal estates which can exist in land.

Interests in land

Rights in land other than the two legal estates (freehold and leasehold) are called interests in land. Basically, there are two types of interests in land, legal interests and equitable interests. S.1 of the Law of Property Act 1925 provides that a limited number of rights in land may, if created by deed, be legal interests. The most common of these are easements and profits *a prendre* (provided that their duration is either that of a fee simple absolute in possession or that of a term of years absolute), and charges by way of legal mortgage. The nature of each of these interests is considered below.

Any interest in land which is not a legal interest must be an equitable interest. Examples of equitable interests are: interests under restrictive covenants; interests (including fee simples, leases, easements or profits *a prendre*) granted for life; interests (including fee simples, leases, easements and profits *a prendre*) created without the use of a deed; interests under contracts to create legal estates or interests; and interests under trusts.

Traditionally, the essential difference between a legal interest and an equitable interest is that a purchaser of land is bound by all legal interests in the land but will not necessarily be bound by all equitable interests in it. The position has been modified by statute, however (Land Charges Acts 1925 and 1972; Land Registration Act 1925).

The extent to which a purchaser of land can discover and is bound by interests in the land which he purchases, and the extent to which the owner of an interest in land can ensure that his interest will bind a purchaser of the land, are considered in the course of the following section.

The conveyancing process

The conveyancing process is the process of transferring ownership of land, i.e. selling land. At present there are two systems of conveyancing in England and Wales. Unregistered conveyancing is the transfer of unregistered land, and registered conveyancing is the transfer of registered land. The traditional system is unregistered conveyancing, but as all land in England and Wales is now subject to compulsory registration, under the provisions of the Land Registration Acts, registered conveyancing is now replacing it.

In order to determine which system of conveyancing is appropriate, it is necessary to determine whether the land in question is registered or unregistered. This requires a search of the Land Register, which is made by sending a form to the District Land Registry in whose area the relevant land is situated.

Unregistered conveyancing

[For the purposes of this section the terms purchaser and vendor will normally imply the solicitors acting on behalf of the purchaser and vendor.]

The stages of the process of conveying unregistered land are as follows.

1 The contract is drafted by the vendor and two copies are sent to the purchaser.
2 The purchaser searches the local land charges register, which is kept by the local authority, in order to determine whether any local land charges are registered in relation to the land. It is usual to send in a search form, requesting the Registrar's staff to perform the search on behalf of the purchaser.

There are two types of local land charges: restrictions imposed on the use of land, such as planning permission restricting the number of houses to be built on the land; and financial charges imposed on the land in respect of work carried out by the local authority, such as the imposition amongst frontagers of the cost of making up a road. If a local land charge is correctly registered then the purchaser is bound by it even if he does not search the register. If a search by the Registrar's staff fails to reveal a local land charge which has been correctly registered, or the local authority has failed to register an existing local land charge, the purchaser is still bound by the local land charge but is entitled to compensation from the local authority.

Using the standard search form, to which additional enquiries may be added, the purchaser makes a variety of other enquiries of the local authority in relation to matters not registerable as local land charges. These may include issues relating to planning, schemes for new roads and slum clearance, and whether the local authority is obliged to maintain roads adjacent to the property. If the local authority's answers are incorrect, the authority may be liable in negligence or for breach of contract.

3 The purchaser makes a variety of enquiries of the vendor, usually by sending a standard form, which may be customized by adding or deleting enquiries as appropriate. Enquiries normally relate to matters such as easements and covenants, services enjoyed, fixtures and boundaries. The vendor is not obliged to reply, but if he does, incorrect information may amount to a misrepresentation.

4 If the purchaser is satisfied by his preliminary enquiries and searches, he may wish to have a survey of the property carried out. This is not essential, but may be a prudent step. If the purchaser needs to raise money by mortgage, the building society will usually conduct its own survey. Where this is the case, it may still be sensible for the purchaser to have his own survey carried out as the building society is only concerned to know whether the property is good security for the mortgage, not whether it is worth the purchase price. (See *Smith v. Eric S. Bush* on page 230 for an example involving a negligent survey.)

5 If the survey is satisfactory, and raising any necessary finance presents no problems, it is then safe for the purchaser to make a legally binding contract with the vendor.

The purchaser may agree the draft which has been sent by the vendor, or he may want amendments. When both parties agree to a final draft of the contract, the purchaser sends one copy to the vendor and keeps the other. The vendor then produces two final copies of the contract and sends them to the purchaser. The purchaser checks them and returns one to the vendor. The vendor and purchaser each sign their copy and then exchange copies (exchange of contracts), the purchaser giving a deposit if required. The contract becomes binding at the time of exchange.

S.2 of the Law of Property (Miscellaneous Provisions) Act 1989 provides that a contract for the sale or other disposition of any interest in land is only valid if made in writing. The writing must embody all the terms agreed, though it may incorporate terms by reference to another document. The contract must be signed by or on behalf of both parties. If there is an exchange of contracts (see above), at least one of the documents, but not necessarily the same one, must be signed by or on behalf of each of them. (These rules are considered in more detail on page 90. See also 'Subject to contract' on p. 92.)

The existence of the contract does not prevent the vendor selling the land to a third party, unless the contract is registered. (See registration of land charges and registered land below.) However, if he does so, he will be acting in breach of contract. Exchange of contracts does not transfer legal title to the property from the purchaser to the vendor.

6 After exchange of contracts, the vendor is required to show that he has the necessary title to the land which he has undertaken by the contract to transfer. The vendor must produce an abstract of title and send it to the purchaser within a reasonable time, unless the contract lays down a specific time limit. The abstract of title is a summary of the documents which prove the vendor's title to the land. In practice, a chronological list of the relevant documents together with photocopies of them are provided by the vendor.

7 The purchaser examines the abstract of title. If he is not satisfied that the vendor has the necessary title, he sends a written requisition to the vendor within a reasonable time (unless the contract specifies a specific time limit). The vendor must reply within a reasonable time (unless the contract specifies a specific time limit). If the vendor does not reply in time or fails to show that he has the necessary title, the purchaser may be entitled to rescind the contract.

8 Immediately before completion (the transfer of legal title in the property from purchaser to vendor), it is necessary for the purchaser to search the registers kept by the Land Charges Department of the Land Registry. In practice these searches are normally carried out by officials of the department on the purchaser's behalf.

Of the five registers kept by the Department, the most significant is the register of land charges. Certain interests in land are registrable as land charges. Examples are: mortgages not protected by deposit of title deeds; contracts to convey or create legal estates; options to purchase land; restrictive covenants created after 1925 and not in a lease; equitable easements or profits *a prendre*; and equitable interests not protected by deposit of title deeds, not arising under a trust and not otherwise being registrable as land charges.

Where a valid registrable interest is registered then, subject to certain exceptions, a subsequent purchaser of an estate or interest in the land will be bound by it, whether he searched in the register or not.

Where such an interest is not registered then, basically, a purchaser for value of a legal estate in the land will not be bound by it. This rule is well illustrated by *Midland Bank v. Green*. In that case, a farmer had made a legally binding contract to sell his farm to his son for a very favourable price. He then fell out with his son, and sold the farm instead to his wife. She knew about the earlier contract between father and son, but the son had never registered it as an estate contract. She paid a very low price for the farm. The son sued on the binding contract he had made with his father, but the court held that his mother could keep the farm, as she was a purchaser for value of a legal estate, and she was not bound by the son's interest because he had never registered it. The case is particularly powerful in reinforcing the rules about registration because the mother had actual knowledge of the estate contract, despite its lack of registration. Nevertheless, the non-registration was the paramount factor.

However, in some cases of non-registration, including contracts to convey or create legal estates, options to purchase land, restrictive covenants and equitable easements and profits, purchasers of interests in land, or purchasers of legal estates other than for money or money's worth, may be bound by the interest even though it was not registered.

Where a legal interest is not registrable as a land charge (for example, a legal easement or profit *a prendre*), a purchaser of an estate or interest in the land will automatically be bound by it. Where an equitable interest is not registrable as a land charge (for example the interest of a beneficiary under a trust) a bona fide purchaser for value of a legal estate without notice of the equitable interest will not be bound by it, but other purchasers of estates or interests in the land will be. Thus, a purchaser of a legal estate who knows of the existence of an equitable interest will be bound by it. Further, a purchaser of a legal estate who would have discovered the existence of an equitable interest if he had properly examined both the title deeds and the vendor's land is deemed to have notice of it and is therefore bound by it. Moreover, a purchaser is deemed to have notice of, and is therefore bound by, interests which his agent either has or is deemed to have notice of.

9   If the purchaser is satisfied that the vendor has the necessary title, the transfer of legal title takes place by deed, the deed being

called a conveyance. (The requirement to use a deed is contained in s.52 of the Law of Property Act 1925, but s.1 of the 1989 Act creates some new rules about the nature of a deed. Most significantly, not only must the deed be signed, but also the signature must be witnessed, and the witness must attest the signature, i.e. sign as witness to it. Two witnesses are required in any case where the executor of the deed needs someone to sign on his behalf, e.g. because he is too ill to sign himself. Such formalities are imposed by law in circumstances where it may be necessary to prove the validity of documents.)

Two copies of the draft conveyance are prepared by the purchaser and sent to the vendor. The vendor returns one copy to the purchaser, either approved or with amendments marked on it. When the final form of the conveyance is agreed, the purchaser makes a final copy, signs it and returns it to the vendor. On the date specified by the contract for completion, the purchase money is paid, the remaining legal work is completed (including handing over of the conveyance, which has been signed by the vendor, to the purchaser) and legal title to the property passes from the vendor to the purchaser. (Note: due to compulsory registration, title must be registered within two months of completion, otherwise the legal estate will not pass.)

Registered conveyancing

[For the purposes of this section the terms purchaser and vendor will normally imply the solicitors acting on behalf of the purchaser and the vendor.]

The stages in the process of conveying registered land are as follows.

1 Local land charges and enquiries of local authorities – see unregistered conveyancing.
2 Enquiries of the vendor/surveys – see unregistered conveyancing.
3 Formation and exchange of contracts – see unregistered conveyancing.
4 After exchange of contracts the vendor is required to show that he has the necessary title to the land which he has undertaken by the contract to transfer. In relation to registered land this may be determined by a search of the register kept by the appropriate District Land Registry.

The register comprises:

(a) The property register, which contains a description of the property.
(b) The proprietorship register, which contains the names of the registered proprietors, states the class of title, and indicates whether the registered proprietor's ability to transfer to land is limited, for example because he holds the property in trust for a beneficiary.

The registered estate will either be freehold or leasehold for a term exceeding 21 years. Different classes of title may be registered, the best and most common of which is absolute title. Absolute freehold title is equivalent to a fee simple absolute in possession. Absolute leasehold title guarantees that the lease is valid and is owned by the registered proprietor.

Where the Registrar is not satisfied that the registered proprietor owns the freehold or possesses a validly granted lease, he may register an inferior class of title which does not guarantee that the registered proprietor owns, respectively, the freehold or a validly granted lease. The inferior classes of registered title are good title (leases only), possessory title (freehold or leasehold) and qualified title (freehold or leasehold).

(c) The charges register: this contains details of certain interests in the property which are owned by third parties. Important examples are 'registrable interests', such as mortgages (legal charges), and 'minor interests', such as contracts to convey or create legal estates, restrictive covenants, equitable easements and profits *a prendre* and interests of beneficiaries under trusts.

Minor interests may be entered on the register by the Registrar as a notice. They may also appear on the register as restrictions, cautions or inhibitions. As a general rule, a purchaser is bound by minor interests which are on the register, and takes the land free of any minor interests which are not registered.

Certain interests are 'overriding interests'. These bind a person who purchases or otherwise acquires registered title to property even though they are not entered on the register. Important examples are leases for a term not exceeding 21 years, local land charges, and legal easements and profits *a prendre*. Further, an interest in land, such as an equitable easement, which would normally only bind a person who purchased or otherwise acquired registered title to the land if entered as a minor interest, will be an overriding interest if, at the time when the land was purchased or acquired, the owner

of the interest was either in occupation of the land or was in receipt of the rents or profits of the land.
5 If the purchaser is satisfied that the vendor has the necessary title, completion takes place by deed, the deed being called a transfer. Two copies of the draft transfer are prepared by the purchaser and sent to the vendor (a standard form is often used). The vendor returns one copy to the purchaser, either approved or with amendments marked on it. When the final form of the transfer is agreed, the purchaser makes a final copy, signs it and returns it to the vendor. On the date specified by the contract for completion, the purchase money is paid and the remaining legal work is completed (including transfer of the deed, which has been signed by the vendor, to the purchaser). Legal title to the property does not, however, pass from the vendor to the purchaser until the transfer has been registered.

Trusts

Basically, a trust exists when one person, the trustee, holds property for the benefit of another person, the beneficiary. Thus, for example, X may hold the freehold of a property on trust for Y. X holds the legal estate of the property but Y has an equitable interest which entitles him to the benefits of the property.

Trusts may be created in a number of ways and for a number of reasons. A landowner may expressly create a trust of his legal estate, either during his lifetime or in his will. For example, assume that X, a widow who owns a house, has a child, Y, who is 8 years old. The terms of X's will may provide that if X dies the house is to be held on trust for Y by two trustees, A and B, until Y reaches the age of 21.

Sometimes, when a landowner does not expressly create a trust of his legal estate, the courts may presume or imply that a trust has been created. For example, suppose that X, the purchaser of a house, agrees that Y, who lives with him, will acquire an interest in the house if Y contributes to the cost of purchasing the property by assisting with the mortgage repayments. Whilst a trust is not expressly created, the courts may be prepared to rule that X holds the legal estate on trust for himself and Y, their relative equitable interests being commensurate with their relative contributions.

Joint ownership

It is quite common for one person to share the ownership of land with another. Obvious examples are a husband and wife who buy a house, or parties who buy premises from which to run their business. The law recognizes two types of co-ownership:

- Joint tenancy.
- Tenancy in common.

With joint tenancy, the co-owners are regarded as an indivisible whole, owning the entire property as one, whereas in tenancy in common, the persons involved each own a share. This can be important if one of the co-owners dies. In a joint tenancy, the effect of death is to vest the deceased person's rights in the land in the surviving owners. If a house is owned in joint tenancy by a husband and wife (i.e. as joint tenants), and the husband dies, the wife is then absolutely entitled to the house. If a property is owned equally by four partners (i.e. as tenants in common), each is at liberty to dispose of his share, and if one dies that share may be inherited by someone other than the surviving partners.

One problem with a tenancy in common is that it could result in large numbers of people owning a share in the same piece of land. This would greatly complicate transfers or other transactions, so it is now provided that tenancies in common can only exist within a trust. The legal estate in the land must be vested in trustees, of whom there can be no more than four. Each of the persons entitled to a share will then hold an equitable interest.

If co-owners have not made it clear exactly how they intend to hold the land, rules exist to determine whether there is joint tenancy or a tenancy in common.

1. For a joint tenancy, the co-owners must be granted the same interest in the land. So if the land is given to A and B, they could be joint tenants, but if the land is given to A and B in shares of two-thirds and one-third respectively, they cannot be joint tenants.
2. For a joint tenancy, all co-owners must acquire their title under the same document.
3. For a joint tenancy, all co-owners must acquire their interest at the same time.
4. For either type of co-ownership each co-owner must be as much entitled as the others to possession of any part of the land.
5. If no answer is provided by these various tests, then it becomes necessary to look at the deed creating the co-ownership, to see if it makes clear which type of co-ownership is intended. A grant to A and B jointly is construed as a joint tenancy; a grant to A and B in equal shares is construed as a tenancy in common. Words such as 'in equal shares', which indicate that owners are to take distinct shares in the property, are said to be words of severance.

6 Where none of the tests is conclusive there are some factual circumstances in which equity will favour a tenancy in common, because it is fairer. These includes cases where land is purchased by co-owners who have contributed in unequal shares, or as partnership property.

Even where these tests indicate that land is held as a joint tenancy it is possible for the parties to alter the arrangement in equity, by severing the joint tenancy. The simplest way to do this is for one tenant to serve notice in writing on the other joint tenants indicating that the server of the notice wishes to sever his share (s.36 Law of Property Act 1925). One example where this could be necessary is where the husband and wife have purchased property as joint tenants. If the marriage breaks down and one party leaves the other, the right of survivorship rule would continue to apply unless the joint tenancy is severed. If a spouse wants to preserve a 'share' in the matrimonial home, severance is a sensible step.

Rights and duties of owners of land

It is difficult to summarize the rights of a landowner. They certainly include the right to occupy the land without interference from others. (Interferences which the law will control include trespass and nuisance.) One important right of every landowner is a right of support for his land from adjacent pieces of land. Any withdrawal of such support, which causes the adjacent land to slip e.g. excavation during building work, may result in a claim for damages. The right of support is a natural right, existing because of the nature of the land itself. A right of support may, however, be acquired, in which case the right is called an easement.

An owner's duties in relation to his land include not only using it in a way which will not constitute a nuisance, but also complying with all the various statutory controls which now exist concerning land use. Significant amongst these are the planning laws contained principally in the Town and Country Planning Act 1990.

The controls on a landowner by statute range from restrictions on killing certain kinds of birds and animals on his land to positive obligations to repair dilapidated property. If an authority wishes to take an owner's land from him, it may be possible to do so under the rules of compulsory purchase. And rights of access may be reserved by statute e.g. for the creation of a long distance footpath. When the range of controls on a landowner is fully examined it is clear that the number of restrictions is far greater then the number of rights.

Access to Neighbouring Land Act 1992

This Act enables a person to apply to the court for an order giving him access to neighbouring land (the servient land) in order to carry out work reasonably necessary for the preservation of his land (the dominant land). Examples of such work include repairing buildings, clearing drains and cutting hedges.

An order may be necessary where the owner or a tenant of the neighbouring land refuses consent to entry, the order being obtained against the person who refuses consent. The court will not make an order if this would interfere with or disturb the use or enjoyment of the servient land, or cause hardship to an occupier of it.

An order specifies a variety of matters, including the nature of the work which can be done, the areas of the neighbouring land which can be visited, the date of entry, terms to avoid or restrict damage, loss or injury and loss of privacy, the manner of the work, the days and hours during which the work may be carried out, and the persons who may enter the land to carry out the work. An order may impose compensation requirements and/or provision for payment to the person against whom the order is obtained, in consideration of entering the land. Breach of the terms of an order may give rise to an action for damages.

An order binds the person against whom it was made, and may be protected against successors and purchasers by registration. This may be necessary where an order permits entry during the term of a lease or in fee simple. Either party to an order may apply to the court for a variation.

Easements

An easement is an interest in land which gives the owner of one piece of land the right to use or restrict the use of land belonging to another. Common examples of easements are:

- Easement of way, which gives a landowner the right to cross land belonging to another.
- Easement of storage, which gives a landowner the right to store goods on land belonging to another.
- Easement of light, which gives a landowner the right to the flow of light to a building on his land over land owned by another.
- Easement of support, which gives one landowner the right to have his land supported by land belonging to another.

However, the courts are not prepared to recognize every right as capable in law of being an easement. For example, a right to a view is incapable of being an easement. Equally, a right of total occupation or control of land cannot be an easement.

An easement must burden one piece of land (servient land) for the benefit of another piece of land (dominant land). For example, in the case of an easement of way, the owner of the dominant land has the right to cross the servient land for the benefit of the dominant land. An easement cannot exist unless it is created for the benefit of dominant land. Moreover, the easement must be of actual benefit to the dominant land, such as increasing its value. A personal benefit to the owner of the dominant land which does not benefit the land itself is not sufficient. For example, a right to cross land in Durham for the benefit of land in Somerset could not be an easement as the right could not be of benefit to the dominant land.

An easement cannot exist if the same person owns and occupies both the dominant land and the servient land. However, where both pieces of land are owned by the same person but one is occupied by a tenant, an easement can exist. Further, an easement cannot be created, whether expressly, impliedly or by prescription, unless at the time of its creation there is a person capable of creating it and a person capable of receiving it.

Easements are normally created expressly. A legal easement is created by deed, its duration being that of a fee simple absolute in possession or a term of years absolute. A landowner may grant an easement over his land to a neighbouring landowner (express grant). Alternatively, when a landowner sells part of his land he may reserve an easement over the land which he sells (express reservation) or may grant the purchaser an easement over the land which he retains (express grant).

Further, a conveyance of dominant land may automatically convert a licence enjoyed with the land into an easement, as s.62 of the Law of Property Act 1925 deems this to be an express grant. For example, if X, a tenant, has permission (a licence – see page 299) to store coal in a shed on land retained by her landlord, the grant of a new lease to X may, unless the new lease specifies otherwise, automatically convert the permission into an easement of storage (*Wright v. Macadam*).

Where easements are not created expressly, they may be created by implication (implied grant or implied reservation). Thus, where a landowner sells part of his land and does not grant the purchaser an easement which is necessary for its enjoyment, or fails to grant to the purchaser an easement which both he and the purchaser intended would exist, the court may recognize as implied into the sale, respectively, the grant of an easement of necessity or the grant of an intended easement. For example, the court might be prepared to recognize the implied grant of an easement of way over a vendor's land where, following a sale, the purchaser discovers that he has no means of access to the land which he has purchased.

The courts are more reluctant to recognize implied reservation than they are to recognize implied grant. Implied grant may also occur where a landowner, prior to selling part of his land, enjoyed over it a right which could have been an easement, except that the dominant and servient lands were both owned and occupied by him (a quasi easement). If he sells the dominant land and retains the servient land, the effect of the rule in *Wheeldon v. Burrows* is that the court may recognize the implied grant of the easement to the purchaser provided that the easement was in use when the land was sold and either was necessary to its reasonable enjoyment or was used regularly and could be detected by visual examination of the land.

As well as being created by express or implied grant or reservation, easements may also be acquired by prescription. Basically, this means that if a landowner, for the benefit of his land, uses or derives a benefit from land belonging to another for twenty years or more, he may acquire an easement. Thus, if X crosses a path over Y's land for twenty years or more, or if the windows of X's house derive a flow of light across Y's land for twenty years or more, X may acquire, respectively, an easement of way or an easement of light. An easement can be acquired by prescription by a freehold owner or his tenant against a freehold owner or tenant of other land, but cannot be acquired by a tenant against his own landlord.

In order to acquire an easement in this way, however, it is necessary that X used Y's land as though he had a right to do so. Thus, if X had Y's permission to cross his land, if X crossed Y's land in secret, so that Y would not discover him, or if X used force in order to cross Y's land, such as damaging fences, he would not acquire an easement in this way. It is also necessary that the use of the servient land be regular. (It is an over-simplification to suggest that twenty years' use will necessarily be sufficient to support a claim of an easement by prescription. In practice it is necessary to satisfy the requirements of at least one of three methods of prescription, the technical details of which fall outside the scope of this book.)

Basically, an easement may come to an end in any of three ways. First, if the owner of the dominant land agrees to 'release' it, normally by deed (express release). Secondly, if the owner of the dominant land abandons the easement (implied release). Thirdly, if the dominant and servient lands become owned and occupied by the same person.

Profits *a prendre*

A profit *a prendre* is an interest in land in the form of a right to remove property from someone else's land. Common examples are rights to fish and to graze cattle. Basically, the legal rules which relate to profits *a prendre* are identical to those which relate to easements (see above) with two major differences. First, in relation to a profit *a prendre* there is no requirement of dominant land. Thus, whilst the benefit of a profit *a prendre* may be attached to dominant land this is not required. Secondly, the only form of implied grant or reservation of a profit *a prendre* is the implied grant or reservation of an intended profit *a prendre* (equivalent to the acquisition of an intended easement).

Licences

Basically, a licence is a permission to use land. It confers no legal estate or interest in land but merely prevents the licensee (the person to whom permission is given) from being a trespasser. An example is provided by permission to cross land belonging to another which either is not intended to create an easement or which cannot create an easement (for example, because there is no dominant land).

Since a licence is not an interest in land, it may be revoked by the licensor (the person who gave permission). What is more, it will not bind subsequent owners of the licensor's land. In contrast, the grant of a legal interest (such as a legal easement of way) confers a legal right to use land belonging to the grantor. Being a legal interest in land, it cannot be revoked by the grantor and will bind subsequent owners of the servient land. Even an equitable easement, which is an equitable interest in land, will bind all subsequent owners of the servient land if it has been appropriately registered.

A *contractual licence* is a licence in the form of a contract. A mere permission which is not contractual is called a *bare licence*. The courts may be prepared to enforce a contractual licence against the licensor, preventing him from revoking it until the contract comes to an end. They will not normally be prepared to enforce a contractual licence against a subsequent owner of the licensor's land. However, in exceptional circumstances, such as where the subsequent owner has behaved unfairly in relation to the licensee, they may be prepared to accept that a contractual licence has given rise to an equitable interest in land.

Sometimes a licence may be coupled to the grant of a legal interest in land, such as a licence to cross land in order to reach a stream and exercise a right to fish (a profit *a prendre*). Where this is a case, as long as the interest (the profit *a prendre*) continues to bind owners of the servient land, so will the licence.

Restrictive covenants

A restrictive covenant is an undertaking which a landowner (the covenantor) makes, usually by deed, not to use his land in certain ways. For example, X might covenant with Y (the covenantee) only to use his house as a single private dwelling. Thus, X would be contractually bound not to use his house for business purposes and not to divide it into flats. However, as a consequence of the principle of privity of contract (see page 88), if the covenantor X sells his land to Z, Y cannot enforce it against the new owner Z.

In certain circumstances, however, a restrictive covenant may give rise to an equitable interest in land, making its burden pass to a purchaser of the covenantor's land, and thus enforceable against such a purchaser. This can only happen if a number of requirements are satisfied.

First, the covenant must be negative. This means that it must not require the covenantor to spend money. Thus, a covenant only to use premises as a single private dwelling would be negative.

Secondly, as in the case of easements (see above), the covenant must have been made for the benefit of the covenantee's land and not merely for the covenantee's personal benefit. (Certain bodies, such as the National Trust, are not required to own benefited land in order to enforce covenants against purchasers of the covenantor's land.) Thus, the land must receive a benefit from the covenant. This could be an increase in or preservation of its value in consequence of the restriction. Or it might be that the covenant prevents undesirable development or activities on the covenantor's neighbouring land.

Thirdly, the covenantor and covenantee must have intended that the burden of the covenant would pass to subsequent owners of the covenantor's land. The effect of s.79 of the Law of Property Act 1925 is that the parties are deemed to intend the burden of the covenant to pass in this way unless the agreement expressly indicates otherwise.

Finally, to be enforceable against all purchasers and subsequent owners of the covenantor's land, it is necessary for the covenantor to register the covenant as a land charge (unregistered land) or have it entered on the register as a minor interest (registered land).

Sometimes, when it becomes necessary to enforce a covenant,

both the covenantor and the covenantee will have sold their land. Where this is the case, as well as establishing that the covenant has given rise to an equitable interest in land which makes its burden enforceable against purchasers of the covenantor's land, it will also be necessary to satisfy equitable requirements relating to the passing of the benefit of the covenant to the purchasers of the covenantee's land. The wording of the covenant may itself 'annex' the benefit of the covenant to the covenantee's land (express annexation), by indicating that the covenantor and covenantee intended the benefit of the covenant to pass to subsequent owners of the covenantee's land. In the absence of express wording to this effect, s.78 of the Law of Property Act 1925 appears to produce the same result provided that it is clear which land was intended to receive the benefit of the covenant. Alternatively, if the covenant was not annexed to the covenantee's land, its benefit may be transferred each time the covenantee's land is sold by the owner of the land to the purchaser (express assignment).

In relation to the enforcement of covenants, special rules apply to 'building schemes'. Basically, a building scheme exists when a developer of land intends to sell the land in plots, each plot being subject to the same covenants for the benefit of all the plots and all purchasers of plots appreciating that this is so. Where a building scheme exists, the benefit of the covenants automatically passes to purchasers of plots. Further, the burden of the covenants will pass to purchasers of plots even though the covenants were not made for the benefit of land possessed by the covenantee (the developer, who presumably will not retain any land when the final plot is sold).

In some cases, when it becomes necessary to enforce a covenant, the original covenantee may have sold the benefited land, but the covenantor has retained the burdened land. In this case, in order to enforce the covenant, it is only necessary to show that the benefit of the covenant has passed to the purchaser of the covenantee's land. This may be achieved in either of two ways. First, by showing that the benefit of the covenant has passed in equity, by annexation or assignment or under a building scheme. Secondly, by showing that the benefit of the covenant has passed at common law. In order to do this, the person who has purchased from the covenantee must satisfy a number of common law requirements.

A covenant may come to an end in a number of ways. First, an application may be made to the Lands Tribunal, who may discharge or modify restrictive covenants. They have power to do so if the covenants have become obsolete, or if they prevent reasonable use of the burdened land and are either against the public interest or provide no practical benefit of substantial value to anyone. The Lands Tribunal can also intervene if the owner of the benefited land has expressly or impliedly agreed that the covenant should be modified or discharged, or in circumstances where discharge or modification of the covenant would not injure the owner of the benefited land.

Secondly, the owners of the land benefited and burdened by a covenant may agree to terminate it (discharge it), usually by deed.

Thirdly, if the two pieces of land become owned by the same person, the covenant will come to an end (extinguish) unless the covenant exists under a building scheme.

Mortgages of land

The enduring quality of land makes it a valuable asset and a superior form of security when needing to raise money. The owner of the land who borrows against it will naturally want to continue to use and enjoy it, but the person lending the money must be given certain rights over the land in the event of the borrower defaulting on payment. These requirements are provided for in the rules governing legal mortgages.

The method of creating a mortgage depends on the type of interest which the landowner holds. If he has a freehold interest there are two possible ways:

1 By a charge by deed, expressed to be by way of legal mortgage.
2 By creating a lease (usually of 3000 years) in favour of the lender (the mortgagee), which provides for the lease to end once the mortgage is redeemed (i.e. the money is repaid), and which permits the borrower (mortgagor) to retain possession, allowing the mortgagee to take possession only on default.

Where the landowner has a leasehold interest (e.g. a lease for ninety-nine years), he can mortgage either by a charge by deed (as in 1 above) or by creating a sub-lease in favour of the mortgagee which is for a period shorter than the unexpired term of his own lease. He usually makes a sub-lease ten days shorter, which leaves it open to create second and subsequent mortgages. This possibility of other mortgages presupposes that the first mortgage is for a sum less than the full value of the property. No one is likely to grant a second or subsequent mortgage if the security provided by the property is exhausted, usually signified in layman's terms by the expression 'mortgaged up to the hilt'.

The rules ensure that a mortgagee has the necessary interest in the land for it to constitute effective security for the loan. They are reinforced by the rights granted to a lender if the borrower defaults. In extreme cases, a mortgagee may need to sell the property in order to realize his security. To enable him to do so, he may rely on:

1. *The right to take possession* – naturally, if a mortgagee has to sell, obtaining vacant possession of the property is a crucial first stage. He is effectively entitled to possession by virtue of the interest in the land granted to him by the mortgage itself, but he can only exercise this right through the court, i.e. a mortgagee must apply to the court for an order for possession.
2. *The right to foreclose the mortgage* – again this right is exercisable through the court. If an order is granted, its effect is somewhat drastic. It removes the right of a mortgagor to redeem his mortgage and vests the legal interest in the land in the mortgagee. In other words, the mortgagor loses all rights to the land. This is so extreme a result that it is unlikely that a court would make such an order. A mortgagee would normally be satisfied to exercise his right to sell.
3. *The right to sell* – the duty of a mortgagee when selling is to obtain the best possible price. He then pays himself what he is due and any surplus is used to discharge subsequent mortgages. If there is any money left, it is paid over to the mortgagor.
4. *The right to appoint a receiver* – a receiver may be appointed to manage or sell the mortgaged property, paying off the loan and paying any surplus profits over to the mortgagor. This is common in relation to mortgages of commercial property.

Naturally a purchaser of land must be careful to ensure that what he is buying is not the subject of an unredeemed mortgage. A legal mortgage, protected by depositing the title deeds to the land with the mortgagee, is not a registrable interest under the land charges system. However, the purchaser always needs to see the title deeds, and if they are not available, this will obviously suggest that they are being held by someone as security to protect his loan.

Adverse possession

Basically, if a squatter (or a succession of squatters) occupies land for twelve years or more, the owner of the land loses his estate in the land and cannot bring an action to recover the land (Limitation Act 1980). Thus, the squatter may become the owner of the land. However, where land is let, whilst the squatter may acquire the tenant's lease, time does not begin to run against the landlord (i.e. the 12 year period does not commence in relation to the landlord's estate in the land) until the lease comes to an end.

A landowner can only lose his estate in land in this way if the squatter's possession of the land is adverse, in the sense that it is inconsistent with the landowner's enjoyment of his property. This will be so if the squatter exercises a sufficient degree of physical control over the land. What is sufficient may vary with the individual facts of each case. For example, fencing a piece of wasteland and cultivating it as a garden, or renovating a derelict building and occupying it as a house, might well be sufficient to amount to adverse possession.

12
Landlord and tenant

The relationship of landord and tenant arises from a lease, which is a grant of exclusive possession of land for a fixed period. Exclusive possession is the most important feature of a lease, as 'The tenant armed with exclusive possession can keep out strangers and keep out the landlord' (*Street v. Mountford*). Leases can be legal estates under s.1 of the Law of Property Act 1925, in which they are described as 'terms of years absolute'.

There are two types of lease, fixed term leases and periodic tenancies. A fixed term lease, such as a 7-year lease or a 99-year lease, expires when the term comes to an end (i.e. after respectively 7 or 99 years). A periodic tenancy, such as a weekly, monthly or yearly tenancy, may be terminated at the end of a period. If not terminated, the tenancy continues for another period. Thus, a weekly tenancy might end after one week or might continue for several years.

Normally, a legal lease must be created by deed (s.52 LPA 1925), but if made in writing, the lease may take effect as an equitable interest. However, a legal lease for three years or less can be made in writing, and a deed is not required, provided that the lease commences immediately and the rent is the best that can reasonably be obtained without taking a premium (s.54).

Once a lease expires, and is not renewed, the leaseholder (sometimes called the lessee or, more commonly, the tenant) is left with nothing. This may raise the question why anyone would want to take a lease rather than buy a freehold estate. Commercial and personal considerations will inevitably play a part in such a decision. A businessman may need offices or premises in a specific area where the only property available is leasehold. The need for the property may be very short term, or a person may lack the funds or the means of raising the funds to buy a freehold property.

From the point of view of the freeholder, there may be significant advantages in granting leases rather than selling land outright. Take the example of a person developing land as a shopping arcade. If he sells all the units within the arcade as freeholds, it will be quite difficult to impose positive obligations on freeholders so that the properties can be maintained in a uniform way and common areas kept in order. If he grants leases, he can impose conditions more easily, and maintain a greater measure of control.

Leases and licences

As is seen below at page 319, a number of Acts of Parliament (such as the Rent Act 1977, the Housing Act 1988 and the Landlord and Tenant Act 1954, Part II) give security of tenure to tenants. That is, they protect tenants by imposing restrictions on the ability of landlords to recover possession of property which they have let. For present purposes, the significant feature of these provisions is that they apply to leases but not to licences.

A licence is a permission to do something in relation to land which would otherwise be a trespass. A licence does not create any interest in the land itself. So, for example, if you visit the cinema, you have a contractual licence to be on the premises. A contractor taking over a site is usually there by virtue of a licence, and not because he has any interest in the land itself. (See page 299 for a further discussion on licences.)

A landlord may wish to retain greater flexibility in the management of his property by creating licences rather than leases, and thus avoiding the application of legislation which might give his tenants security of tenure. Traditionally, an agreement for occupancy of premises which was called a licence and/or which purported not to give the occupiers exclusive possession of the premises would be classified by the courts as a licence. For example, in *Somma v. Hazelhurst* a single room was let to a man and woman who were living together. It was let under two separate agreements which gave the landlady the right to share the room with them and to nominate replacement occupiers to share with them. The Court of Appeal held that the agreement could not be a lease, as it did not give the occupiers exclusive possession. Rather, it was a licence, to which the protection given to tenants by the Rent Act 1977 Act did not apply.

In *Street v. Mountford*, however, the House of Lords held that the decision in *Somma v. Hazelhurst* was incorrect. *Street v. Mountford* concerned an agreement, described as a licence, under which two rooms of a house were let, the occupier agreeing that the agreement did not give her a tenancy protected by the Rent Act 1977. The House of Lords held that the agreement gave the occupier exclusive possession of the property, and consequently, it was a lease to which the Rent Act 1977 Act applied.

In relation to *Somma v. Hazelhurst*, their Lordships held that the true effect of the agreements in that case was to grant exclusive possession of the premises to the occupiers. The fact that two separate agreements were given rather than one and that the landlady reserved the right to share the room or nominate replacement occupiers was a mere sham, an attempt to disguise the true nature of the agreement. Thus, in reality, the agreement was a lease, not a licence, the two agreements being read together as one.

Basically, the effect of the decision of the House of Lords in *Street v. Mountford* is as follows. If the courts believe that the true effect of an agreement is to give the occupier exclusive possession of premises then, in the absence of special circumstances, the agreement will be treated as a lease. This will be so even though the agreement is described as a licence and even though the agreement purports not to give the occupier exclusive possession of the premises.

The House of Lords in *Street v. Mountford* accepted that there may be special circumstances in which an agreement which gives or appears to give exclusive possession is only a licence. Some examples which occur commonly include:

- Lodgers and hotel guests normally only have a licence. They do not have exclusive possession of the premises as they receive services which require the licensor or his agents to have unrestricted access to the room.
- An employee who is required to occupy premises as a necessary part of the proper performance of his contractual duties only has a licence, even though he has exclusive possession of the premises.
- An agreement made between relatives or friends or with charitable intent which gives exclusive possession may only amount to a licence if creation of a lease was not intended.

A significant number of cases involving the distinction between a lease and a licence have come before the courts in recent years. Each tends to turn on its own individual set of facts, and may have limited value as a precedent. In every case, the court is seeking to determine the intention of the parties, particularly whether the tenant truly has exclusive possession, however the landlord may have tried to disguise it.

Assignment and sub-letting

Under a lease, both the landlord and the tenant possess interests in land. The tenant possesses his lease. The landlord possesses his reversion, the right to resume occupation of the premises when the tenancy ends.

A landlord may own the freehold of the premises let or may himself be a tenant. For example, L, who owns the freehold of a property, may grant a 99-year lease to T who grants a 7-year lease to S. L is the head landlord, T is L's tenant S's landlord, and S is T's tenant and L's sub-tenant.

A landlord who owns the freehold of the let premises may sell his interest. The purchaser becomes the owner of the reversion and thus becomes the tenant's new landlord. A tenant (including a landlord who does not own the freehold) may sell his lease. This is technically called an assignment. The purchaser becomes the new tenant (and also becomes a landlord if the vendor had a sub-tenant).

It may be, however, that the lease prohibits assignment by the tenant, either absolutely or without the landlord's consent. Where the lease prohibits assignment without the landlord's consent, s.19 of the Landlord and Tenant Act 1927 provides that consent must not be unreasonably refused. Further, undue delay in giving consent or notifying the tenant of a refusal of consent may render the landlord liable in damages to the tenant (Landlord and Tenant Act 1988). If the tenant assigns without consent, he has a defence to an action brought against him by the landlord, if the landlord's refusal was unreasonable.

As an alternative to assigning the lease, the tenant may choose to sub-let and receive rent from his own tenant. The lease may prohibit sub-letting, either absolutely or without the landlord's consent. Again, where sub-letting is prohibited without the landlord's consent, consent must not be unreasonably refused (s.19 Landlord and Tenant Act 1927). Likewise, the landlord may be liable in damages to the tenant where there is undue delay in giving consent, or notice of refusal.

The original parties to a lease can enforce its covenants against each other, because the lease is a contract, and both landlord and tenant are privy to the contract. Where either or both the landlord and tenant are not the original parties to the lease, because the original landlord sold his reversion and/or the original tenant assigned his lease, there is no privity of contract between them. There is, however, privity of estate, due to the existence between them of the relationship of landlord and tenant.

Where there is privity of estate but no privity of contract between a landlord and tenant, only covenants which 'touch and concern the land' are enforceable between the parties. A covenant does not touch and concern the land if it benefits only the parties to the lease and not the land itself. Examples of covenants which touch and concern the land are covenants to repair, covenants to pay rent and covenants for quiet enjoyment. Examples of covenants which do not touch and concern the land are options to purchase and obligations to repair other premises.

Regardless of whether a covenant touches and concerns the land, the original tenant may himself remain liable for its breach under the contract. Thus, when a tenant assigns, it may be advisable for him to require on the part of the assignee a covenant indemnifying the assignor in respect of any breaches of covenant by the assignee. Such a covenant is implied by s.77 of the Law of Property Act 1925 into assignments of leases.

Where a tenant sub-lets, there is neither privity of contract nor privity of estate between the head landlord and the sub-tenant. Thus, covenants of the head lease are only enforceable between the head landlord and the sub-tenant if the requirements for enforcement of restrictive covenants (see page 300) are satisfied. In order to enforce a restrictive covenant against a sub-tenant, the landlord is not required to possess neighbouring land which is benefited by the covenant, however, as his reversion counts as land which can receive the benefit of a restrictive covenant.

Regardless of whether a covenant of the head lease is enforceable between head landlord and sub-tenant, the original tenant may remain liable under the contract for its breach. Thus, when a tenant sub-lets it may be advisable to require the sub-tenant to covenant to comply with the covenants of the head lease and to indemnify the tenant in respect of any breaches of those covenants.

Covenants in leases

Covenants in a lease lay down the obligations of the landlord and tenant. There are two types of covenants, express covenants and implied covenants. Express covenants are expressly agreed upon by the parties to the lease. Implied covenants do not result from express agreement by the parties to the lease but are implied into it by common law or statute, unless the parties agree otherwise. However, certain covenants implied by statute into a lease cannot be excluded by agreement of the parties.

Implied covenants (other than repairing covenants)

Tenant's implied covenants. Other than repairing covenants (which are examined below), two major examples of tenant's implied covenants (i.e. implied covenants imposing an obligation upon the tenant) are the implied covenant to pay rent and the implied covenant to pay taxes relating to the property. In relation to rent, in the absence of an express covenant or contrary agreement, a covenant may be implied if the wording of the lease implies that the payment of rent is intended.

In relation to taxes, in the absence of an express covenant or contrary agreement, the tenant is under an implied obligation to pay rates and taxes relating to the property for which the landlord is not personally liable.

Landlord's implied covenants. Other than repairing covenants, two major examples of landlord's implied covenants are the implied covenant not to derogate from his grant and the implied covenant for quiet enjoyment. In relation to non-derogation, in the absence of an express covenant or contrary agreement, the landlord is under an implied obligation not to do things which make the leased premises less suitable for the purpose for which they were let, assuming that the landlord knows the purpose, e.g. blocking the flow of air to timber drying sheds.

Quiet enjoyment, despite its name, is not really about noise or lack of it. It covers any actions of the landlord which would stop the tenant from being able to use the premises. In the absence of an express covenant or contrary agreement, the landlord is under an implied obligation not to interfere with the tenant's quiet enjoyment of the premises (for example, by unlawful entry or eviction). With regard to both of these implied covenants, the landlord may also be liable in respect of acts done in breach of the covenants by persons claiming under him, such as tenants of other neighbouring property which he owns.

Express covenants (other than repairing covenants)

Tenant's express covenants. Examples of express covenants which may be made by a tenant are a covenant to insure, a covenant to pay taxes relating to the property, a covenant only to use the leased premises for certain purposes, a covenant not to assign or sublet, and a covenant to pay rent. In relation to insurance, a tenant may agree to insure the premises, the extent and nature of the policy being determined by the covenant.

In relation to taxes, an express covenant may impose upon the tenant an obligation more or less extensive than, or identical to, that which would otherwise be implied.

In relation to use of the premises, the tenant may agree to restrictions on his use of the premises. For example, he may agree only to use the premises as a single private dwelling, which would prevent both use for business purposes and sub-letting as flats. The significance of covenants not to assign or sub-let is considered on page 308.

In relation to rent, the tenant will normally expressly covenant to pay a specified amount of rent on a specified day. Rent is normally payable in arrear unless the parties have agreed to payment in advance. A tenant may be entitled to deduct sums from the rent before paying it to the landlord. The lease may specifically permit or prevent the making of such deductions. Where the lease is silent, the basic principle is that the landlord is entitled to the entire rent. The lease may contain a term which suspends the tenant's obligation to pay rent in circumstances such as damage or destruction of the premises by fire.

The lease may contain a rent review clause, which permits regular rent reviews throughout the course of the tenancy, say once every three years. The lease will normally contain a formula to be used in determining the revised rent, perhaps an open market rent to be fixed by agreement or arbitration or a form of index-linking, possibly to the retail prices index.

Landlord's express covenants. Examples of express covenants which may be made by a landlord are a covenant to insure, a covenant to pay taxes relating to the property and a covenant for quiet enjoyment. In relation to insurance, if the landlord covenants to take out an insurance policy, it may be that the tenant agrees to pay the insurance premiums.

In relation to such taxes, the landlord may covenant to undertake liability which would otherwise fall within the tenant's implied obligations.

In relation to quiet enjoyment, the requirements of an express covenant may be wider in some respects but more limited in others than those of an implied covenant. For example, an express covenant for quiet enjoyment continues to bind the party who made it throughout the duration of the lease whereas an implied covenant for quiet enjoyment only binds a landlord whilst he remains landlord.

Provisos and options

A lease may contain a number of provisos and options. Examples of options are options to purchase and options to renew. An option to purchase may entitle the tenant to purchase the landlord's reversion, which may be the freehold or may be a lease depending upon whether his landlord owns the freehold or is himself a tenant. The tenant may be entitled to purchase at a fixed price or, alternatively, the lease may lay down a mechanism and formula to be applied in determining the purchase price.

An option to renew may entitle the tenant to renew the lease by serving notice on the landlord on a specified day. The tenant will be entitled to require a new lease for a term specified by the option. The lease may specify a fixed rent or, alternatively, may lay down a mechanism and formula to be applied in determining the new rent. Options should be registered in order to bind purchasers of the landlord's estate (see page 289).

Examples of provisos are provisos for forfeiture and break clauses. A landlord cannot forfeit a lease unless the lease contains a proviso which gives the landlord a right of re-entry on breach by the tenant of conditions or covenants of the lease (see page 319).

A break clause is a proviso which empowers one or both parties to a lease to terminate the lease at certain specified dates by serving notice on the other (see page 319).

The obligation to repair

When negotiations for a lease are taking place, it is important to establish the responsibilities with regard to repair of the property. The parties may expressly agree on responsibility for repairs, such agreement forming the 'express repairing covenants' of the lease. Where no such express agreement has been reached, the parties may find that there are repairing obligations implied by common law or by statute – 'implied repairing covenants'.

Express repairing covenants

A lease may expressly impose repairing obligations upon landlord or tenant in the form of express repairing covenants. For example, a lease might require the tenant:

(a) to put premises in repair at or within a reasonable time of the commencement of the tenancy; or
(b) to leave premises in repair (i.e. to hand them over in repair) at the end of the tenancy; or
(c) to keep premises in repair throughout the tenancy (i.e. to put the premises in repair at the commencement of the tenancy, keep the premises in repair throughout the tenancy and hand them over in repair at its end).

Where the landlord covenants to repair, he is not obliged to repair until he has notice of the disrepair, i.e. from the tenant or from his own inspection of the premises.

Once repairs are undertaken, the standard required is 'such repair as, having regard to the age, character and locality of the premises, would make them reasonably fit for the occupation of a reasonably-minded tenant of the class who would be likely to take them'. This was established by Lord Esher in *Proudfoot v. Hart*. The criteria are applied by looking at the age, locality etc. as they would have been at the start of the tenancy. This can be important where, for example, the locality becomes less desirable during the course of a long lease. The standard of repair required is not then reduced because the locality would no longer attract the same sort of tenant.

A covenant to repair requires painting and decorating in order to prevent the property falling into disrepair. A lease may also contain an express covenant to paint, which may specify the materials to be used and the frequency of the work.

Express repairing covenants may specifically exclude liability to repair in respect of disrepair resulting from 'fair wear and tear', whether caused by normal and reasonable use by the tenant for proper purposes or by the normal action of time and weather. The tenant must, however, undertake such repairs as are necessary to prevent the premises falling into further disrepair.

A covenant to repair does not normally impose an obligation to make improvements, or to renew the premises in their entirety. However, such a covenant will normally impose an obligation to rebuild in the case of destruction of premises by fire, unless the lease expressly excludes such liability.

Remedies for breach of repairing covenants

1 *Damages*. Where a party to a lease fails to perform his repairing obligations he may be liable in damages.

   Where the landlord is in breach of his repairing obligations, the tenant cannot recover damages in respect of the period prior to the date when the landlord was notified of the disrepair. If the landlord fails to repair within a reasonable time of receiving such notice, the tenant may sue for damages or carry out the repairs and recover their cost from the landlord.

   Where the tenant is in breach of a covenant to put, keep or leave the premises in repair, s.18(1) of the Landlord and Tenant Act 1927 provides that the damages cannot exceed the amount by which the value of the landlord's reversion was diminished in consequence of the breach. The section also provides that, where the tenant is obliged to leave the premises in repair, no damages are recoverable if, at or shortly after the termination of the tenancy, the premises are to be demolished or structural alterations which would render the repairs valueless are to be made.

In the case of a lease for a fixed term of or exceeding 7 years, of which at least three years remain unexpired, s.1 of the Leasehold Property (Repairs) Act 1938 provides that, before bringing an action for damages or forfeiting the lease in respect of the breach by the tenant of a covenant to keep or leave the premises in repair, the landlord must first serve notice on the tenant under s.146 of the Law of Property Act 1925 (see page 320). This notice must, amongst other matters, inform the tenant of his right to serve a counter-notice within 28 days of the service of the s.146 notice. If the tenant serves a counter-notice, the landlord cannot sue for damages or forfeit the lease without the leave of the court. The court may give leave if:

(a) the value of the landlord's reversion has been substantially diminished by the breach or will be so diminished unless the breach is immediately remedied; or
(b) the breach must be immediately remedied in order to give effect to the provisions of a statute, a bye-law or of delegated legislation; or
(c) in those cases where the tenant is not in occupation of the whole of the premises, the immediate remedying of the breach is required in the interests of the occupier; or
(d) the expense required in order to remedy the breach is small compared to the much greater expense which would probably be required if the work required was postponed; or
(e) there are special circumstances which make the giving of leave just and equitable.

S.1 of the 1938 Act does not apply in respect of the breach of a covenant to put premises in repair upon or within a reasonable time of taking possession (s.3). Further, s.1 does not apply to breach of a covenant to carry out internal decorative repairs, though s.147 of the Law of Property Act 1925 empowers the court to grant the tenant relief from such repairs.

2 *Forfeiture.* A landlord can forfeit a lease if the tenant is in breach of his repairing obligations and the lease contains a proviso which gives the landlord a right of re-entry on breach by the tenant of those obligations (see page 319).

3 *Landlord's right to repair at the tenant's expense.* Where the lease empowers him to do so, the landlord may enter the premises, carry out repairs and recover their cost from the tenant.

4 *Specific performance.* A tenant may obtain an order of specific performance against his landlord compelling the landlord to carry out his repairing obligations. It appears that a landlord cannot obtain such an order against his tenant.
5 *Tenant's set-off of repair costs against rent.* If the landlord fails to repair within a reasonable time of receiving notice of disrepair and the tenant carries out the necessary repairs, the tenant may be entitled to deduct their cost from rent payments.
6 *Receiver.* Upon application by the tenant, the court may appoint a receiver to manage the property if it is just and convenient to do so. This may be the case where the landlord refuses to perform his repairing obligations. (The general power of the High Court to appoint a receiver is contained in s.37 Supreme Court Act 1981. A specific power relating to buildings containing two or more residential flats is given to the County Court by Part II and s.52 of the Landlord and Tenant Act 1987.)

Implied repairing covenants

Where a lease fails to lay down express repairing obligations, repairing obligations may be implied at common law. Further, repairing obligations may be imposed by statute, regardless of the terms of a lease.

Relating to the landlord

1 *Fitness for habitation of houses* (S.8 Landlord and Tenant Act 1985). Where s.8 applies, it implies, regardless of any agreement or term of the lease to the contrary, a condition that the house is fit for human habitation at the commencement of the tenancy and an undertaking that the landlord will keep the house so fit throughout the tenancy.

S.9 provides that a house is only unfit for human habitation for the purposes of s.8 if it is not reasonably suitable for occupation due to defects in its condition with regard to one or more of the following circumstances: state of repair; stability; freedom from damp; internal arrangement; natural lighting; ventilation; water supply; drainage and sanitary conveniences; facilities for preparation and cooking of food and for disposal of waste water.

It appears that s.8 only requires the landlord to repair where it is possible to make the premises fit for habitation at reasonable expense. The section entitles the landlord to enter the premises at reasonable times of the day to view their state and condition on giving 24 hours' notice to the tenant or occupier. The landlord's liability under s.8 does not arise until he has received notice of the want of repair. Breach by a landlord of the requirements of s.8 may entitle the tenant to vacate the premises without paying rent and to recover damages.

S.8 applies to lettings of houses for human habitation provided that the rent does not exceed £80 per year if the house is located in London or £52 per year if the house is located elsewhere. Whilst these financial limits remain so low, the section remains of extremely limited application.

S.8 does not apply to leases for a term of three years or more which require the tenant to put the premises into a condition reasonably fit for human habitation provided that neither party has the option to determine the lease within three years.

2 *Fitness for habitation of furnished houses.* Subject to express provision to the contrary in the lease, there is an implied condition at common law that a furnished house will be fit for habitation when the tenancy begins. If this is not the case, for example because of defective drains, the tenant can either repudiate the lease or sue for damages. The landlord is not required to keep the premises fit for habitation throughout the tenancy as long as they are fit for habitation at its start.

3 *Repair of the structure of dwelling-houses* (s.11 Landlord and Tenant Act 1985). Where s.11 applies to a lease of a dwelling-house, its effect is to imply into the lease covenants on the part of the landlord:

(a) to keep in repair the structure and exterior (including drains, gutters and external pipes);
(b) to keep in repair and proper working order the installations for supply of water, gas and electricity and for sanitation (including basins, sinks, baths and sanitary conveniences, but not other fixtures, fittings and appliances for making use of the supply of water, gas or electricity); and
(c) to keep in repair and proper working order the installations for space heating and heating water.

The covenants implied by s.11 do not require the tenant:

(a) to do repairs for which the tenant is liable either under his duty to use the premises in a tenant-like manner (see page 318) or under an express covenant to the same effect; or
(b) to rebuild or reinstate the premises in the case of destruction or damage by fire, or by tempest, flood or other inevitable accident; or
(c) to repair tenant's fixtures.

The standard of repair required by the covenant implied by s.11 is to be assessed having regard to the age, character and prospective life of the dwelling-house and the locality in which it is situated.

S.11 entitles the landlord to enter the premises at reasonable times of the day to view their state and condition on giving 24 hours' written notice to the occupier. The landlord is not liable under s.11 until he has notice of the disrepair and is not in breach of covenant unless he fails to carry out the required repairs within a reasonable time of receiving such notice.

As a general rule, s.11 applies to leases of dwelling-houses granted on or after 24 October 1961 for a term of less than 7 years.

S.14 provides that s.11 does not apply to:

(a) leases granted to existing tenants, or to former tenants remaining in possession, where s.11 did not apply to the previous lease;
(b) tenancies of agricultural holdings;
(c) leases granted on or after 3 October 1980 to various specified bodies, such as local authorities;
(d) leases granted to a government department or to the Crown (unless the lease is managed by the Crown Estate Commissioners).

The landlord is not merely obliged to carry out repairs on structure, installations etc. which form part of the property let to the tenant. Rather, he is required to repair the structure of all parts of the building in which he has an estate or interest. Equally, he is required to maintain installations which serve the dwelling-house, provided that they form part of the building in which he has an estate or interest or are owned by him or under his control. For example, where a tenant's flat is heated by radiators, but the boiler is in the basement, which does not form part of the property let to the tenant, the landlord is obliged to maintain both the radiators and the boiler.

S.12 provides that an agreement or term of a lease is void to the extent to which its effect is to modify or exclude the provisions of s.11 unless it was authorized by the County Court. The County Court may authorize such an agreement or term with the consent of the parties if it is reasonable to do so. Unless such an agreement or term is authorized, any covenant made by the tenant to carry out repairs is of no effect to the extent that it requires the tenant to carry out repairs which the landlord is obliged to carry out under s.11.

4 *Repair and maintenance of common parts of blocks of flats.* Subject to express provision to the contrary in the lease, there is an implied condition at common law that landlords will take reasonable steps to repair and maintain common parts of the premises which remain in their control, such as lifts and stairs. This implied condition basically appears only to apply in relation to residential blocks of high-rise flats, though it may apply in other special situations in which maintenance of the common parts by the landlord is essential in order to make the tenancy agreement effective.

Relating to the tenant

1 *Obligation not to commit waste.* The law of tort imposes upon tenants an obligation not to commit waste.

Voluntary waste comprises damage to or destruction of premises resulting from deliberate or negligent acts (such as demolition or the making of alterations), but does not encompass damage or destruction which results from their reasonable use.

Ameliorating waste comprises acts which improve the value of land or premises (such as constructing an extension of existing premises). The courts will not normally be prepared to award any remedy in respect of ameliorating waste.

Permissive waste comprises damage to or destruction of premises resulting from omission to undertake necessary repairs (such as failure to replace slates resulting in the loss of a roof). Periodic tenants (except, perhaps, yearly tenants) are not liable for permissive waste.

Where a tenant commits waste, the landlord may bring an action for damages and (except in the case of permissive waste) may claim an injunction to prevent continuing or repeated acts of waste.

2 *The tenant's duty to use the premises in a tenant-like manner.* Where the lease does not expressly impose repairing obligations, it is implied that the tenant will use the premises in a tenant-like manner.

A fixed term tenant, being liable for permissive waste, should hand the premises over in repair at the end of the tenancy, though the tenant is not required to carry out repairs in respect of 'fair wear and tear'. A yearly tenant should carry out minor repairs necessary to keep the premises wind and water tight, but is not required to carry out repairs in respect of 'fair wear and tear'. A periodic tenant for a period of less than a year should perform small tasks which any reasonable tenant would perform (such as cleaning the windows).

Whatever the length of the tenancy, the tenant should carry out repairs required in consequence of voluntary waste.

Security of tenure: termination of a tenancy at common law

At common law, a tenancy may terminate in a variety of ways. Until it terminates, both parties are bound by its terms.

The purpose of the present section is to examine briefly these common law methods of termination. It should be noted at this point, however, that the termination of business tenancies is regulated by Part II of the Landlord and Tenant Act 1954. Similarly, the ability of a landlord to obtain possession of a dwelling-house let under a residential tenancy is restricted by the Rent Act 1977 and Part I of the Housing Act 1988. These statutory provisions are examined later, after a discussion of the common law principles.

Methods of termination at common law

Expiration of a fixed term

A fixed term tenancy terminates when the term expires. Thus, for example, a 7-year lease expires after seven years have passed.

Service of notice under a break clause

A break clause is a proviso which empowers one or both parties to a lease to terminate the lease at certain specified dates by serving notice on the other. For example, a 14-year lease might contain a proviso empowering the tenant to terminate the lease at the end of the seventh year by prior service of notice.

Forfeiture

Where a tenant is in breach of a covenant of his lease, the landlord can only forfeit the lease if it contains a proviso which gives him a right of re-entry on breach of that covenant. The option of forfeiture might, typically, be available to a landlord where a tenant has been in arrears of rent for 21 days.

If a landlord is aware of a breach which entitles him to forfeit the lease, and then expressly or by implication affirms the continued existence of the lease, he waives his right to forfeit it, unless it is a continuing breach (such as a state of disrepair). Acceptance of, or demand for, rent due in respect of a period following the breach will, for example, give rise to such waiver.

Other than in the case of forfeiture for non-payment of rent, s.146 of the Law of Property Act 1925 requires the landlord, before forfeiting the lease, to serve on the tenant a notice which:

(a) specifies the breach;
(b) if it can be remedied, requires the tenant to remedy it; and
(c) requires the tenant to make monetary compensation for it.

Following service of the s.146 notice, the landlord can proceed to enforce the forfeiture unless the tenant remedies the breach and provides reasonable compensation within a reasonable time. Other than where the tenant peaceably vacates the premises, the landlord will normally be required to enforce his right of re-entry by bringing proceedings in court.

Where s.146 applies to a forfeiture, it entitles the tenant to apply to the court for relief against forfeiture. Taking into account both the conduct of the parties and all other relevant circumstances, including the fact that the tenant has remedied the breach or has promised to do so, the court can grant or refuse relief as it thinks fit. Special rules apply under s.146 to sub-tenants, including mortgagees. (See also page 314 for consideration of the Leasehold Property (Repairs) Act 1938.)

In the case of forfeiture in respect of non-payment of rent, whilst s.146 does not apply, a right of re-entry will not arise until the landlord has formally demanded the rent unless, as is normally the case, this requirement is excluded by a proviso of the lease or by operation of s.210 of the Common Law Procedure Act 1852. The tenant may apply for relief from forfeiture. The extent of the tenant's entitlement to relief, and the powers of the court in respect of an application for relief, vary depending upon whether the action is brought in the High Court or the County Court.

Merger

Merger occurs when a tenant acquires his landlord's reversion and intends the two estates to merge into one, terminating the tenancy. Such intention will normally be expressed in the assignment of the landlord's reversion to the tenant.

Surrender

Surrender may occur in either of two situations.

1 *Surrender by express agreement of the parties.* Express surrender occurs when landlord and tenant agree to end a lease before it is due to expire. Express surrender will normally be by deed, though an agreement in writing is sufficient in the case of terms of three years or less.
2 *Surrender by operation of law.* Surrender by operation of law occurs when both parties to a lease act in such a way that the surrender of the lease before it is due to expire is implied. This is the case where the conduct of the parties is such as permits of no explanation other than that the lease has been surrendered.

For example, surrender may be implied where a tenant abandons possession and his landlord re-lets the premises to a new tenant. Again, it may be implied when a landlord grants his tenant a new lease before the old one has expired. It will probably not be implied where a tenant abandons possession and his landlord leaves the premises vacant in anticipation of the tenant's return.

Surrender, whether express or by operation of law, does not invalidate a lawful sub-lease.

Service of notice to quit

A periodic tenancy may be terminated by service of notice to quit. The terms of a tenancy may expressly indicate the length of notice required in order to terminate it. Where this is not the case, the basic rule is that a tenancy may be terminated by service of one period's notice and, consequently:

- to terminate a weekly tenancy, one week's notice is required;
- to terminate a monthly tenancy, one month's notice is required;
- to terminate a quarterly tenancy, one quarter's notice is required.

In the case of a yearly tenancy, or a periodic tenancy the period of which exceeds one year, six months' notice is required.

In general, notice must be served so as to terminate a tenancy on an anniversary of its creation. In the case of a yearly tenancy, this would be one or more years from the date of grant. In the case of a weekly or monthly tenancy, this would be, respectively, one or more weeks or months from the date of grant. A quarterly tenancy must normally be terminated on a quarter day (25 March, 24 June, 29 September or 25 December).

Notice to quit a dwelling-house must be in writing and must be served at least four weeks before the date of termination (Protection from Eviction Act 1977, s.5 — though s.5 does not apply to certain tenancies which are excluded by s.3A).

Frustration

The doctrine of frustration is applicable, though in limited circumstances, to leases (see *National Carriers v. Panalpina Ltd.*, page 117).

Offences related to the recovery of possession of premises

S.6 of the Criminal Law Act 1977 provides that it is an offence, without lawful authority, to use or threaten violence for the purpose of securing entry into premises, knowing that someone opposed to the entry is present on the premises.

S.1(2) of the Protection from Eviction Act 1977 provides that it is an offence to unlawfully deprive or attempt to deprive the residential occupier of premises (namely, the person who occupies those premises as a residence under contract, statute or rule of law) of his occupation of all or part of the premises in the absence of belief, with reasonable cause, that the residential occupier has ceased to reside there.

The only lawful method of obtaining possession of premises occupied by a residential tenant is by obtaining a court order, unless either the tenant peaceably vacates the premises or the tenancy is a Housing Act 1988 excluded tenancy.

S.1(3) provides that it is an offence for any person to do acts likely to interfere with the peace or comfort of a residential occupier of premises or members of his household or to persistently withdraw or withhold services reasonably required for the occupation of the premises as a residence, intending in either case to cause the residential occupier to give up possession of the premises or to cause him to refrain from exercising rights or pursuing remedies in respect of the premises.

S.1(3) (A) provides that it is an offence for the landlord of a residential occupier or his agent to do such acts or withdraw or withhold such services, knowing or having reasonable cause to believe that the conduct is likely to result in one of those consequences, unless the person who did the acts or withdrew the services had reasonable grounds for so doing.

A tenant who gives up possession of premises in consequence of conduct falling within s.1(2) or (3) may be entitled to recover damages from the landlord.

S.2 provides that where a person resides in premises which are let as a dwelling, a right of re-entry or forfeiture can only be lawfully enforced by proceedings in court. S.3 provides that where a lease of a dwelling-house comes to an end and the occupier continues to reside in it, the landlord can only lawfully regain possession by court proceedings unless the tenancy was a protected tenancy or an excluded tenancy.

Business tenancies (Landlord and Tenant Act 1954, Part II)

Where the provisions of Part II of the Landlord and Tenant Act 1954 apply to a tenancy, the Act may entitle the tenant to a new lease or to the continuation of his existing tenancy when the contractual term expires or the landlord serves notice to quit.

The Act only applies to leases, not to licences (s.23). It should be noted, however, that the principles stated by the House of Lords in *Street v. Mountford* (see page 307) are applicable to business leases. Thus, an agreement which is described as a licence but gives exclusive possession of premises for a fixed or periodic term will amount to a lease, not a licence, unless the agreement is one of the exceptional types recognised by the House of Lords in *Street v. Mountford*.

Under s.23, the tenant must occupy at least part of the premises. This does not require the tenant to occupy the premises personally. Rather, it may be sufficient if the premises are occupied by the tenant's agents or employees. The tenant need not be in sole occupation of the premises. The tenant must, however, retain a sufficient degree of control over the premises. For example, where a business tenant sublets the premises and acts as manager, providing services to the sub-tenants, this may be sufficient to amount to occupation of the premises by the tenant. The premises may include land without a building, such as an undeveloped lot used as a car park.

Occupation of the premises must, at least in part, be for the purposes of a business carried on by the tenant (s.23). Business includes a trade, profession or employment. The expression thus covers a wide variety of enterprises, from shops and offices to hospitals and clubs.

The Act does not apply to a tenancy if the lease prohibits business use by the tenant unless the landlord has agreed to the business use. The Act also does not apply where premises are mainly used for residential purposes and any business use is subsidiary to this. This may be so, for example, where a residential tenant takes in a lodger.

Where the purpose of a letting is a mixed business and residential purpose, the business purpose not being subsidiary to the residential purpose, then the Act may apply to the tenancy.

Types of tenancy to which the Act does not apply

- Agricultural holdings (s.43).
- Mining leases (s.43).
- Service tenancies (s.43) – tenancies granted to the tenant for the purpose of the tenant's employment or an office or appointment which the tenant holds.
- Leases of licensed premises (s.43). The Act applies to off-licensed premises but not to most on-licensed premises.
- Residential tenancies (s.23)
- Leases of premises required for public purposes (s.57) or regional development (s.60).
- Leases of premises required for national security purposes (s.58).
- Fixed term tenancies not exceeding six months (s.43). If the tenant has been in occupation for more than twelve months or the lease contains a term permitting extension or renewal of the tenancy beyond six months then the Act does apply.
- Extended tenancies granted under the Leasehold Reform Act 1967 (s.16 Leasehold Reform Act 1967).
- Tenancies of premises where the parties have agreed to a new tenancy of the premises (s.28).

Contracting out of the Act

Section 38 of the Act empowers the court, on an application by both landlord and tenant, to authorize an agreement between the landlord and tenant that the provisions of the Act which relate to termination of the tenancy and entitle the tenant to a new tenancy will not apply to the tenancy. The court will normally authorize such an agreement if both parties have received proper legal advice. If such an agreement has not been authorized by the court, it will be void.

Protection under the Act for the business tenant

S.24 provides that a business tenancy to which the Act applies can only be terminated by forfeiture, by surrender, by merger, by service of notice to quit by the tenant, or by satisfying special requirements laid down in the Act.

Thus, where a fixed term tenancy expires or the landlord serves notice to quit (in relation to a periodic tenancy or under a break clause), then, unless the special requirements laid down by the Act have been satisfied, the tenancy does not terminate. Rather, the tenancy continues until terminated in one of the five ways referred to above.

A tenancy which continues in this way is described as a continuation tenancy. Basically, it amounts to a continuation of the contractual tenancy and consequently, it is possible for both landlord and tenant to assign their interests. The terms of the continuation tenancy are the same as the terms of the contractual tenancy, with the exception of any terms permitting the landlord to terminate the tenancy or creating personal obligations which do not run with the land. Whilst the tenancy continues in this way, the tenant continues to have the right to remove tenant's fixtures and to use easements given to him by the lease.

Under a continuation tenancy, the landlord cannot increase the rent except by applying to the court (s.24A). He can only do this if either he has commenced the special procedure which the Act lays down for determining the tenancy, or the tenant has commenced the special procedure which the Act lays down for applying for a new tenancy. The court may be prepared to grant an interim rent, which will remain in force until the continuation tenancy ends.

The court is not bound to grant such a rent but if it chooses to do so the rent must be an open market rent fixed in accordance with principles laid down by the Act. The principles are basically those which the Act lays down in relation to determining the rent payable under a new tenancy (s.34). Additionally, the court is required to take into account the rent payable under the terms of the tenancy and to determine the rent upon the basis of the assumption that the tenancy is a yearly tenancy (s.24A).

If a continuation tenancy ceases to be a business tenancy, the landlord may terminate it by giving the tenant between three and six months' notice or in accordance with the terms of the tenancy (s.24).

Termination by the tenant

The Act permits termination by surrender and merger. The Act also allows the tenant to terminate the tenancy by serving notice to quit (in relation to a periodic tenancy or under a break clause), although he may not do so until he has been in occupation for at least one month under the tenancy (s.24).

In relation to the expiration of a fixed term tenancy, the tenant must give his landlord written notice that he does not wish the tenancy to continue either at least three months before the date when the tenancy would expire if the Act did not apply or at least three months before any quarter day following that date (s.27).

With regard to termination of a tenancy by notice to quit (in relation to a periodic tenancy or under a break clause), the usual common law notice requirements (see pages 319–21) apply. In order to terminate a continuation tenancy, the tenant must give his landlord written notice at least three months before any quarter day (s.27).

Termination by the landlord

Subject to the discretion of the court to grant relief, the Act allows the landlord to forfeit the tenancy where breach of a condition or covenant gives rise to a right of re-entry (s.24). Other than this, the landlord can only terminate the tenancy if he satisfies special requirements which the Act lays down (see below). In order to rely upon expiration of a fixed term or service of notice to quit (in relation to a periodic tenancy or under a break clause), the landlord must satisfy these special requirements. If he does not do so, then a continuation tenancy will result (s.24). Similarly, where a landlord wishes to terminate a continuation tenancy, his options are either to forfeit, if appropriate, or to satisfy the special requirements laid down by the Act.

The *special requirements* which the Act lays down for termination of a tenancy by the landlord are as follows:

1 The landlord must serve written notice (a 's.25 notice') on the tenant in the form laid down by regulations made under the Act. The notice must specify the date of termination and must be served on the tenant between six and twelve months before the date which it specifies (s.25). If it has not already terminated, the contractual tenancy will terminate on this date (though the current tenancy may continue as a continuation tenancy beyond this date).

In relation to the expiration of a fixed term tenancy, the date of termination cannot be earlier than the date when the tenancy would have expired had the Act not applied to it.

With regard to termination of a tenancy by notice to quit (in relation to a periodic tenancy or under a break clause) the date of termination cannot be earlier than the earliest date upon which the tenancy could have been brought to an end, had the Act not applied to it, by notice to quit served by the landlord. Provided that the s.25 notice conforms with the requirements of the tenancy agreement, the landlord need not serve a separate notice to quit on the tenant.

In relation to the termination of a continuation tenancy, the date of termination can be any date chosen by the landlord.

The notice must require the tenant, within two months of its service, to notify the landlord in writing whether he will be willing to give up possession on the date of termination specified by the notice. If the landlord does not wish the tenant to be granted a new tenancy, the notice must state this, and must also indicate the grounds upon which the landlord objects to the grant of a new tenancy. The landlord may only oppose the grant of a new tenancy on grounds laid down by s.30. These grounds are outlined at page 328.

2 The landlord who serves the notice will be the tenant's immediate landlord if he either owns the freehold or has a tenancy which will not expire within fourteen months (s.44). If this is not the case, the landlord competent to serve the notice will be the first landlord in the chain above the tenant's immediate landlord who either owns the freehold or has a tenancy which will not expire within fourteen months.

3 If the tenant does not want a new tenancy, he need not respond to the s.25 notice served on him by the landlord. The tenancy will terminate on the date specified by the notice.

If the tenant does want a new tenancy, he must serve a written counter-notice on the landlord indicating that he refuses to give up possession (s.29). He must serve this counter notice on the landlord within two months of the serving of the s.25 notice on him by the landlord, otherwise he loses his right to apply for a new tenancy.

4 Once the tenant serves the counter-notice he becomes entitled to apply to the court for a new tenancy if the parties cannot agree upon one themselves (s.29). If the tenant wishes to apply to the court he must do so not less than two nor more than four months after the service of the s.25 notice by the landlord, otherwise he loses the right to apply. The relevant court may be the County Court or the High Court, depending upon the rateable value of the premises (s.63).

Once the tenant applies to the court for a new tenancy, s.64 provides that the current tenancy cannot terminate within three months of the disposal of the tenant's application by the court. This is so whether or not the court grants a new tenancy. Thus, if the termination date specified by the landlord's s.25 notice falls within this three month period, the current tenancy will not terminate until the three months have passed.

The court will order the grant of a new tenancy unless the landlord establishes one or more of the grounds upon proof of which the court may or must refuse the tenant's application (s.29). If the tenant changes his mind he may apply to the court for revocation of the order within fourteen days of its making (s.36). If the tenant does this the current tenancy will continue for a period agreed by the parties or determined by the court, thus giving the landlord a reasonable opportunity to re-let or dispose of the premises (s.36).

If the court grants a new tenancy it will determine the terms of the new lease if the parties cannot reach agreement upon them. In the absence of agreement, the Act specifies in ss.32–35 rules with regard to: the length of the tenancy, up to a maximum of 14 years; the property to be included in the tenancy; the rent payable, which will be an open market rent fixed in accordance with statutory principles; and other terms of the new tenancy.

5 Under s.30 there are statutory grounds on which landlords can object to the grant of a new tenancy. Proof of certain of these grounds means that a court *must* refuse the tenant's application for a new tenancy. In other cases, the court *may* refuse to grant a new tenancy.

Mandatory grounds for refusing the tenant's application

There are three grounds upon proof of which by the landlord the court must refuse the tenant's application.

Ground d – the landlord has offered and is willing to provide the tenant with alternative accommodation which is suitable for the tenant's requirements on terms which are reasonable.

In assessing whether the alternative accommodation is suitable for the tenant's requirements (including the preservation of goodwill), the court should take into account the time when the new accommodation will be available for the tenant, the nature and class of the tenant's business and the situation, extent and facilities of the accommodation provided by the current tenancy.

Ground f – the landlord intends to demolish or reconstruct the premises or to undertake substantial construction work, and could not reasonably do so without obtaining possession of the premises. The landlord must prove at the time of the court hearing that he has the necessary intention. He will do so if, on the evidence, a reasonable man would think that the landlord had a reasonable prospect of carrying out his intention. (Note, the tenant may be entitled to compensation under s.37 where the landlord relies on this ground — see page 332.)

Ground g — the landlord intends to occupy the premises either as his residence or for the purposes of a business to be carried on by him. This ground is not normally available to a landlord who acquired the freehold or a tenancy of the premises by purchase within five years of the date of termination specified by his s.25 notice. Again, the landlord must prove at the time of the court hearing that he has the necessary intention. (Note, the tenant may be entitled to compensation under s.37 where the landlord relies on this ground -- see page 332).

Discretionary grounds for refusing the tenant's application

There are four grounds upon proof of which by the landlord the court may, but is not obliged to, refuse the tenant's application.

Ground a — the tenant ought not to be granted a new tenancy in consequence of the state of repair of the premises resulting from the tenant's failure to comply with his obligations under the current tenancy to repair and maintain them. The breach must be serious enough to persuade the court not to grant a new tenancy, an undertaking made by the tenant to remedy the breach being a relevant factor.

Ground b — the tenant ought not to be granted a new tenancy in consequence of his persistent delay in paying rent due under the current tenancy. This requires either a failure to pay rent due on a number of occasions or a very long delay in paying rent due. The court will take into account factors such as the reason for the delay and how, if a new tenancy is granted, the landlord may be protected against breaches of the tenant's covenant to pay rent.

Ground c — the tenant ought not to be granted a new tenancy in consequence of other substantial breaches of his obligations under the current tenancy, or for any other reason connected to his use or management of the premises. The court will take into account factors such as the seriousness of the breach and whether the tenant has remedied or proposes to remedy it.

Ground e — the tenant ought not to be granted a new tenancy where: the tenant is a sub-tenant of part of the property let by the landlord under a superior tenancy; and the landlord might reasonably expect to achieve a substantially better rent by letting the property as a whole; the landlord requires possession of the premises in order to let or dispose of the property let under the superior tenancy as a whole. The court will take into account factors such as the fact that the landlord consented to the sub-letting. The ground is only of value to the landlord where the superior tenancy is due to terminate shortly after the sub-tenancy. [Note, the tenant may be entitled to compensation under s.37 where the landlord relies on this ground – see page 332.]

Tenant's request for a new tenancy

A tenant who has not been served a s.25 notice by his landlord may wish to apply to the court for a new tenancy. The requirements which must be satisfied are as follows:

1. The current tenancy must have been granted for a term exceeding one year. This includes a tenancy granted for a fixed term and thereafter from year to year.
2. The tenant must serve a s.26 notice, requesting a new tenancy, on the 'competent landlord' (who will not necessarily be the tenant's immediate landlord (s.44) – see page 327). A tenant cannot serve a s.26 notice if his landlord has served a s.25 notice on him, or if he has given his landlord notice to quit, or if he has notified his landlord (under s.27) that he does not wish his fixed term tenancy to continue when its expiration date arrives.

    The notice must be in writing and in the form laid down by regulations made under the Act. It must state the date upon which the new tenancy is to commence. The current tenancy will automatically terminate on this date. The date must be between six and twelve months after the making of the s.26 request to the landlord, and cannot be earlier than the date on which, had the Act not applied to it, the tenancy would have expired or could have been terminated by notice to quit served by the tenant. The notice must contain the tenant's proposed terms of the new tenancy, including the property to be comprised in the tenancy and the rent to be paid.
3. If the landlord does not wish to oppose the tenant's application for a new tenancy, he need not respond to the s.26 notice served on him by the tenant. He thus loses his right to oppose the tenant's application.

    If the landlord does wish to oppose the tenant's application, he must serve a written counter-notice on the tenant indicating that he opposes the application. He must serve this counter notice on the tenant within two months of the serving of the s.26 notice on him by the tenant, otherwise he loses his right to oppose the tenant's application.

The notice must indicate the grounds upon which the landlord objects to the grant of a new tenancy. The grounds upon which the landlord may so object, laid down by s.30, were outlined on page 328.

4 Once the tenant serves the s.26 notice he becomes entitled to apply to the court for a new tenancy if the parties cannot agree upon one themselves (s.29). If the tenant wishes to apply to the court, he must do so not less than two nor more than four months after the service of the s.26 notice on the landlord, otherwise he loses the right to apply. The relevant court may be the County Court or the High Court, depending upon the rateable value of the premises (s.63).

Once the tenant applies to the court for a new tenancy, s.64 provides that the current tenancy cannot terminate within three months of the disposal of the tenant's application by the court. This is so whether or not the court grants a new tenancy. Thus, if the termination date specified by the tenant's s.26 notice falls within this three-month period, the current tenancy will not terminate until the three months have passed.

The court will order the grant of a new tenancy, unless the landlord establishes one or more of the grounds upon proof of which the court may or must refuse the tenant's application (s.29). If the tenant changes his mind he may apply to the court for revocation of the order within fourteen days of its making (s.36). If the tenant does this, the current tenancy will continue for a period agreed by the parties or determined by the court, thus giving the landlord a reasonable opportunity to re-let or dispose of the premises (s.36).

If the court grants a new tenancy it will determine the terms of the new lease if the parties cannot reach agreement upon them. In the absence of agreement: the length of the tenancy will be that which the court considers to be reasonable in the circumstances, to a maximum of fourteen years (s.33); the property included in the new tenancy will be that part of the property included in the current tenancy which the court designates in the circumstances (s.32): the rent payable will be an open market rent, fixed in accordance with principles laid down by the Act (s.34); the other terms will be determined by the court having regard to the terms of the current tenancy and all other relevant circumstances (s.35).

5 If the landlord establishes one of the statutory grounds upon

which he is entitled to object to the grant of a new tenancy (see page 328), it may be that the tenant will not be granted a new tenancy. Upon proof of certain grounds by the landlord, the court must refuse the tenant's application for a new tenancy. Upon proof of other grounds, the court has discretion to grant or refuse to grant a new tenancy.

Tenant's compensation

Under s.37 Landlord and Tenant Act 1954, Part II

Compensation under this section is available in two cases:

1 Where the tenant's application to the court for a new tenancy is refused because the landlord intends to demolish the premises, or because the landlord intends to occupy the premises himself, or because the landlord wants to amalgamate a sub-tenant's property with other property in order to get a better rent by letting as a whole (s.30).
2 Where either the landlord's s.25 notice or his counter notice served in response to the tenant's s.26 notice states any or all of these three grounds of opposition (but no others), and the tenant either fails to apply for a new tenancy or makes and then withdraws an application. (Note, the leave of the court is required to withdraw an application.)

The amount of the compensation is determined by reference to a formula laid down from time to time by statutory instrument.

It may be, however, that landlord and tenant have agreed in writing to exclude the tenant's right to compensation under s.37. Such an agreement may be valid if the tenant and his predecessors in the business occupied the premises for business purposes for less than five years by the time he leaves. Similarly, such an agreement may be valid if made after the right to compensation has accrued. Otherwise, however, s.37 provides that such an agreement will be void.

Compensation for improvements under s.1 Landlord and Tenant Act 1927, Part I

Entitlement to compensation. Part I of the 1927 Act applies to premises held under a lease which are used for trade or business purposes, but does not apply to mining leases, leases of agricultural holdings or leases which express in writing that the premises are let to the tenant as holder of an office, appointment or employment which he continues to hold (s.17). An agreement depriving a tenant of his right to compensation under Part I is invalid unless made for valuable consideration before 10 December 1953 (s.9 as amended by s.49 Landlord and Tenant Act 1954).

Where Part I of the 1927 Act applies, it entitles the tenant, on vacating his holding at the end of the tenancy, to compensation from his landlord in respect of improvements made by him or his predecessors in title (s.1). Where premises are used only partly for trade or business purposes, compensation is only payable in relation to improvements if and to the extent to which they relate to the trade or business (s.17). It is the tenant's immediate landlord who is required to pay compensation. If he is not the freeholder, then when his tenancy comes to an end and he vacates the premises, he is entitled to claim compensation from his landlord (s.8).

In order to give rise to this right to compensation, an improvement must not be a fixture which the tenant is entitled to remove, and it must add to the letting value of the holding at the end of the tenancy (s.1). Moreover, the improvement must not have been made for valuable consideration (s.2).

Further, a tenant is not entitled to compensation under Part I unless the requirements laid down by s.3 have been satisfied. Prior to making the improvement, the tenant is required to notify the landlord of his intention, providing the landlord with a specification and plan of the improvement. Unless the making of the improvement is required by statute, the landlord is entitled to serve notice of objection on the tenant within three months of receiving the tenant's notice.

If the landlord does not serve a notice of objection within three months, he loses his right to do so. The tenant can lawfully go ahead with the improvement even if it is prohibited by the terms of his lease.

If the landlord does serve a notice of objection within three months, the tenant can apply to the court for a certificate that the improvement is proper. When such an application is made, all superior landlords must be notified of the tenant's intention and have a right to be heard before the court. The court must give the tenant a certificate if it is satisfied that the improvement will add to the letting value of the holding at the end of the tenancy, is reasonable and suitable to the character of the property, and will not diminish the value of any other property belonging to the landlord.

The landlord is entitled to offer to make the proposed improvement himself in return for a reasonable increase in rent or an increase determined by the court. In this case the court cannot grant a certificate unless it is subsequently shown to the court's satisfaction that the landlord has failed to do so.

Before giving the tenant a certificate, the court is entitled to make such modifications to the specification or plan as it thinks fit and to impose such conditions as it thinks reasonable. Once the certificate has been issued, the tenant can lawfully carry out the improvement, even if it is prohibited by the terms of his lease. He must comply with conditions imposed by the court and must complete the work within time limits agreed with the landlord or fixed with the court.

Claiming compensation

The tenant's claim for compensation, made in the form laid down by rules of court, must be served on the landlord within time limits laid down by s.47 of the Landlord and Tenant Act 1954. Where the claim relates to the giving of notice to quit, whether by the landlord or the tenant, the claim must be made within three months of the giving of the notice.

If the claim for compensation is in relation to termination following the service of a s.26 notice by the tenant, it must be made within three months of the date upon which the landlord served his counter-notice. If the claim is in relation to termination by expiration of time, it must be made between three and six months before the end of the tenancy. Where the claim is in relation to forfeiture, it must be made within three months of the effective date of a court order for possession or, in the case of re-entry without a court order, within three months of the date of re-entry.

Amount of compensation

In the absence of agreement between the parties to the lease, the amount of compensation is determined by the court (s.1). The amount of compensation cannot exceed either the net addition to the value of the holding which directly results from the improvement, or the reasonable cost of carrying out the improvement at the end of the tenancy less the cost of putting the improvement into a reasonable state of repair (unless or to the extent to which the lease obliges the tenant to carry out such repairs).

In determining the net addition to the holding's value, the intended use of the premises after the end of the lease, including an intention to demolish or reconstruct, and the likely time lapse between the date of termination and any change of use, demolition or reconstruction must be taken into account. If the landlord does not carry out his intentions within the time specified, the tenant is entitled to make a further application for compensation. The amount of compensation will be reduced in accordance with the value of any benefits received by the tenant or his predecessors from the landlord or his predecessors in consideration of the making of the improvement (s.2(3)).

Residential tenancies

In relation to a residential tenancy, where the provisions of the Housing Act 1988 or the Rent Act 1977 apply, the tenant may be provided with security of tenure.

Housing Act 1988

A tenancy to which the Act applies is known as an *assured tenancy*. The Act only applies to leases, and not to licences. However, as was noted on page 307, an agreement which is described as a licence, but which gives exclusive possession of premises for a fixed or periodic term may in fact create a tenancy (*Street v. Mountford*).

The tenancy must be of a dwelling-house, the purpose of the letting being residential. A dwelling-house may be a house or part of a house (s.45) and thus can include a flat. It must be let as a separate dwelling, which must provide the tenant or joint tenants with exclusive possession of rooms in which the tenant can conduct the main activities of living, especially cooking, eating and sleeping. Other rooms, such as bathrooms, may be shared (s.3).

In order to be within the protection of the Act, the tenant or one of the joint tenants must be an individual (i.e. not a company) and must occupy the dwelling-house as his only or principal home. Furthermore, the tenancy *must not* be one of the following, which are examples of tenancies incapable of being an assured tenancy:

- A tenancy entered into before 15 January 1989.
- A tenancy of a dwelling-house the rent for which (or rateable value if the tenancy was entered into prior to 1 April 1990) exceeds specified limits.
- A tenancy under which either no rent is payable or the rent payable falls below specified limits.
- A business tenancy within Part II of the Landlord and Tenant Act 1954.
- A tenancy of premises with an 'on-licence' for the sale of intoxicating liquors.

- A tenancy under which agricultural land exceeding two acres is let with the dwelling house.
- A tenancy under which the dwelling-house is comprised in an agricultural holding (see Agricultural Holdings Act 1986) and is occupied by the person who controls the farming of the holding.
- A tenancy granted to a student at a specified educational institution by a specified body.
- A tenancy the purpose of which is to let the dwelling-house to the tenant for a holiday.
- A tenancy of a dwelling-house which forms part of a building, where the landlord is an individual who occupies another dwelling-house in the same building as his only or principal home. The tenancy is not necessarily rendered incapable of being an assured tenancy by the presence of a resident landlord if the dwelling-house forms a flat in a purpose-built block of flats. (Note — the complex provisions of Sched. I para. 10, relating to resident landlords, fall outside the scope of this book.)
- A tenancy under which the interest of the landlord belongs to a local authority or other specified body.
- A protected tenancy (see discussion of the Rent Act 1977 below); a housing association tenancy (see Rent Act 1977); a secure tenancy (see Housing Act 1985).

Security of tenure under the 1988 Act

1 *Periodic tenancies.* A landlord who wishes to terminate a periodic tenancy which falls within the protection of the Act (a periodic assured tenancy) may only do so by obtaining a court order (s.5).

Before taking proceedings in court, the landlord must normally serve on the tenant a notice which satisfies the requirements of s.8. The notice must state the landlord's statutory grounds for possession (see page 338), indicate the earliest date on which the proceedings can begin (which must be at least two weeks from the date of service of the notice) and, indicate that the proceedings will not begin later than twelve months from the date of service of the notice.

Where the notice specifies, with or without other grounds, any of statutory grounds 1, 2, 5, 6, 7, 9 or 16 (see page 338), the proceedings cannot begin earlier than the date at which the landlord could have terminated the tenancy at common law by serving notice to quit and, in any event, not earlier than two months from the date of service of the notice.

The court can dispense with the s.8 notice requirement where

it is just and equitable to do so, except where the landlord seeks to recover possession under statutory ground 8.
2 *Fixed-term tenancies.* Unless the tenant surrenders his tenancy, a landlord who wishes to terminate a fixed term tenancy which falls within the protection of the Act (a fixed term assured tenancy) may only do so either by obtaining a court order or by exercising his power to terminate the tenancy under a break clause (s.5). In order to obtain a court order, the landlord must comply with the s.8 notice procedure (outlined above).

If a fixed term assured tenancy comes to an end other than by surrender or by court order, the tenant is entitled to remain in possession under a periodic assured tenancy which arises under s.5 (a statutory periodic tenancy), the period of which is the same as that for which rent was payable under the fixed term tenancy. Basically, the terms of the statutory periodic tenancy are the same as those of the fixed term tenancy except that terms permitting termination of the tenancy cease to have effect so long as the tenancy is assured.

At the end of a fixed term tenancy, if the landlord grants the tenant a new fixed term or periodic tenancy of substantially the same dwelling-house, a statutory periodic tenancy will not arise.
3 *Possession by court order.* The court may only grant the landlord an order for possession of a dwelling-house let on an assured tenancy if the landlord establishes one or more of the statutory grounds for possession (s.7). In relation to a fixed term tenancy, the court cannot make an order if the tenancy has not expired or otherwise been terminated unless at least one of grounds 2, 8, 10, 11, 12, 13, 14 or 15 is established and the lease permits termination on the ground in question.

Where an order is made in relation to a fixed term tenancy which has come to an end, this is sufficient to terminate a statutory periodic tenancy which has arisen under s.5 without further notice being required (ss.7 & 8).

There are two types of grounds for possession, mandatory grounds and discretionary grounds. Upon proof of a mandatory ground, the court is required to make an order for possession (subject both to what was said immediately above with regard to the making of orders in relation to fixed term tenancies and to satisfaction of the s.8 notice requirements examined above).

Upon proof of a discretionary ground, the court may make an order for possession if it considers it reasonable to do so. The court possesses discretion to stay or suspend the execution of the order or postpone the date of possession for such period as it thinks fit.

Mandatory grounds (Part I of Schedule 2, Housing Act 1988)

In order to obtain possession on grounds 1–5, the landlord must have notified the tenant not later than the beginning of the tenancy that possession might be recovered on the ground in question. In relation to grounds 1 and 2, the court can dispense with the notice requirement where it is just and equitable to do so.

Ground 1 applies either where the landlord at some time prior to the beginning of the tenancy occupied the dwelling-house as his only or principal home or where the landlord requires the dwelling-house as his or his spouse's only or principal home.

Ground 2 applies where the dwelling-house is subject to a mortgage granted before the tenancy began and the mortgagee requires vacant possession in order to exercise a power of sale.

Ground 3 applies to fixed term tenancies for a term not exceeding eight months if at some time within the twelve months prior to commencement of the tenancy the dwelling-house was let as a holiday home.

Ground 4 applies to fixed term tenancies for a term not exceeding twelve months if at some time within the twelve months prior to commencement of the tenancy the dwelling house was let to a person pursuing or intending to pursue a course of study provided by an educational institution specified by the Secretary of State and the tenancy was granted by that institution or by another specified institution or body of persons.

Ground 5 applies where a dwelling-house is held for the purpose of being available for occupation by a minister of religion as a residence from which to perform his duties and the court is satisfied that the dwelling house is required for this purpose.

Ground 6 applies where the landlord intends to demolish or reconstruct the whole or a substantial part of the dwelling-house or to carry out substantial works. In order to obtain possession on this ground, it must be established that:

(a) the work cannot reasonably be carried out without the tenant giving up possession because: the tenant will not agree to such a variation of the terms as would permit the work to be carried out; or the nature of the work is such that no such variation is practicable; or the tenant is not willing to accept an assured tenancy of such part of the dwelling house as would leave in the landlord's possession so much of the dwelling-house as would reasonably enable the work to be carried out; or the nature of the work is such that such a tenancy is not practicable; and
(b) the landlord seeking possession either acquired his interest in

the dwelling-house before the grant of the tenancy or the interest was in existence at the time of the grant and neither the landlord nor any other person who has acquired the interest since that time acquired it for money or money's worth: and
(c) the assured tenancy on which the dwelling-house is let did not come into being in succession to a Rent Act 1977 or Rent (Agriculture) Act 1976 statutory tenancy (see page 343).

S.11 provides that where the court makes an order for possession on ground 6 or ground 9 (but not any other ground) the landlord must pay to the tenant the reasonable expenses likely to be incurred by the tenant in moving from the dwelling-house. The amount will either be determined by agreement between landlord and tenant or by the court.

Ground 7 applies to a periodic tenancy which has devolved under the will or intestacy of the former tenant (but not by statutory succession) if proceedings for recovery of possession are commenced either not later than 12 months after the death of the former tenant or, if the court so directs, after the date on which the landlord became aware of the former tenant's death. For the purpose of this ground, acceptance by the landlord of rent from a new tenant after the former tenant's death does not create a new periodic tenancy unless the landlord agrees in writing to change a term or terms of the tenancy. (For statutory succession, see page 342.)

Ground 8 applies when both at the date of service of the s.8 notice and at the date of the court hearing:

- At least thirteen weeks' rent is unpaid in cases where the rent is paid weekly or fortnightly.
- At least three months' rent is unpaid in cases where the rent is paid monthly.
- At least one quarter's rent is more than three months in arrears in cases where the rent is paid quarterly.
- At least three months' rent is more than three months in arrears in cases where the rent is paid yearly.

Discretionary grounds (Part II of Schedule 2, Housing Act 1988)

Ground 9 applies where suitable alternative accommodation is available for the tenant or will be available when the order for possession takes effect.

Part III of Schedule 2 provides that a certificate of the local housing authority for the district in which the dwelling-house is located, certifying that the authority will provide suitable alternative accommodation for the tenant by a specified date, is conclusive evidence that suitable alternative accommodation will be available by that date.

Otherwise, Part III of Schedule 2 provides that alternative accommodation is suitable if:

(a) it will either be let on an assured tenancy (other than an assured shorthold tenancy or an assured tenancy in relation to which notice is given that possession may be required on any of grounds 1 to 5), or the terms of the letting will give the tenant security of tenure reasonably equivalent to that provided by such an assured tenancy; and
(b) the accommodation is reasonably suited to the needs of the tenant and his family in relation to proximity to place of work; and
(c) the accommodation is similar in relation to rent and extent to that provided in the neighbourhood by any local housing authority for persons whose needs in relation to extent are similar to those of the tenant and his family or is reasonably suited to the means of the tenant and to the needs of the tenant and his family in relation to extent and character; and
(d) if any furniture was provided under the assured tenancy, furniture is provided which is either similar to that formerly provided or is reasonably suited to the needs of the tenant and his family.

Accommodation is not suited to the needs of the tenant and his family if their occupation of it would render it an overcrowded dwelling within the meaning of Part X of the Housing Act 1985.

S.11 provides that where the court makes an order for possession on ground 6 or ground 9 (but not any other ground) the landlord must pay to the tenant the reasonable expenses likely to be incurred by the tenant in moving from the dwelling-house. The amount will either be determined by agreement between landlord and tenant or by the court.

Ground 10 applies where rent is unpaid on the date on which the proceedings for possession are begun and (except where the court considers it to be just and equitable to dispense with the requirement that notice be served under s.8) was in arrears at the date of the service of the section 8 notice.

Ground 11 applies where (whether or not rent is in arrears on the date on which proceedings for possession are begun) the tenant has persistently delayed in paying rent.

Ground 12 applies where any obligation of the tenancy, other than one related to the payment of rent, has been broken or has not been performed.

Ground 13 applies where the condition of the dwelling-house or common parts has deteriorated due to acts of waste by, or the neglect or default of, the tenant or any other person residing in the dwelling-house. Where the waste, neglect or default is that of a lodger or sub-tenant, it must be shown that the tenant has not taken such steps as he ought reasonably to have taken in order to remove the lodger of sub-tenant.

Ground 14 applies where the tenant or any other person residing in the dwelling-house has been guilty of conduct which is a nuisance or annoyance to adjoining occupiers, or has been convicted of using the dwelling-house or allowing its use for immoral or illegal purposes.

Ground 15 applies where the condition of any furniture provided under the tenancy has deteriorated owing to ill-treatment by the tenant or by any other person residing in the dwelling-house. In the case of ill-treatment by a lodger or sub-tenant, it must be shown that the tenant has not taken such steps as he ought reasonably to have taken in order to remove the lodger or sub-tenant.

Ground 16 applies where the dwelling-house was let to the tenant in consequence of his employment by the landlord (or a previous landlord under the tenancy) and the tenant is no longer in that employment.

4 *Possession under an assured shorthold tenancy.* To be an assured shorthold tenancy, a tenancy must be a fixed term assured tenancy, granted for a term of not less than six months, which the landlord has no power to determine within six months of its commencement (s.20). Further, before the assured tenancy was entered into, the landlord must have served on the tenant a notice which stated that the tenancy was to be a shorthold tenancy.

Where an assured shorthold tenancy comes to an end, a statutory periodic tenancy arises and any new assured tenancy of the same premises with the same landlord and tenant will be a shorthold tenancy, even though it does not satisfy these requirements, unless the landlord serves notice on the tenant before the tenancy is entered into that it is not to be a shorthold tenancy. Thus, a periodic tenancy may be an assured shorthold tenancy if it follows a fixed term assured shorthold tenancy.

A tenancy cannot be an assured shorthold tenancy if, immediately before the tenancy was granted, one of the tenants was tenant under a tenancy granted by the landlord which was an assured tenancy but was not a shorthold tenancy.

Where a tenancy is an assured shorthold tenancy, the Act gives the landlord a specific right to recover possession in addition to the grounds for possession which have already been examined. In relation to a fixed term assured shorthold tenancy, s.21 provides that the court must order possession when the tenancy comes to an end, even if a statutory periodic tenancy has come into existence, provided that the landlord has given the tenant at least two months' notice stating that he requires possession and no further assured tenancy (other than a statutory periodic tenancy) is in existence. Notice must be given to the tenant before or on the day when the tenancy comes to an end.

In relation to a periodic assured tenancy, s.21 provides that the court must order possession if the landlord has notified the tenant that after a specified date, possession of the dwelling-house is required under s.21. The date cannot be earlier than the earliest day on which the tenancy could have been terminated at common law by notice to quit given on the same day as the s.21 notice.

Rent control under the Housing Act 1988

Fixed term assured tenancies. The Act does not regulate the amount of rent which may be charged or the extent of rent increases under a fixed term tenancy.

Periodic assured tenancies. S.13 provides for rent increases under assured periodic tenancies (including statutory periodic tenancies). The minimum period between such increases is one year. The s.13 procedure is not applicable where the lease contains a rent review clause which is binding upon the tenant, and the section does not prevent the parties from reaching an agreement of their own. In practice, it appears that the s.13 procedure is not widely used.

Assured shorthold tenancies. S.22 provides that a tenant under an assured shorthold tenancy may apply to a rent assessment committee for determination of the rent which the landlord might reasonably be expected to obtain under the tenancy, provided that the requirements of the section are satisfied.

Succession by tenant's spouse to an assured periodic tenancy

Where the tenant under an assured periodic tenancy dies, the spouse or partner of the tenant can take over the tenancy under s.17, provided that the spouse or partner occupied the dwelling-house as his or her only or principal home immediately before the death of the tenant. This rule can only operate where the deceased tenant was not already a successor, e.g. by virtue of being the survivor of joint tenants, or by inheriting under the will or intestacy of a previous tenant.

Rent Act 1977

Basically, the Rent Act 1977 applies to private residential tenancies entered into before 15 January 1989, though certain tenancies created after that date do still fall within the provisions of the 1977 Act. In relation to tenancies created after that date, the appropriate legislation is normally the Housing Act 1988.

A tenancy to which the Rent Act applies is known as a *protected tenancy*. Like the Housing Act 1988, the Rent Act applies to tenancies but not to licences. The tenancy must be of a dwelling-house, which may be a house or part of a house, and thus can include a flat. The dwelling-house must be let as a separate dwelling, which must provide the tenant or joint tenants with exclusive possession of rooms in which the tenant can conduct the main activities of living, especially, cooking, eating and sleeping. Other rooms, such as bathrooms, may be shared (s.22).

The tenancy *must not* be one of the types which are incapable of being protected tenancies. The following are examples of types of tenancy which are incapable of being protected tenancies:

- Certain tenancies where the rent/rateable value exceeds specified limits (ss.4 and 25).
- A tenancy under which either no rent is payable or the rent payable falls below specified limits.
- A shared ownership lease granted under Part V of the Housing Act 1985 (s.5A), or a lease granted by a housing association which satisfies the requirements of s.5A(2).
- A dwelling-house let with land other than the site of the dwelling-house (s.6) − though where land is let with a dwelling-house and the primary purpose of the agreement is the letting of the dwelling-house, the Act may apply.
- A dwelling-house let at a rent which includes payments for genuine board and attendance (s.7).
- In relation to a student studying a course provided by a specified educational institution, a tenancy granted to the student by a specified body (s.8).
- A tenancy the purpose of which is to let the dwelling-house to the tenant for a holiday (s.9).
- A tenancy under which the dwelling-house is comprised in an agricultural holding (see Agricultural Holdings Act 1986) and is occupied by the person who controls the farming of the holding (s.10).
- A tenancy of premises with an 'on-licence' for the sale of intoxicating liquors (s.11).

- A tenancy of a dwelling-house which forms part of a building, where the landlord from the time the tenancy was granted has occupied another dwelling-house in the same building as his residence (s.12). The tenancy is not rendered incapable of being a protected tenancy by the presence of a resident landlord if the dwelling-house forms a flat in a purpose-built block of flats. (Note, the complex provisions of s.12, relating to resident landlords, fall outside the scope of this book.)
- A tenancy under which the interest of the landlord belongs to certain types of public body such as a county council, development corporation or housing action trust (s.14).
- Business tenancies which fall within Part II of the Landlord and Tenant Act 1954 (including premises with mixed business/residential use) (s.24(3)).

Security of tenure under the Rent Act

A protected tenancy comes to an end when the term expires (if the lease is for a fixed term) or when the tenancy is otherwise terminated at common law. Following the termination of the protected tenancy, s.2 provides that the former protected tenant becomes a statutory tenant of the dwelling-house provided that and for as long as the dwelling-house remains his residence.

A statutory tenancy is not an interest in land; rather, it is a statutory right to remain in possession of a dwelling-house. The terms of the statutory tenancy are those of the protected tenancy except to the extent to which they are inconsistent with provisions of the Act (s.3). Thus, terms of the protected tenancy entitling the landlord to recover possession of the dwelling-house will not become terms of the statutory tenancy.

1 *Termination of a statutory tenancy by act of the tenant.* A statutory tenancy will not arise, or will come to an end, if the dwelling-house ceases to be the tenant's residence (s.2). The fact that the tenant is frequently absent from the dwelling-house for lengthy periods does not necessarily mean that the dwelling-house has ceased to be his residence.

Similarly, where a tenant has two houses, the fact that he occupies one house more regularly than the other will not necessarily make the former house his residence for this purpose. In each case, the primary consideration in determining whether a dwelling-house is the tenant's residence is his intent.

A statutory tenant may terminate his statutory tenancy by

serving notice on the landlord (s.3). If the protected tenancy was a fixed term tenancy, the tenant must give the landlord at least three months' notice for this purpose. If the protected tenancy was a periodic tenancy, the length of notice required is either that which was required under the protected tenancy or four weeks, whichever is the greater.

A statutory tenant may surrender his statutory tenancy, expressly or impliedly (see page 321).

2   *Termination of a statutory tenancy by the landlord.* A landlord can only obtain possession of a dwelling-house which is subject to a statutory tenancy by obtaining a court order (s.98). In order to obtain a court order, the landlord must prove one or more of the discretionary or mandatory grounds for possession which are laid down by the Act. If the tenant can prove a discretionary ground, the court may then order possession if it considers it reasonable to do so. The court possesses discretion to stay or suspend the execution of the order or postpone the date of possession for such period as it thinks fit. On proof of one or more of the mandatory grounds, the court must order possession.

The landlord also needs a court order in order to obtain possession of a dwelling-house which is subject to a protected tenancy. Even though the landlord can establish grounds for possession under the Act, the court will not grant such an order in respect of a protected tenancy unless the landlord is entitled to possession of the dwelling-house at common law.

Discretionary grounds for possession (Rent Act 1977)

S.98(1) (a) applies where suitable alternative accommodation is available for the tenant. The phrase 'suitable alternative accommodation' is extensively defined in Part IV of Schedule 15 to the Act in terms similar to those already examined in the context of the Housing Act 1988 at page 339.

Schedule 15, Part I of the Rent Act 1977 sets out the following discretionary grounds for a landlord obtaining possession.

Case 1 applies where the tenant has failed to pay rent due or has broken or failed to perform an obligation of the tenancy.

Case 2 applies where the tenant has been guilty of conduct which is a nuisance or annoyance to adjoining occupiers, or has been convicted of using the dwelling-house or allowing its use for immoral or illegal purposes.

Case 3 applies where the condition of the dwelling-house or common parts has deteriorated due to acts of waste by, or the neglect or default of, the tenant, his lodger or a sub-tenant.

Case 4 applies where the condition of any furniture provided under the tenancy has deteriorated owing to ill-treatment by the tenant, his lodger or a sub-tenant.

Case 5 applies where the tenant has given his landlord notice to quit and, consequently, the landlord has contracted to sell or let the dwelling-house.

Case 6 applies where the tenant has, without the landlord's consent, either assigned or sublet the whole of the dwelling-house.

Case 7 was repealed by the Housing Act 1980.

Case 8 applies where the dwelling-house was let to the tenant in consequence of his employment by the landlord or a former landlord and the tenant is no longer in that employment, provided that the dwelling-house is reasonably required by the landlord for a person engaged in full-time employment by him or by one of his tenants.

Case 9 applies where the dwelling-house is reasonably required by the landlord as a residence for himself, for a child of his over 18 years old, for his father or mother or for his father-in-law or mother-in-law. This case does not apply if the landlord became landlord by purchasing the dwelling-house or an interest in it after 23 March 1965.

Part III of Schedule 15 provides that the court shall not make an order under case 9 if greater hardship would be caused by granting the order than by refusing to grant it.

Case 10 applies where the tenant has sublet part of the dwelling-house under a protected or statutory tenancy at a rent in excess of the maximum rent recoverable under Part III of the Rent Act 1977.

Mandatory grounds for possession under Part II of Schedule 15, Rent Act 1977

In order to obtain possession under cases 11–15 and 20, the landlord must have notified the tenant not later than the beginning of the tenancy that possession might need to be recovered under the appropriate case (though, in relation to cases 11, 12 and 20, the court may dispense with these requirements where it is just and equitable to do so).

Case 11 applies where the owner, prior to letting the dwelling-house, occupied it as his residence and:

(a) the house is required as a residence by him or by a member of his family who lived with him when he last occupied the dwelling-house; or
(b) he has died and the house is required as a residence for a member of his family who was living with him at the time; or

(c) he has died and the house is required by a successor in title (but not a purchaser) either as his residence or in order to dispose of it with vacant possession; or
(d) the house is subject to a mortgage granted before the tenancy and the mortgagee requires vacant possession in order to exercise a power of sale; or
(e) the house is not reasonably suited to his needs, having regard to his place of work, and he requires it in order to dispose of it with vacant possession, the proceeds to be used to acquire a house as his residence which is better suited to his needs.

Case 12 applies where the owner intends to occupy the dwelling-house as his residence when he retires from regular employment and:

(a) he has retired and requires the house as his residence; or
(b) he has died and the house is required as a residence for a member of his family who was living with him at the time; or
(c) he has died and the house is required by a successor in title (but not a purchaser) either as his residence or in order to dispose of it with vacant possession; or
(d) the house is subject to a mortgage granted before the tenancy and the mortgagee requires vacant possession in order to exercise a power of sale.

Case 13 applies to fixed term tenancies for a term not exceeding eight months if at some time within the twelve months prior to commencement of the tenancy the dwelling-house was let as a holiday home.

Case 14 applies to fixed term tenancies for a term not exceeding twelve months if at some time within the twelve months prior to commencement of the tenancy the dwelling-house was let to a person pursuing or intending to pursue a course of study provided by a specified educational institution and the tenancy was granted by a specified body.

Case 15 applies where a dwelling-house is held for the purpose of being available for occupation by a minister of religion as a residence from which to perform his duties and the court is satisfied that the dwelling-house is required for this purpose.

Cases 16, 17 and 18 relate to premises once occupied by agricultural workers which are required for agricultural employees, provided that notice was served prior to commencement of the tenancy that possession might be required under the relevant case, and that the other requirements of the relevant case are satisfied.

Case 19 may be applicable where a dwelling-house was let under a protected shorthold tenancy (see consideration of the similar provisions of the Housing Act 1988 at page 341).

Case 20 applies where a dwelling-house was let after 28 November 1980 by a person who was a member of the regular armed forces when he acquired it, provided that he can satisfy requirements similar to those in case 12.

Succession to a statutory tenancy (S.2 and Part I of Schedule 1, Rent Act 1977 as amended by s.39 and Schedule 4, Housing Act 1988)

Where a statutory tenant who was the original tenant under the protected tenancy dies, the tenant's spouse (including a common law spouse) succeeds to the statutory tenancy provided that the spouse resided in the dwelling-house immediately prior to the tenant's death. If there is no resident spouse, then a family member who lived with the tenant in the dwelling-house immediately before the tenant's death and for the two years prior to it succeeds to an assured periodic tenancy under the Housing Act 1988.

If a surviving spouse has succeeded to a statutory tenancy, then on the death of that person, any member of his or her family who was also a member of the original tenant's family may be able to succeed to an assured periodic tenancy.

Premiums (Rent Act 1977, Part IX)

It is an offence to require a premium in addition to rent in relation to the grant of a protected tenancy. Any part of a contract requiring the payment of such a premium is void.

Where an excessive price is charged for the purchase of furniture, and the grant of the tenancy is conditional upon such payment, the excess part of the payment above that which is reasonable is treated as a premium.

A requirement that rent be paid prior to commencement of the period to which it relates is void, and it is an offence to require or receive such payments.

Taking a deposit is allowed provided that it does not exceed one sixth of the annual rent and is reasonable.

Protected shorthold tenancies (Housing Act 1980 ss.51–55 and Housing Act 1988 s.34 and Schedule 18)

A tenancy may only be a protected shorthold tenancy if it is a fixed term tenancy for a term of between one and five years which cannot be brought to an end by the landlord before expiration of the term other than in consequence of breach by the tenant of his obligations under the lease. Further, prior to commencement of the tenancy, the landlord must have notified the tenant that the tenancy was to be a protected shorthold tenancy. If a tenant was a protected or statutory tenant of a dwelling-house immediately prior to the grant of a new protected tenancy, the new tenancy cannot be a protected shorthold tenancy.

Where a shorthold tenancy is created after 15 January 1989 it will be an assured shorthold tenancy under the Housing Act 1988 unless granted under a contract made before that date.

Rent control (Rent Act 1977, Part IV)

Either or both parties to a protected tenancy may apply to the rent officer to have a fair rent registered for two years. Usually, no further application can be made within the two-year period. This registered rent is the maximum which the landlord can recover though if it exceeds the current rent no increase in rent is authorized unless permitted by the terms of the tenancy. If no such application is made, the rent payable is that which is authorized by the lease or agreed by the parties.

Similarly, either or both parties to a statutory tenancy may apply to the rent officer to have a fair rent registered for two years. This is the maximum which the landlord can recover though if it exceeds the current rent no increase in rent is authorized unless permitted by the terms of the tenancy. If no such application is made, the maximum rent payable is that which was payable under the last rental period of the protected tenancy (or, if a fair rent was registered under the protected tenancy, that which was so registered).

In determining a fair rent, regard is to be had to all relevant non-personal circumstances, such as: the age, character, locality and state of repair of the dwelling-house; the quantity, quality and condition of furniture; and any premium which has lawfully been required or received on the grant or renewal of the tenancy.

In determining a fair rent, certain matters must be disregarded, namely: scarcity in the neighbourhood; disrepair or defects attributable to the failure of the tenant or his predecessor in title to comply with his obligations under the tenancy; improvements carried out by the tenant or his predecessors in title other than under the terms of the tenancy; and improvement of or deterioration of furniture due to the conduct of the tenant or a person residing or lodging with him or his sub-tenant.

In practice, the best method of assessment is normally comparison with registered fair rents of comparable dwelling-houses.

Long leases of residential premises: the Leasehold Reform Act 1967

The Act applies to tenancies which satisfy the following conditions, laid down by s.1.

1   The tenancy must be a long tenancy. This basically means that the tenancy must be for a fixed term exceeding 21 years (s.3).
2   The tenancy must be at a low rent. This means that the rent must fall within limits laid down by the Act.
3   The tenancy must be a tenancy of a house (but does not include a flat created by horizontal division of a building).
4   The rateable value of the house must fall within specified limits (if the tenancy was granted after 1 April 1990 a statutory formula is applicable). (Note, this requirement only applies where the tenant wants an extended lease, not where he wishes to purchase the freehold, Housing and Urban Development Bill 1993.)
5   The tenant must satisfy the residence requirement. This basically means that when the tenant makes a claim under the Act, he must have occupied the house as his main residence for the preceding three years. Neither a company nor a tenant of an agricultural holding may be in occupation for the purposes of the Act.

Where the requirements of the Act are satisfied, the tenant may claim the freehold of the house (s.8) or a 50-year extension to the term of the lease (s.14) by serving notice on his landlord. A tenant may, however, lose his right to make a claim in a number of ways, for example by forfeiture, or by failing to make a claim within two months of receiving notice to terminate the tenancy from the landlord, or by giving the landlord notice to terminate the tenancy or being granted a new tenancy under Part II of the Landlord and Tenant Act 1954 (business tenancies).

The landlord may apply to the court for possession. The court may grant possession if the landlord can prove statutory grounds, for example, that he or one of his family wishes to occupy the house.

Where the freehold is claimed, the purchase price will be calculated upon the basis of factors laid down by the Act (s.9). If the tenant decides not to go ahead with the purchase when the price is calculated, he loses his right to make another claim for three years.

Where an extension to the term is claimed, the rent will be calculated upon the basis of factors laid down by the Act (s.14). The other terms of the extended lease will basically be those of the original lease, subject to modifications relating to options and rights of termination (s.15).

A tenant who claims an extended tenancy under the Act is not subsequently entitled to claim a further extension. He may purchase the freehold before the date when his original tenancy would have ended. After that date, he has no rights to enfranchisement under the Act. (Note: where the 1967 Act is not applicable to a long tenancy, or the tenant does not claim the freehold or an extended tenancy under it, he may be entitled to security of tenure under Part I of the Landlord and Tenant Act 1954, or Sched. 10 of the Local Government and Housing Act 1989 which gives the tenant an assured tenancy.)

Residential tenancies of flats: Landlord and Tenant Act 1987

Tenants' right to purchase interest of landlord where landlord is in breach of repairing covenants

Part III of the Act applies to qualifying residential tenants of flats under leases for terms exceeding 21 years. Its effect is that a majority of the qualifying tenants are entitled to compulsorily purchase the landlord's interest in premises to which Part III applies, if he fails to remedy the breach of his repairing or management covenants within a reasonable period after receiving notice from the tenants and is unlikely to do so, or a manager has been in place for three years (s.29). The landlord's interest is transferred to a nominated person on terms either agreed by the parties or determined by a rent assessment committee.

Tenants' right to purchase interest of which landlord intends to dispose

Part I of the Act applies to qualifying residential tenants of flats. Its effect is that if the landlord decides to dispose of an interest in premises to which Part I applies, qualifying tenants are given the right to purchase the interest. Part I applies to disposals such as the sale or lease of the landlord's freehold reversion or, if he is a tenant, the assignment of his lease. It does not apply to disposals such as disposals by mortgage, or compulsory purchase, or family or charitable gifts, or where required by an option or under a will (s.4).

Where the landlord proposes to dispose of an interest in the premises, he must notify the qualifying tenants of his intention, offering the interest to them (s.5). The tenants must be given at least two months to accept the offer (s.6) or serve a counter-notice on the landlord making a counter-offer (s.7), either of which requires a majority of the qualifying tenants.

If the tenants accept the offer, or the landlord accepts their counter-offer, they must be given another two months to nominate a person to take the landlord's interest. If they fail to nominate such a person within the two-month period, the landlord has 12 months to dispose of the premises for consideration not less than that specified in the notice.

If they do not accept the offer within the specified period, or the landlord rejects their counter-offer, the landlord has 12 months to dispose of the premises, though the consideration must not be less than that specified in the notice. Alternatively, in response to a counter-offer, the landlord may make a further offer.

Tenants' right to apply to the court for variation of the terms of the lease

Part VI of the Act permits a residential tenant (or residential tenants) of a flat (or flats) under a lease (or leases) for a term exceeding 21 years to apply to the court for variation of the terms of the lease or leases. The court may order variation if grounds are established (s.35). Basically, variation may be ordered where the terms of the lease are inadequate with regard to matters such as repair and maintenance of the building, of the flat or of installations, provision of insurance or computation of service charges.

(Note: the Housing and Urban Development Bill 1993 introduces two new rights for tenants of flats with long leases: first, a collective right to purchase the freehold; secondly, where this is not available, an individual right to an extended lease.)

Council and other public sector tenants: Housing Act 1985

The Housing Act 1985 applies to a tenancy of a house let as a separate dwelling (s.79), provided that the tenant occupies the dwelling-house as his only or principal home (s.81) and the landlord is a local authority or other specified public body (s.80). It also applies to a licence to occupy a dwelling-house (s.79), other than a temporary licence given to an occupier who entered the premises as a trespasser.

Tenancies to which the Act applies are known as *secure tenancies*. The Act does not apply to tenancies for terms exceeding 21 years, nor where the tenant occupies the premises for the purposes of his employment where the local authority (or another specified public body) is his employer, nor where the house is used for temporary accommodation, because it is on land acquired for the purposes of development. The Act is also inapplicable where the house is let to the tenant as temporary accommodation whilst work is carried out on his home if the tenant is not a secure tenant of his home. There are also exclusions from the protection in respect of tenants of agricultural holdings and on-licensed premises, and in the case of certain student lettings, and in relation to the provision of housing for homeless persons and persons moving into the district in order to take up employment. The Act is also inapplicable where Part II of the Landlord and Tenant Act 1954 applies to the tenancy.

A tenancy ceases to be secure if the tenant gives up possession of the house or sub-lets the entire house (s.93) or (in the case of a fixed term tenant) if he dies (s.90). Secure tenancies cannot normally be assigned (s.91).

Succession

When a secure periodic tenant dies, his spouse or a member of the tenant's family who had resided with the tenant for the 12 months before his death and had occupied the house as his principal home, may succeed to the tenancy (s.89). Only one succession is allowed.

Possession

A secure fixed term tenancy may end by expiration of the term or by court order in certain circumstances (s.86). Once the fixed term ends (even if by forfeiture), a periodic tenancy will automatically arise (s.86) unless the tenant has been granted another tenancy of the dwelling-house which commences as soon as the original tenancy ends. The period of the periodic tenancy will be determined by reference to how the rent was paid under the fixed term tenancy. The terms will be a modified form of those of the original tenancy.

A secure periodic tenancy may only be terminated by court order. The court can only make an order if grounds for possession are established by the landlord (s.84). The grounds relied upon must be contained in a notice which the landlord serves on the tenant prior to bringing proceedings. The date specified by the notice for commencement of possession proceedings cannot be earlier than the first possible date of common law termination. The grounds are contained in Schedule 2.

Grounds 1–8. On grounds 1–8, the court will only make an order for possession if it is reasonable to do so, the court possessing the power to adjourn proceedings, postpone or suspend an order, or impose conditions in relation to rent. The grounds are as follows:

1. The tenant has failed to pay rent or is in breach of covenant.
2. The tenant or persons residing with him have caused a nuisance or annoyance to neighbours or have been convicted of using the house for illegal or immoral purposes.
3. The house has deteriorated in consequence of acts of waste by the tenant or persons residing with him. If acts of a lodger or sub-tenant, the tenant must have failed to take reasonable steps to remove him.
4. Furniture provided by the landlord has deteriorated in consequence of ill-treatment by the tenant or persons residing with him. If acts of a lodger or sub-tenant, the tenant must have failed to take reasonable steps to remove him.
5. The tenant was granted the tenancy in consequence of a false statement made by the tenant intentionally or recklessly.
6. The tenant acquired the tenancy by exchange and either paid or received a premium.
7. The building which contains the house is mainly used for non-residential purposes, the tenancy was granted because the tenant is employed by the landlord and the nature of the tenant's behaviour or that of a person who resides with him has been such that, taking into consideration the use of the rest of the building, it is not proper for him to remain in occupation.
8. The tenant was a secure tenant of a different house and accepted the present house on the basis that he would leave when the work on his original house was complete, that work now being complete.

Grounds 9–11. On grounds 9–11, suitable alternative accommodation must be available at the time when the order is to take effect. The grounds are as follows:

9. The occupier is liable for an offence in respect of the overcrowding of the house.
10. The landlord intends to demolish, reconstruct or carry work out on the building and cannot reasonably do so without possession.
11. The landlord is a charity and the tenant's continued occupation of the house would conflict with its objectives.

Grounds 12–16. On grounds 12–16, the court will only make an order for possession if it is reasonable to do so. The court has the power to adjourn proceedings, postpone or suspend the order, or impose conditions in relation to rent. Further, suitable alternative accommodation must be available at the time when the order is to take effect. The grounds are as follows:

12 The building which contains the house is mainly used for non-residential purposes, the tenancy was granted because the tenant is employed by the landlord, the employment has ended and the landlord needs the house for a new employee.
13 The house is adapted for the disabled, no disabled person is living there and the landlord needs it for a disabled person.
14 The landlord is a housing association or trust which only lets houses to people who are difficult to house, no such person is living in the house and the landlord needs it for such a person.
15 The house is part of a group for people with special needs, there being special facilities in the area for such persons, no such person is living in the house and the landlord needs it for such a person.
16 The house is more extensive than the tenant needs, the tenant succeeded to the tenancy, not being the deceased tenant's spouse, and notice was served on the tenant between 6 and 12 months after the former tenant's death.

Right to buy

A person who has been a secure tenant (not necessarily of the same house or under the same landlord) for two years or more at any time may be entitled to purchase his house or take a long lease (125 years) of his flat (s.119). A tenant cannot claim this right if an order for possession of the house or flat has been made (s.121). Certain dwellings (such as dwellings particularly suitable for the elderly or disabled) are excluded (Schedule 5). The price is calculated taking account of factors specified by the Act (s.127), subject to a discount based upon the length of the period during which the tenant has been a secure tenant. The tenant also has the right to a mortgage from his landlord (s.132).

13

Planning law

Central and local planning responsibilities

If people are to enjoy and maintain a reasonable quality of life in the place where they live, then issues relating to land use planning should obviously concern them. The very word 'planning' suggests change, and the policies developed should aim to maintain and improve the environment as an attractive place in which to live and work.

Any coherent system of planning control is necessarily dependent on a proper administrative structure. The local government structure was established by the 1890s and the first Planning Act followed in 1909. The emphasis in that early Act was directed towards planning being used as a tool to secure better sanitary conditions but it introduced for the first time the important issues of 'amenity and convenience'. Early planning Acts were based on the idea of local councils preparing planning schemes to control land use in suburban areas. General control of land use was introduced in 1947, creating the modern system of planning law. The main principles are now contained in the Town and Country Planning Act 1990.

The overall structure for administering the rules on planning is headed by the Secretary of State for the Environment who has power over local planning authorities – county councils, district councils, metropolitan authorities and other bodies vested with planning responsibilities. County councils must draw up a 'structure plan', which is a policy statement concerning land use and the environmental implications for their area.

Structure plans deal with strategic issues relating to matters such as population and its distribution, industrial and business location, and transport and communications. Once a structure plan is in place for an area it is then necessary for the district councils to prepare their local plans.

A local plan will be much more detailed than the structure plan and may relate to one small town or a small geographically identifiable area. It will contain proposals for development and land use, and measures for controlling traffic and improving the environment. It will seek to provide for acceptable levels of change, and will set down allocations of land for particular types of development. In terms of timescale, a local plan is probably geared to a ten-year span. Inevitably, any place or area cannot function in isolation from its surrounding area, and the cumulative effect of decisions taken by private individuals, local and national government, businesses and recreational organizations will all affect how a town, village or area develops.

Take the example of a small country town, supporting a local auction mart with a regular market day, where the town centre is congested and full of buildings of historic significance. A new by-pass opens and the town's one major industrial employer is booming and seeking to expand. It is obvious in such a case that a local plan will need to give thought to issues such as the availability of land for industrial purposes, provision of housing for key workers, expansion of parking facilities within the town and the growth of any tourist industry with attendant demands for accommodation and catering, while still adequately catering for long-established businesses like the auction mart.

Local plans should aim:

- To apply the strategic objectives of the structure plan and relate them to precise areas of land.
- To provide a well-reasoned framework for development control.
- To provide a detailed basis for coordinating development by various agencies.
- To bring detailed planning issues before the public.
- To ensure that planning decisions can be taken in full knowledge of local needs and possibilities.
- To focus opportunities for public and private investment.
- To examine particular social, economic and land use problems and to propose solutions.

The local plan will be especially useful when applications for planning permission are being dealt with.

In metropolitan areas, the two-tier county/district hierarchy does not exist. Rather, all planning activities are dealt with by a single authority. Thus, in such areas, structure and local plans are being replaced by unitary plans.

Development

The key concept in planning control is development. It is defined by the 1990 Act, s. 55 as: 'The carrying out of building, engineering, mining or other operations in, on, over or under the land, or the making of any material change in the use of any building or other land'. If an activity comes within this definition, then as a broad general rule, planning permission is required. The usual way to seek planning permission is to apply to the local planning authority. However, there are instances where no individual planning permission is necessary. This may be because:

1 The operation planned is not within the meaning of the word 'development'.
2 The development is permitted within the terms of the General Development Order, so that no specific application need be made.

The definition of the term 'development' has two aspects – 'operations' and 'uses'. An operation results in changes to the physical form of the land which will have some degree of permanence. Use consists of the activities done in, or on, the land which do not change its physical form. Inevitably, the definition of development has given rise to much dispute and case law. This sometimes happens because a person goes ahead and does something which the local planning authority finds out about, and for which it considers that planning permission should have been obtained, e.g. the erection of a building. The planning authority may then serve an enforcement notice, ordering the building to be removed. The costs involved will usually spur the person who has erected the building to appeal against the enforcement notice to the Secretary of State.

The s.55 definition of development is further extended in the Act itself. 'Building' is defined to include any structure or erection, but does not include the plant and machinery within the building. 'Building operations' are defined to include: 'Rebuilding operations, structural alterations of, or additions to, buildings, and other operations normally undertaken by a person carrying on business as a builder'. This at first sight seems to be virtually all-embracing, but s.55 of the Act does say that development does *not* include: 'The carrying out of work for the maintenance, improvement or other alteration of any building, being works which affect only the interior of the building or which do not materially affect the external appearance of the building.'

For many years, it was not altogether clear where demolition work stood within the s.55 definition of development. The position has now been resolved by the Planning and Compensation Act 1991, which provides that demolition of a building is a 'building operation' and will normally amount to development. As such, it will usually be permitted under the General Development Order.

Change of use

Only those changes of use which are 'material' require planning permission. Inevitably, what is material is a question of degree, but the Act aims to give some assistance by laying down certain specific instances of material change of use. These are:

1 Where a single dwelling house is converted into two or more separate dwellings.
2 Where there is to be a display of advertisements on any external part of a building not previously used for that purpose.
3 Where refuse or waste materials are to be deposited, including extending an existing tip.

If the change of use proposed is within the scope of what is permitted by the Use Classes Order, then there is no requirement to seek planning permission. The latest order is the Town and Country Planning (Use Classes) Order 1987. Where a person has obtained permission for one use and implemented that permission, he can then change the use of the land and buildings to any of the similar uses provided for by the order. So, for example, if permission was initially obtained to use premises as an art gallery, and that use was implemented, then those same premises can subsequently be used as a museum, public library, reading room, public hall, exhibition hall, day centre, educational institution or creche without the need to seek further permission. Previous permission to use premises as a casino would allow change of use to a cinema, concert hall, bingo hall, dance hall, swimming bath, skating rink or gymnasium, without further permission.

Of course, if the use for which permission was originally granted is not listed in the Use Classes Order, then its rules cannot assist and a normal planning application will be necessary. Again planning permission for the original use may have been subject to conditions which over-ride the provisions of the Use Classes Order. The order may permit a change of *use* without further permission but if the change of use requires new buildings, or alterations or extension to existing buildings, an individual planning permission is still likely to be required.

Permission under the General Development Order 1988

Instead of seeking permission for development by applying to the local planning authority, there are cases, covered by the General Development Order made by the Secretary of State, where permission is automatically given for all classes of development set out in the order. Some commonly encountered examples of development which are permitted by the order include:

1 Development within the curtilage of a dwelling house. Where it is proposed to extend a dwelling house, then provided that the extension does not increase the overall size of the house by more than 15 per cent (10 per cent in the case of a terraced house) calculated by reference to cubic capacity, such development is permitted. There are other conditions to be satisfied, e.g. that the height of the building when extended must not exceed the height of the original dwelling house.
2 Construction of porches and garages, subject to size, location and height limitations.
3 Minor operations, e.g. gates, fences, walls, hard-standing for cars, installation of oil storage tanks.
4 Temporary buildings. Where building operations for which planning permission has been granted are in progress, this rule allows temporary buildings needed in connection with the building work such as site huts or workmen's lavatories to be erected. But they must be removed at the end of the building operations.
5 Temporary uses. Land can be used for most temporary purposes for up to twenty-eight days each year. This rule allows land to be used for an occasional agricultural show, or point-to-point, or as a site for a summer fete. The permission also covers erection of movable structures (e.g. marquees) for these purposes. If the purpose is to hold a market, or motor racing or speed trials, or clay-pigeon shooting, temporary use is limited to fourteen days each year.
6 Building or engineering operations required for agricultural purposes may be carried out on agricultural land, e.g. erection of barns and steadings. This permission is restricted in terms of size and height limitations, and distances from highways.
7 Additions to industrial buildings, limited in size and height.

It should be noted that permitted development under the General

Development Order can only operate once. Other extensions to a property which have already occurred, even if validly covered by planning permission, may reduce or remove permitted development rights. Even if work looks as though it is covered by the General Development Order, the local planning authority may have made an 'Article 4 Direction'. Such directions are made in the interests of good planning in a particular area. The local planning authority will impose restrictions on development in that area, so that an application for permission must be made, even though it appears that permission already exists under the Order. There is likely to be an 'Article 4 Direction' if the place is a conservation area, or if a particular street has a special character or appearance. Landowners in such areas should have been served with a copy of the direction.

A further situation in which planning consent may be automatically granted is where a 'simplified planning zone' is established by a local planning authority. This is a new idea created by the Housing and Planning Act 1986 and as yet, it is impossible to judge what its impact will be. Supplementary provisions are contained in the Planning and Compensation Act 1991. Within such zones, designated by a local planning authority only after extensive consultative processes, certain developments or classes of development will be permitted without the need for formal application. These simplified planning zones are likely to have particular significance in run-down industrial areas, but it could be argued that the greatest incentive for development in such areas is the type of financial inducement which was previously available in enterprise zones. The mere relaxation of planning controls is not, in itself, likely to act as a significant inducement. The fact that no formal application need be made, and that no fee is payable is likely to be an insignificant detail to large developers.

It must be borne in mind that planning permission can do nothing to help in those cases where the type of development is prevented by a restrictive covenant, or where a tenant is precluded from carrying out the development in question by the express terms of his lease.

Making a planning application

One immediate advantage of development which is permitted under the General Development Order is that it costs the developer nothing to get his permission. Since 1980, fees have been payable in respect of planning applications, and the level of these can be varied from time to time. The application will not be considered until the fee is paid. In the longer term, the Government intends to increase the fees to cover all of a local authority's development control costs.

Applications are made to the local planning authority, and may be made by anyone, whether or not he owns the land or any interest in it. The actual owner does not need to give his consent, but the applicant has to give notice of his application to anyone with a material interest in the land. The necessary forms are supplied by the local planning authority, and it may be sensible to appoint a surveyor or architect to make the application. A full application will require plans so that the site can be adequately indentified and the development proposals fully understood. Enagaging professional assistance may be costly, and at an early stage a developer may simply want to discover the attitude of the planning authority in principle to what is proposed. In such a case, an outline planning application is appropriate. A developer may still have to acquire the land and commission detailed plans, but he can then go ahead with more confidence in his business venture, armed with his outline permission. This is not usually the best way for private individuals to proceed, as the outline permission, if granted, must be followed by a second application for 'approval of reserved matters'.

Once an application has been submitted, with the correct number of copies and the appropriate fee, one copy will be placed in the planning register, which is a public document available for inspection by anyone.

All planning applications must be publicized, usually by a site notice or a notification to neighbours. In some instances, a newspaper advertisement and a site notice may be required, for example, where an 'environmental impact statement' accompanies the application, or where there is a 'major' development such as mineral extraction. 'Bad neighbour' development is no longer given separate treatment.

The planning authority may need to consult with the highway authority or the housing authority or a parish council or the Ministry of Agriculture or the Department of the Environment. This latter consultation may be required if the development proposed would involve some radical departure from the local plan. The planning authority will have to refer the application direct to the Secretary of State if he has used his power to 'call in' the application. This is likely to happen only in the case of very large-scale controversial development proposals.

The 1991 Act introduced the concept of environmental assessment into the planning process, to implement a European Community Directive. Circular 15/88 states that 'Formal environmental assessment is essentially a technique for drawing together, in a systematic way, expert quantitative analysis and qualitative assessment of a project's environmental effects, and presenting the results in a way which enables the importance of the predicted effects...to be properly evaluated by the decision making body before a decision is given'.

An environmental assessment is required in the case of developments such as oil refineries, or thermal or nuclear power stations. An assessment may be required in relation to developments for industries such as rubber, paper and textiles, if the development is likely to have significant environmental effects because of its nature, size or location.

The planning authority must now make a decision on the application before it. It will take account of structure and local plans, the views of any organizations it has consulted, objections received, and any other material considerations, which may in appropriate cases include the need within the area for development of this type. The authority should keep in mind the central government guidance, which is that a positive approach to planning applications should be adopted, and that permission should only be refused in order to protect the public interest. It has been said that planning laws are not concerned with the protection of the private interests of one individual against another.

A decision by the planning authority should be taken within eight weeks or within sixteen weeks if an environmental impact statement is required. That decision may be:

- Unconditional permission or approval.
- Conditional permission.
- Refusal of permission.

In the vast majority of cases permission is conditional. The conditions imposed can be whatever the planning authority 'thinks fit', so long as the authority acts reasonably and the conditions themselves are within the general spirit and objects of planning law. For instance, permission may be given only for a limited period, or subject to work being commenced within a particular period. Conditions may relate to the materials to be used, or the location of a building on the site.

If planning permission is refused, or the applicant is aggrieved by the conditions imposed, he may appeal to the Secretary of State. Before considering an appeal it may be worthwhile negotiating with the planning authority. Its objections may possibly be overcome by design changes or the resiting of the proposed development. In such cases the renewed planning application will be exempt from a fee, as long as it is made within twelve months.

Where the planning authority fails to make a decision within the eight or sixteen week period (as appropriate), the applicant is entitled to treat this as a refusal, and appeal to the Secretary of State.

The planning authority may refuse to determine an application, if a similar application has been refused by the Secretary of State within the previous two years (Planning and Compensation Act 1991).

Appeals

An appeal must be lodged within six months of receiving the planning authority's decision. It may be an appeal against refusal of permission or an appeal against the imposition of conditions on the permission. There are three methods for dealing with appeals – written representations or public inquiry:

1 *Written representations* – this procedure can be used where both parties consent. It avoids the expense of an inquiry. Each side must provide the other with a full written statement of its case. Each then has the right to comment on the other's case. All the written material is put before an inspector, who may decide to make a site visit. This written representations procedure is used in the majority of cases.
2 *Hearings* – these are in private, and used only in a few exceptional situations.
3 *Public inquiry* – this will be conducted by an Inspector appointed by the Secretary of State. The Inspector makes the decision in the case, unless it is one of a limited class of appeals where only the Secretary of State may give the decision. The Inspector hears the appeal as if it had come to him at first instance. This can result in a case where permission has been granted subject to conditions now being refused permission altogether. There is now a procedure under which only the conditions are subjected to review, thus avoiding jeopardy to the whole permission. Inquiries tend to be formal and legalistic. The Inspector is required to observe the rules of natural justice. He must conduct the inquiry fairly, giving both sides an adequate chance to present their case. The proceedings will be held in public, and they give an opportunity to persons closely affected by the development to put their objections. Once the Inspector (or the Secretary of State in appropriate cases) has made his decision, he must communicate it, together with the reasons for the decision. His decision can only be challenged by applying to the High Court on a point of law to have the decision quashed.

Planning enforcement

The planning rules need to provide for those cases where development is carried without without valid permission, or where a developer breaches the conditions imposed on the permission. Breaches of the planning rules can come to light in a number of ways. Some are reported, often anonymously, by neighbours or persons affected by the development. Others become obvious when a subsequent permission is applied for. The planning officers are regularly out and about making site inspections and they tend to know what is going on and what has been authorized within their 'patch'. They liaise with the building inspectors, who have the opportunity to observe 'unauthorized' development in the course of their work. Breach of the rules is not a criminal offence as such. The planning authority uses an enforcement notice where some action is necessary.

It must be emphasized that where possible, the planning authority will seek to resolve matters without the need to resort to enforcement procedures. For example, the development may be of a type which would clearly have been permitted, if permission had been sought at the appropriate time. In such a case, the developer is encouraged to make a late application, which will be heard in the ordinary way by the planning authority. Steps short of enforcement procedures were introduced by the Planning and Compensation Act 1991. Two new kinds of notice, a planning contravention order and a breach of condition notice, assist planning officers to seek information and ensure compliance. Only as a last resort, when persuasion has failed, will an enforcement notice be issued. The notice must specify the alleged breach of the planning laws, and the steps which must now be taken to remedy the breach, together with the time limit for taking the necessary steps. The measures to be taken can vary considerably. For example, where a building has been erected without permission, at one extreme the step may consist of ordering it to be demolished, or at the other, may simply require some tree planting or landscaping to disguise the building.

A developer may appeal to the Secretary of State against an enforcement notice on a number of grounds, e.g. the matters specified in the notice do not constitute a breach of planning laws, or the time specified to take the necessary steps is unreasonably short. The appeal has the effect of suspending the enforcement notice until the appeal has been resolved. This means that the operation or change of use can go on as before until the appeal is resolved. That could be disastrous where, by allowing the operation to continue, it would be impossible to subsequently correct matters. In such circumstances it may be appropriate to issue a stop notice. Failure to comply with a stop notice is a criminal offence. If the enforcement notice is quashed on appeal, the local planning authority must compensate an owner or occupier of land for losses occasioned by the stop notice. This can include a contractor who has been held up on a site. In addition to the stop notice, the new rules in the Planning and Compensation Act 1991 provide for a local planning authority to apply for an injunction to restrain any actual or apprehended breach of planning control.

If an enforcement notice is upheld on appeal, or if no appeal is made against it, then it must be complied with within the prescribed time limits. Non-compliance with an enforcement notice is a criminal offence, punishable by a fine. Moreover, in cases of non-compliance, the local planning authority may enter the land and carry out the necessary steps at the expense of the owner.

Planning and the preservation of amenity

A vital part of the planning process is the preservation and enhancement of amenity. That term is not defined by the planning legislation, but the dictionary defines it as 'pleasantness, as in situation and characteristics'. The ordinary planning rules can do much to preserve amenity, by refusing permission in appropriate cases, or attaching conditions to preserve or improve amenity in others. Additionally, certain specific forms of control are particularly geared to protect amenity.

Trees

Local planning authorities may make tree preservation orders in the interests of amenity. These may be designed to protect individual trees or areas of woodland. In such cases, once an order is made, permission must be sought to fell or lop a tree. If such permission is granted, the local planning authority will often require replacement planting. Ordinary planning permission may be granted subject to the preservation of trees on the site, or subject to the planting of additional trees.

Interference with protected trees is a criminal offence, punishable by a fine.

Buildings of special interest and conservation areas

Buildings of special interest are usually refered to as 'listed buildings'. They become listed by the Secretary of State after consultation with experts if they are of special architectural or historic interest. The buildings listed are often proposed to the Secretary of State by the local planning authority. The owner of the building cannot object to its listing. He may know nothing of a decision to list the building until he is notified by the local planning authority. The listing will also be registered in the local land charges register. The local planning authority may give six months protection to a building under a 'building preservation notice' while they initiate the procedures for listing with the Secretary of State.

If a building is listed, any work to alter or demolish it must have 'listed building consent'. If such consent is refused, the owner can then appeal and argue that the building should never have been listed. This argument is significant because the owner had no right initially to object to the listing. Many owners do not welcome listing of their property. Not only is the planning control much more rigorous, but it may also cause considerable loss if it was hoped to sell the building to realize its development value. This was the problem facing the parties in the case of *Amalgamated Investment Co Ltd v. John Walker and Sons Ltd* where a property with development potential was agreed to be sold for £1.7 million. The building on the land was listed before the sale went through, and the value of the land then slumped to £250,000! In exceptional cases, an owner may serve a 'listed building purchase notice' on the planning authority requiring them to purchase the listed building, but generally there is no compensation if listed building consent for development is withheld.

The concept of conservation areas stems from the Civic Amenities Act 1967. Prior to 1967 the emphasis was on the preservation of individual buildings, not areas. Local planning authorities are now required to determine those areas which are of special historical or architectural interest, where it is desirable to preserve or enhance their appearance. Effectively these are areas which are rich, or potentially rich, in listed buildings, or of scenic beauty. Where an area is to be designated as a conservation area, the local planning authority must give notice in the local press, and enter appropriate notices in the land charges register.

When the local planning authority is considering designating a conservation area it will take account not only of individual buildings but groupings of buildings, their relationship to each other, and the quality and character of the space between buildings. The group of houses and shops in a marketplace of a small town may not be particularly worthy of note individually, but the overall impact of the area may be sufficiently pleasing to warrant creating a conservation area. Naturally, owners within such an area might consider themselves to be weighed down by the extra obligations relating to development. District councils are empowered to give grants to owners of listed buildings for their repair and maintenance, which may go some way to redress the balance. New development within a conservation area is not forbidden, but it must be sympathetic to the general character of the area.

Advertisements

The Secretary of State can, by regulations, control the display of advertisements so far as it appears to be expedient in the interests of amenity or public safety. This is generally designed to control the display of outdoor advertising, which includes any model, sign, placard, board, notice, device or representation, whether illuminated or not. The definition is wide enough to cover nameplates on shops and professional premises, as well as road signs. The power of the planning authority relates to issues of amenity and public safety; it could not therefore refuse permission on grounds of social desirability.

Under Regulation 6 of the Control of Advertisements Regulations 1989 a number of advertisements may be displayed without express consent. These include functional advertisements of local authorities and statutory undertakers; nameplates for businesses or companies; and advertisements of local religious, cultural, political, social or recreational events. If there is no deemed consent under Regulation 6 planning permission is required.

Wasteland

Where a local planning authority considers that the amenity of an area is adversely affected by any vacant site or wasteland, it can serve a notice on the owner or occupier requiring him to take steps to remedy the condition on the land. Failure to comply is punishable by a fine and the local authority can then enter, carry out the necessary work, and charge the owner.

The Refuse Disposal (Amenity) Act 1978 also gives local planning authorities powers in relation to abandoned vehicles and dumped refuse, which may not strictly be within the law of planning but which are certainly important in the preservation of amenity.

Financial problems – compensation and betterment

When a planning decision has been made (e.g. to create a new motorway, or build a new airport runway) there will be owners of property adjacent to the development whose property needs to be acquired by compulsory purchase for the purposes of the development. For some owners the value of their property will be significantly reduced by the development, while for others the value of the property will be increased by the planning decision.

Local authorities and development corporations have power to acquire land compulsorily for planning purposes. In such a case the appropriate authority or corporation will have initiated the necessary legal steps to acquire the land. There are two situations, however, where a landowner can initiate proceedings to force the local authority to purchase his land:

1 *Purchase notices* – where planning permission has been refused or made subject to onerous conditions, an owner may argue that his land is incapable of reasonably beneficial use in its existing state. He may then serve a purchase notice on the local authority, requiring it to buy the property. Compensation is payable to the owner in the same way as it would be under an ordinary compulsory purchase order. This will only be effective if the adverse planning decision renders the land virtually useless, not where it merely renders it less valuable.

2 *Blight notices* – where an owner is unable to sell his land at a fair market price because the prospect that it may be compulsorily purchased has rendered it virtually unsaleable, he may serve a blight notice on the appropriate public authority requiring it to buy his land. Compensation is again payable as under a normal compulsory purchase order.

As a broad general rule, no compensation is payable merely because planning permission has been refused. Exceptionally, where a revocation or modification of planning permission results in an existing right being lost (e.g. where the owner of agricultural land could have erected a barn) compensation may be payable.

The problems of taxing betterment is a political issue and approaches to the question have varied according to the government of the day. Most recently this was achieved by the Development Land Tax Act 1976 introduced by a Labour administration, but this was abolished by a later Conservative government.

14
Highways

Introduction

Even a short journey by road will serve to convince the traveller that the law relating to highways must be complex. Workmen are to be seen digging up portions of the road; part of the road is blocked off for repair; notices indicate the building of a new road; trees adjacent to the road have become a hazard with their overhanging branches; the road has a bad surface which needs repair; builders have deposited a skip and piles of sand and stones on the road by a building site; demonstrators have blocked the road with a march; animals have strayed on to the road; snow is making the road dangerous. Many of these incidents can cause annoyance and, more seriously, delay, injury or loss. Who will be responsible? Against whom can a complaint be made? When can damages be claimed? The answer to these questions can be learned from an analysis of highway law.

The first question to be posed is what is a highway? Simply, the answer is any portion of land over which the public has the right to pass. When it is remembered that many highways have existed for over 1000 years, and that the law on this subject has been developing for almost as long, this may explain some of the complexity. Most of the relevant statutory law on the point is now contained in the Highways Act 1980. The main aim of the Act was to bring together many rules previously contained in Highways Acts passed between 1959 and 1971 and in the Private Street Works Act 1961 but account was also taken of recommendations made by the Law Commission. The enormity of the subject can be grasped when it is seen that the Highways Act 1980 runs to 432 pages.

Although the mental picture which most people have of a highway is a busy road full of motor traffic, the term is far wider than that. Highways can be subdivided, and further definitions of some other common expressions will be useful:

1. *Footpath* – a highway over which the public have a right of way on foot only.
2. *Footway* – a part of a highway which is also a carriageway (see below) where the public have a right of way on foot only. A footway is what is commonly called the pavement at the side of a road.
3. *Carriageway* – a highway over which the public have a right of way for the passage of vehicles, i.e. what is commonly called a road.
4. *Bridleway* – a highway over which the public have a right of way on foot or on horseback.

Certain obvious questions arise in connection with highways:

- How does a highway come into existence?
- Who owns the highway?
- Who is responsible for the upkeep of the highway?

Acquisition and ownership of highways

There are two ways in which a highway can be created: by dedication or by statute.

Dedication

A highway can come into existence if the owner of the land concerned dedicates the right to cross it to the general public, which right is then accepted by the public actually using it. Dedication sounds like a very formal act, but in fact it can occur impliedly. Several points about dedication should be noted:

1. The owner of the land must intend to dedicate it as a highway. He shows such intention in the case where he makes a formal dedication, e.g. if he uses a formal document. Often, his intention will be implied from the circumstances, e.g. he knows people are passing and repassing over his land, and he takes none of the steps available to him to stop the creation of a highway (see below).
2. The highway must be dedicated to the public generally.
3. The person dedicating the land as a highway must be granting a right for all time. Effectively, therefore, only an outright owner of land can dedicate.

4 The right to pass and repass must be accepted, by the public actually using the highway.

Some assistance in proving the existence of a highway created by dedication is given by statute. Section 31 of the Highways Act 1980 provides that where a way has been enjoyed by the public as a right, without interruption for twenty years, then it is deemed to be dedicated as a highway unless the owner can prove that he did not intend to dedicate the way. The owner could best prove lack of intention to dedicate by showing that he had displayed notices to that effect, e.g. 'No public right of way', or 'This land is not dedicated as a public highway'.

The rule created by s.31 is important because it places responsibility on the owner to show that there is no highway, rather than on the public to show that there is. Even if a twenty-year period has not elapsed, it would still be open to a member of the public to try to prove the existence of a highway. It would then be necessary to prove that the owner intended to dedicate it. This might be possible, for example, by showing that the owner had allowed repairs to be undertaken by the highway authority. (Merely because a highway is acknowledged to exist by the landowner it does not necessarily follow that it is maintainable at public expense – see below.)

Creation of highways by Act of Parliament

The main rules are now consolidated in the Highways Act 1980, s.24. Under its provisions highway authorities can build highways including particular types of road (e.g. motorways where access and passage are restricted to certain classes of vehicle).

Who owns the highway?

Although the highway authority, when building new roads, may have acquired the land on which the highway is built, it is quite usual for the land under a highway to remain in the ownership of a private person, possibly the person who dedicates the highway. Often the landowners on either side of a highway own the land to the centre line of the road. Naturally, the use to which they can put such land is very limited. At common law, owners did have rights, e.g. to tunnel under the land, but many of the common law rights have been eroded by controls in Acts of Parliament. Ownership of a highway may still be important, however, if it is one which is not maintainable at public expense (see below).

Upkeep of the highway

The position with regard to repair and maintenance of highways is more easily understood if a brief mention is made of the historical development of the rules. At common law (before the introduction of any statutory changes), the duty and expense of repair rested with parishes, of which there were originally more than 10,000. The drawbacks of such a fragmented system were obvious, especially once methods of communication and transport were quicker. Changes made by statutes gradually produced the result that 'highway authorities' were created with responsibility for repair and maintenance. Today the position is that highway authorities are either:

1 The Minister of Transport, or
2 County councils.

The responsibility of the Minister of Transport is for trunk roads, i.e. the principal roads constituting the national system of routes for through traffic in Great Britain. However, the work of repair and maintenance is commonly delegated to the county council as agents. The county councils are the highway authorities responsible for all other roads maintainable at public expense. In turn, repair work is commonly delegated by a county council to district councils.

Section 36 of the 1980 Act deals with the question of which highways are maintainable at public expense. Effectively, these are:

1 Those highways in existence before 1835.
2 Highways created since then which have been adopted by the highway authority. Adoption can occur in a number of ways, e.g. by the dedicator serving notice on the highway authority which is accepted within the terms of s.37, or by agreement (s.38). Most important, however, are the adoption procedures with regard to private streets provided for by Part XI of the 1980 Act.

New Roads and Street Works Act 1991

One important aspect of this new Act is to impose a duty under s.58, where substantial works are to take place on a highway, to give notice to statutory undertakers, such as gas, electricity and water authorities, so that any work due to take place on that section of the road can be coordinated. After completion of work, there would then be a twelve month period of grace before more work could be undertaken. The new rules are being phased in gradually.

Streets and street works

Although the word 'street' may conjure up a mental picture of a road with buildings on either side of it, the definition provided by the 1980 Act is much more vague. 'Street' includes any highway, road, lane, footpath, square, court, alley or passage, whether a thoroughfare or not (s.329). The Act goes on to define a street as one which is not a highway maintainable at public expense. In order for a private street to be 'adopted', i.e. to become maintainable at public expense, street works as specified by the 1980 Act must be executed, and then the adoption rules applied (s.228). These rules are important to landowners who have done the work specified and now want to rid themselves of responsibility for repair and maintenance. This will be particularly significant for an estate developer.

The private street works code

Street Works Authorities (generally outside London these are the county councils) have power to make bye-laws regulating the building of streets (s.186) and power to execute street works in private streets (s.205). Where this power is used, it is generally referred to as the Private Street Works Code. This involves the council preparing:

- A specification of the street works.
- An estimate of the probable expense.
- A provisional apportionment of those probable expenses between the premises liable to be charged – this means premises fronting the street.

Once these details are approved by the Street Works Authority, it must:

1. Publish notices containing relevant details in local newspapers.
2. Post notices in the street affected.
3. Serve individual notices on the owners of the premises liable to be charged for the work, stating the sum provisionally apportioned to their premises. Generally, this will be decided according to the length of frontage of the premises to the street, but s.207 provides for the provisional sum calculated to be increased or decreased, depending on factors such as the degree of benefit to be derived by the owners of the premises, and any work already carried out by the owners.

Once all the required notices have been given, the owners of premises affected may wish to object. Section 208 sets out possible grounds of objection. These include arguments that:

- The street is not a private street.
- The estimated expenses are excessive.
- The proposed works are unreasonable.
- The provisional apportionment is unacceptable.

Hearings of these objections take place before the Magistrates Court, which has authority to quash the proposals in whole or in part, or amend them.

Once any objections are resolved and the work is completed, the Street Works Authority can then recover the costs in the agreed proportions from the owners, who may be given up to thirty years to pay. The actual amount due becomes a charge on the premises. This means that the agreed cost passes as a liability to any new owner of the premises.

Once the street works have been executed, it is then possible to 'adopt' the street and make it a highway maintainable at public expense under s.228. This is done by the Street Works Authority displaying a notice in the street declaring it to be a highway maintainable at public expense. If no objection is made within a month, the street becomes such a highway. But a majority of owners in the street may object and in that case the Street Works Authority must apply to the Magistrates Court for an order overruling their objections, before the street can be turned into a highway maintainable at public expense.

Advance Payments Code

The standards imposed by Street Works Authorities in the making up of streets mean that this work can prove very costly. The authority operating under the Private Street Works Code is responsible for recovering the sums due after completion of the work and money may be outstanding for long periods. The drawback of this system can be avoided in cases where the Advance Payments Code applies. The purpose of this code is to secure payment of the expenses of executing street works in private streets adjacent to new buildings. Section 219 provides that where (a) it is proposed to erect a building for which plans must be deposited with the local authority and (b) the building fronts onto a private street where the Street Works Authority could execute street works under the Private Street Works Code, then no building can take place until the owner of the land has paid or secured the payment of the sums required to the Street Works Authority under the Advance Payments Code. The sanction for disobeying this rule is a fine. However, s.219 is subject to a number of important exceptions, e.g. where an agreement under s.38 has been made with the Street

Works Authority, whereby the person undertaking street works at his own expense will, on completion of the works, dedicate the street as a highway.

If the Advance Payments Code does apply, then the Act provides the machinery for establishing the amount to be deposited – this is 'such sum as, in the opinion of the Street Works Authority, would be recoverable under the Private Street Works Code if the Street Works Authority were to carry out the work'. Provision is made for appeal to the Secretary of State for Transport against the amount proposed. Once an advance payment has been made, it opens the way for the 'adoption' procedure under s.229. As long as one frontager to the street has paid under the Code then a majority of the frontagers (either in terms of numbers of owners, or length of frontage owned) may request the Street Works Authority to secure the carrying out of appropriate street works and then declare the street to be a highway maintainable at public expense.

Two final points with regard to street works should be noted:

1 Where a private street is in need of repair to obviate danger to traffic, the Street Works Authority can require owners fronting onto the street to undertake specified repairs within a given time. If they fail to do so, the Street Works Authority may proceed with the repairs and recover the cost proportionately from the owners. Where a notice is served requiring urgent repairs, it may be in the interests of the owners to ask the authority to undertake the work of properly making up the street under the Private Street Works Code, as it will, thereafter, be a highway maintainable at public expense (s.230).

2 Section 236 gives power to a Street Works Authority to resolve to bear all or any part of the costs of works under the Private Street Works Code. This then discharges or reduces the liability of the owner.

Standard of the highway

When considering the state of repair of a highway, all highway authorities should bear in mind the words of a judge in the House of Lords:

> It is the duty of the road authorities to keep their public highways in a state fit to accommodate the ordinary traffic which passes or may be expected to pass along them. As the ordinary traffic expands or changes in character, so must the nature of the maintenance and repair of the highway alter to suit the change.

In fact, most highway authorities will wish to improve highways under their care in so far as their budget allows them to do so. The Highways Act 1980 makes extensive provision for such improvements. Part V of the Act contains rules, e.g. with regard to dual carriageways and roundabouts, cycle tracks, footways, guardrails, refuges, subways and footbridges, levelling of highways, improvement of corners, fencing and lighting of the highway, and roadside planting.

Where a person is affected by a failure to repair the highway, he may wish to exercise the enforcement procedure provided by s.56 of the 1980 Act. To use this, a notice is served on the highway authority (or other party alleged to be responsible for maintenance) requiring the repair to be done. If the highway authority disputes the notice, the case is referred to the Crown Court, which may order the highway to be put in proper repair within a specified period.

The failure to repair the highway may have led to injury or damage being caused. Until 1961 it was very difficult to bring a civil action for damages against a highway authority, as the rule was that the authority owed no liability for injury caused by failure to repair, only for injury caused by negligent carrying out of the repairs. That rule is now changed; the present law is that the highway authority can be fully liable for injury caused by its failure to maintain or repair the highway. However, by way of defence, the authority may plead that it took such care as was reasonably required to secure that the highway was not dangerous to traffic (s.58). Naturally, this will involve factors such as the type of highway involved, the usual nature of the traffic using it, and the state of knowledge of the highway authority. People frequently seek to bring actions under these rules when they have been injured as pedestrians by a fall on an uneven or broken pavement.

An interesting example of the application of s.56 can be seen in *Wentworth v. Wiltshire County Council*. A farmer had a long running dispute with the council about the status of the road to his farm, which he argued was a highway maintainable at public expense. The council did not agree. After the dispute had raged for a considerable number of years, matters came to a head when the Milk Marketing Board withdrew the tanker to his farm because of the state of the road. On application to the county court, the farmer was successful in obtaining a s.56 order, under which the council were given nine months to effect the repairs. Unfortunately, as a result of the withdrawal of the milk tanker, the farmer had had to abandon his dairy business. He now wanted to seek financial compensation. The High Court refused to make such an order, as he had a remedy under s.56, which he ought to have pursued earlier, and his losses in relation to his dairy business were economic loss.

Heavy traffic

The damage caused to highways is often due to excessively heavy traffic. It seems very unjust that the public should bear the cost of repair and maintenance when the damage is perhaps caused by the vehicles of one person, company or organization. Rules to redress this injustice are contained in s.59. They provide that a highway authority may recover 'excessive expenses' for maintaining the highway from any person causing excessive weight or other extraordinary traffic to pass along the highway. The excess expenses are those incurred over and above the average expenses for maintaining that or a similar highway which have resulted from damage arising from the extraordinary traffic.

Where a person knows that he will be operating traffic likely to cause such damage, he can agree in advance a payment to the highway authority. The advantage of this is that the payment will be a fixed amount, whereas if the operator waits for recovery procedures to be used by the highway authority, the 'excess expenses' are an unknown quantity. There is a significant amount of case law particularly on the meaning of the phrases 'excessive weight' and 'extraordinary traffic' which are not defined in the Act. Given the loads carried by builders and the type of vehicles being brought to sites, it will be appreciated that the section may be relevant to them.

Closing and obstruction of highways

It may be necessary, from time to time, to close or divert a highway e.g. while work is in progress on it. Alternatively, the need for the highway may have disappeared and the authority may wish to close it permanently. The procedure for 'stopping up' is contained in the Highways Act 1980, s.116. Generally, this involves an application to the Magistrates Court by the highway authority. (Where the closure is sought by a person other than the highway authority, he must ask the highway authority to take the necessary steps. There is no appeal if the highway authority refuses to activate the procedure but it must not unreasonably withhold consent.) The magistrates can make an appropriate order once the highway authority has followed through the necessary steps. These include notifying adjoining owners and occupiers, advertising the proposed closure or diversion in the press, and fixing notices at each end of the highway.

Any obstruction of, or interference with, the highway, unless authorized in some way, may constitute a civil and/or criminal offence. The Highways Act 1980, Part IX, contains numerous offences (e.g. in relation to straying animals and unlawful deposits) and also creates various duties and powers.

The highway authority is under a general duty to assert and protect the rights of the public to the use and enjoyment of any highway, and to prevent as far as possible the obstruction of the highway. It may enforce these rules in legal proceedings. Some specific examples of interference with the highway which are relevant to construction work should be noted:

1 Section 133 – if excavation or other work on land adjoining a street causes damage to a footway (pavement), the highway authority can recover the cost of repair from the landowner, or from the person causing the damage.
2 Section 131 – it is an offence, without lawful authority, to make an excavation on the highway, or to deposit anything on a highway that will cause it damage. The penalty imposed in respect of such offences is a fine.
3 Section 137 – it is an offence to wilfully obstruct free passage along the highway.
4 Section 139 – builders' skips may not be deposited on a highway without permission of the highway authority. When granting such a permission, the authority may specify the size of the skip, conditions as to its siting, lighting and guarding, means for making it visible to oncoming traffic, disposal of its contents and its ultimate removal. The owner must make sure that it:
    (a) Is properly lit.
    (b) Is clearly marked with his name and address.
    (c) Is removed as soon as practicable when full.
    (d) In all respects complies with any conditions attached to the permission.

Breaches of these rules are punishable by fines. The section specifically states that 'nothing in this section is to be taken as authorizing the creation of a nuisance'. Where the skip is deposited on a highway, even when permission has been granted, s.140 provides that the highway authority or a police constable can require it to be removed or repositioned, or may themselves remove or reposition it and recover from the owner the expense of doing so.

5 Section 168 deals with building operations affecting public safety. If building operations in, or near, a street cause an accident which gives rise to a risk of injury to persons in the street then the owner

of the land on which the work is taking place or any other person liable for causing such a risk (e.g. the building contractors) is liable to a fine.
6. Section 169 provides that where scaffolding will obstruct the highway, it can only be erected if a licence is obtained from the highway authority. There is a right of appeal against refusal of a licence, or against any conditions attached to it. If a licence is granted, then the scaffolding must be adequately lit during darkness and any necessary traffic signs must be erected and maintained.
7. Section 170 controls the mixing of mortar and cement on the highway in such a way that it may stick to the highway or enter drains and sewers and solidify. The section permits mixing in a receptacle or on a plate, but any breach of the rules is an offence carrying a fine.
8. Section 171 controls deposits of building materials and the making of excavations in streets. Such activities can be carried on with the consent of the highway authority, which may impose conditions, particularly if there is a danger of damage or obstruction to the apparatus of statutory undertakers (e.g. the gas, electricity and water boards). Appeal against refusal of permission, or the conditions imposed, is to the Magistrates Court.
9. Section 172 regulates when building works must be close boarded by a hoarding or fence. This section operates whenever a building is to be erected or taken down in a street, unless the County Council dispenses with the requirement. Where a hoarding is required, conditions are likely to be imposed regarding the creation and lighting of a footway and the maintenance of the hoarding. Under s.173, it is an offence to use such a hoarding which is not securely fixed to the satisfaction of the Council.

It should be noted that some of the foregoing sections apply to highways whereas others apply to streets. Many create statutory offences punishable by a fine. But the behaviour constituting the statutory offence may also cause damage or injury to an individual, who may be able to bring a civil claim for damages in the law of tort. A number of torts may be relevant, e.g. a person falling into an unfenced and unlit excavation on the highway may succeed in negligence, whereas a person injured by the collapse of an unsafe hoarding around building work in a street may succeed in negligence or public nuisance. An accumulation of rubbish in a builder's skip may constitute a nuisance. For the relevant rules in tort see Chapter 9.

The tort of greatest significance in relation to activities on or near the highway is public nuisance. Claimants have succeeded in cases where crowds blocked the highway, where smoke obscured visibility in the highway, where a dangerous lamp overhanging the highway fell on to it, where there was unlit scaffolding, where the danger consisted of broken cellar flaps and where the danger arose from a heap of slates piled at the side of the road.

Building and improvement lines

Part V of the 1980 Act contains a number of rules, powers and duties with regard to the improvement of highways. It may be obvious to the highway authority that, because of changing or increasing traffic volume, it will sooner or later become necessary to widen a particular highway. The need may not yet be urgent or funding may not be available, but it would be absurd if in the meantime the local authority continued to allow building work to take place, thereby making ultimate widening of the highway more difficult and costly. To prevent this situation occurring, s.73 of the Act provides that the highway authority may prescribe an improvement line for the street in question. This is the line to which it is ultimately intended to widen the street. Naturally, such a decision can affect the value of adjacent properties and the development which can be undertaken in relation to them. It would not be surprising, therefore, if the owners wished to object to the prescription of the improvement line. This can be done by any aggrieved person appealing to the Crown Court. If the appeal fails, it should be noted that s.73 (9) provides for compensation to be paid to persons whose property is thus injuriously affected.

Section 74 of the Act provides for the highway authority to prescribe a building line. Once such a building line is operative, no new building shall be erected beyond the building line. Rather like the rules in s.73, provision is made to compensate owners whose property is injuriously affected, but unlike s.73 there is no provision in s.74 for appealing against the imposition of the building line. Schedule 9 merely provides that the highway authority must consider objections.

Once either type of line is properly prescribed, it must be shown on a duly authenticated plan which is available for inspection by interested parties. Both ss. 73 and 74 provide means whereby these lines can be removed if they are subsequently seen to be unnecessary.

15
Building Regulations

Purpose and scope of the Building Regulations

Since 1875, the law has imposed controls to ensure that buildings are constructed in a way which is conducive to good health. Originally, local authorities made bye-laws on matters such as the materials to be used, ventilation and sanitation. Relevant legislation is now consolidated in the Building Act 1984 which empowers the Secretary of State for the Environment to make Regulations. New Building Regulations were made in 1985, replacing the 1976 set. There was a significant change in approach in the 1985 version, in that the Regulations now impose much less detailed control. The latest set of Regulations was issued in 1991. Many specific points previously contained in the Regulations are now covered by guidance documents, a system similar to the codes of practice and guidance issued by the Health and Safety Commission. Section 7 of the 1984 Act provides that a failure to comply with guidance documents does not in itself render a person liable to civil or criminal proceedings, but in any such proceedings a failure to comply may be relied on as tending to establish liability.

Under s.1 of the Building Act 1984, the Secretary of State may make regulations:

1 To secure the health, safety, welfare and convenience of persons in or about buildings and of others who may be affected by buildings.
2 To further the conservation of fuel and power.
3 To prevent waste, undue consumption, misuse or contamination of water.

The regulations may relate to the design and construction of buildings, and the provision of services, fittings and equipment in, or in connection with, buildings. Such regulations are to be known as Building Regulations. The Secretary of State has power to dispense with or relax a Regulation in particular cases where he considers that its operation would be unreasonable (s.8). When making regulations, the Secretary of State is advised by the Building Regulations Advisory Committee, appointed under s.14.

Enforcement procedures

The Regulations apply to any 'building work' or 'material change of use' of a building. The work done must be supervised, either by the local authority or an approved inspector, i.e. a person approved by the Secretary of State. The approved inspector must have no professional or financial interest in the work. At present, only one approved inspector has been approved, which is a wholly owned subsidiary company of NHBC. Its approval is limited to supervision of dwellings up to eight storeys in height.

When work is to be supervised by the local authority, the builder will provide the authority with a building notice, or he will deposit full plans. Building notices are only appropriate for small and minor works. The local authority can ask the builder for such extra detail as it needs, and will make the same site inspections as in other cases. On a deposit of full plans, these must be passed or rejected by the authority within five weeks. Where the plans show that work will be carried out in accordance with the Building Regulations, the authority must approve them. Appropriate fees are paid on deposit of the plans. These are usually worked out on a sliding scale, depending on the job cost.

Where the local authority is supervising, there are prescribed time limits for notifying the authority about aspects of the work:

1 At least forty-eight hours notice of commencement.
2 At least twenty-four hours notice before covering up foundations, or a dampcourse, or any concrete material.
3 At least twenty-four hours notice before covering up any drain.
4 Notice within seven days of completion of the work.

Many local authorities issue pro-forma notices for a builder to post off to them at relevant times.

Under s.35 of the 1984 Act, contraventions of the Building Regulations render a person liable on summary conviction in the Magistrates Court to a fine of up to level five on the standard scale. The scale is laid down in the Criminal Justice Act 1982. A local authority may also serve notice on an owner requiring him to pull down or remove the work, or make the alterations necessary so that it does comply with the

Regulations. If necessary, the local authority may pull down or remove the work themselves, and charge the owner (s.36). No s.36 notice can be served after the expiration of twelve months from the date of completion of the work in question.

Under s.38 (Which is not yet (1991) in force), breach of a duty imposed by Building Regulations which causes death or personal injury is to be actionable in civil proceedings, unless the specific regulation provides otherwise. This rule is in addition to any rights of action in negligence at common law.

Defective and dangerous buildings

Sections 76–83 of the Building Act 1984 create several important controls regarding defective and dangerous premises and demolition. These are in addition to rules which already exist in the Environmental Protection Act 1990 and the Town and Country Planning Act 1990. Where a building is in such a defective state that it is prejudicial to health or a nuisance, the local authority may serve a notice on the owner indicating that they intend to remedy the defective state of affairs. Nine days after serving the notice, the authority may execute the work and recover its expenses from the owner. An owner can serve a counter-notice within seven days, saying that he intends to remedy the situation himself. The local authority can then only intervene if the owner does not make a start within a reasonable time, or does not make reasonable progress. The advantage of using this s.76 procedure is speed. It is only available in cases where it appears to the authority that the procedures under the Environmental Protection Act 1990 would cause unreasonable delay. The drawback of the procedure is that an authority may need to bring proceedings to recover its expenses, and at that point, the court can enquire whether the authority was justified in the assumptions it made and the steps it took. If not justified, the authority cannot recover the expenses or any part of them. Similar rules exist with regard to dangerous buildings (ss.77 and 78), and dilapidated buildings and neglected sites (s.79).

Demolition work is now subject to a measure of control by local authorities by virtue of ss.80–3. Where the whole or part of a building is to be demolished, then no work should commence until a notice of intention to demolish has been given to the local authority and either the local authority has served a s.81 notice, or six weeks have elapsed since they received the notice of intention. If the authority decides to serve a s.81 notice, it may require the demolition contractor to take any, or all, of the following steps:

- Shore up adjacent buildings.
- Weatherproof surfaces of adjacent buildings exposed by demolition.
- Repair and make good any damage to adjacent buildings caused by the demolition.
- Remove rubbish or material resulting from the demolition.
- Disconnect and seal sewers or drains.
- Make good the surface of the ground.
- Make arrangements for disconnection of services.
- Take such steps as the authority considers necessary for the protection of the public and preservation of public amenity.

A contractor can appeal to the Magistrates Court against a s.81 notice, particularly against the shoring-up and weatherproofing requirements, where he may argue that the adjacent owner ought to pay or contribute to the costs.

For the effective operation of many of these rules, it is essential that authorized officers of local authorities have power to enter premises. Powers are granted by s.95, and must sometimes be exercised with the authority of a justice's warrant. Officers must ensure that they have appropriate authority to enter, otherwise their actions constitute the tort of trespass.

Index

Abatement, 220
ACAS, 193
Acceptance, 78
Acts of Parliament, 6
Administration order, 21, 72
Advance Payment Code, 375
Adverse possession, 303
Advertisements, 368
Advisory committees, 262
Affirmation, 124
Agency, 131
  apparent authority, 132
  architect's authority, 133
  authority of agent, 131
  disclosed principal, 136
  duties of agent, 135
  duties of principal, 134
  Estate Agents Act, 137
  estate agents' authority, 131
  indemnity, 134
  lien, 135
  misdescription, 138
  named principal, 136
  necessity, 132
  powers of attorney, 134
  privity of contract, 136
  ratification, 133
  termination, 137
  undisclosed principal, 136
Agreement, 76, 81
Annual General Meeting, 68
Appeals, 23, 28, 48
Arbitration, 21, 29
Architects, 162

Article 4 directions, 361
Articles of Association, 64, 65
Ascertained goods, 144
Assignment, 160, 308
Assured tenancies, 335
Attachment of earnings order, 21, 179
Average clauses, 159

Bailiff, 21, 26
Ballots, 206
Bankruptcy, 22
Barrister, 32
Basic award, 194
'Beauty contests', 201
Betterment, 368
Bills, 6
Blight notices, 369
Breach, 123
Breach of statutory duty, 251
Budget, 41
Builders' skips, 379
Building contracts, 161
  architect, 162
  architect's instructions, 166
  breach, 170
  certification, 169
  clerk of works, 162, 165
  collateral agreements, 172
  extensions of time, 167
  goods on site, 165
  insurance, 168
  letter of intent, 163
  nominated sub-contractors, 171

non-completion, 170
payment, 168
possession of the site, 165
practical completion, 169
progress of work, 166
standard forms, 161, 164
sub-contracts, 171
tenders, 163
variations, 167
Building lines, 381
Building operations, 358
Building Regulations, 382
Business liability, 97
Business organizations, 61
Bye-laws, 10, 46

Cabinet, 35
Causation, 241
Caveat emptor, 3, 83, 103, 108, 141, 151, 153
Certainty of agreement, 80
Certification of trade unions, 198
Certification Officer, 198
Certiorari, 23, 45, 49
Change of use, 359
Charging order, 22
Circuit Judge, 21, 27
Civil liability, 2, 20
Civil Service, 42
Clerk of works, 162, 165
Clerk to the Justices, 25
Codes of practice, 271
Collateral agreements, 172
Collective agreements, 177, 197
Commissioner for Administration, 45
Committal proceedings, 24
Common duty of care, 245
Common law, 16
Companies, 62
  Administration Order, 72
  Annual General Meeting, 68
  Articles of Association, 64, 65
  capital, 65
  company finance, 69
  crystallization, 70
  debentures, 69
  directors, 67
  elective resolutions, 69
  fixed charges, 70
  floating charges, 70
  incorporation, 62

  insolvency practitioners, 71
  legal personality, 63
  limited liability, 62
  liquidation, 71
  management of the company, 67
  Memorandum of Association, 64
  name of the company, 64
  objects clause, 64
  ordinary resolutions, 68
  polls, 69
  preferential creditors, 73
  private companies, 66
  public companies, 66
  registrar, 64
  shares, 66
  special resolutions, 68
  Table A, 66
  ultra vires, 64
Compensatory awards, 194
Compulsory insurance, 156
Conditional sale, 150
Conflict of interest, 181
Consent, 218, 243
Conservation areas, 367
Consideration, 81
Constituencies, 35
Construction regulations, 276
Constructive dismissal, 191
Consumer, 143
Consumer credit, 149
Contempt of court, 22
Continuation tenancy, 325
Continuous service, 185
Contract, 75
  acceptance, 78
  affirming the contract, 124
  agreement, 76, 81
  breach, 123
  business liability, 97
  certainty, 80
  conditions, 92
  consideration, 81
  counter offers, 78
  damages, 126
  damages in lieu of rescission, 107
  dealing as a consumer, 98
  discharge, 114
  duress, 109
  economic duress, 86, 109
  exclusion clauses, 95
  fiduciary relationship, 103
  form of the contract, 90

388 Index

Contract – *continued*
  fraudulent misrepresentation, 104
  frustration, 117
  illegal contracts, 110
  injunction, 126
  innocent misrepresentation, 106
  innominate terms, 94
  insurance, 104
  intention, 88
  invitation to treat, 76
  limitation periods, 130
  measure of damages, 127
  misrepresentation, 101
  mistake, 101, 108
  mitigation, 127
  negligent misrepresentation, 104
  offer, 76
  performance of existing duties, 83
  privity of contract, 88
  promissory estoppel, 86
  public policy, 86, 111
  quantum meruit, 91
  remoteness of damage, 126
  repudiation, 123
  requirement of reasonableness, 99
  rescission, 106
  restraint of trade, 111
  specific performance, 125
  subject to contract, 92
  terms, 92
  undue influence, 110
  utmost good faith, 104
Contract of service, 173
Contribution, 159, 243
Contributory negligence, 243
Control test, 174
Conventions, 36
Conversion, 211
Conveyance, 287, 291
Copyrights, 182
Corporate personality, 43
Counter-claim, 20
County Court, 19, 20
Court of Appeal, 19, 23
Covenants, 308
Credit cards, 150
Criminal liability, 2, 23
Crown, 5
Custom, 3

Damages, 50, 126
Dangerous buildings, 384

Dealing as consumer, 98, 143
Debentures, 69
Deceit, 107
Declaration, 45, 50
Dedication of highways, 371
Deductions from wages, 179
Deeds, 286, 290, 293, 297, 302
Defamation, 181
Defective buildings, 250, 384
Defective products, 252
Degree of risk, 238
Delegated legislation, 10
Demolition, 359
Designs, 183
Development, 358
Development plans, 356
Directors, 67
Discharge of contracts, 114
Disciplinary rules, 192
Discrimination, 186
Dismissal, 188
Distinguishing cases, 15
Distress, 26
Duress, 109
Duty of disclosure, 152
Duty to take care, 224

Easement, 286, 289, 292, 296
Economic duress, 86, 109
Economic loss, 226
Economic torts, 203
Elections, 35, 46
Elective resolutions, 69
Employment, 173
  ACAS, 193
  basic awards, 194
  collective agreements, 177
  compensatory awards, 194
  conflict of interest, 181
  continuous service, 185
  contract of service, 173
  control test, 174
  copyrights, 182
  deductions from wages, 179
  designs, 183
  disciplinary rules, 192
  discrimination, 186
  dismissal, 188
  duties of the parties, 178
  equal pay, 179, 186
  faithful service, 181
  guarantee payments, 180

holiday pay, 178
integration test, 174
itemized pay statement, 177
mixed or multiple test, 175
part-time employees, 178
patents, 182
periods of notice, 188
procedure, 194
reasonable care, 180
redundancy, 195
redundancy payments, 196
re-engagement, 193
references, 180
re-instatement, 193
restraint of trade, 182
secret profit, 181
statutory rights, 184
termination, 188
trade unions, 197
vicarious liability, 255
wages, 179
week's pay, 196
written statements, 177
Enforcement notices, 365
Equal pay, 179, 186
Equitable interests, 286, 289–92, 300
Equity, 1, 86, 110, 125
Escape, 221
Estates in land, 285
Estoppel, 86
Euisdem generis, 12
European Community, 4
Eviction, 322
Exchange of contracts, 289, 291
Exclusion clauses, 95
Execution, 21
Express terms, 92
Expressio unius, 12
Expulsion from trade union, 202
Extensions of time, 167
Extraordinary traffic, 378

Factories, 274
Fair reasons for dismissal, 190
Faithful service, 181
Fee simple, 285
Fiduciary relationship, 103
Financial procedures, 41
Fines, 2, 26
Fire, 222
Fixed charge, 70
Fixed term leases, 305, 319

Fixtures, 283
Floating charge, 70
Forfeiture, 312, 314, 319
Form of the contract, 90
Fraudulent misrepresentation, 104
Freedom of speech, 5
Freeholds, 285
Frustration, 117

General Development Order, 360
Golden formula, 205
Golden rule, 12
Goods, 144
Government, 34
Green form scheme, 32
Guarantee payments, 180
Guidance notes, 271, 382

Health and safety, 259
  advisory committees, 262
  codes of practice, 271
  Commission, 261
  compensation, 279
  construction regulations, 276
  duties:
    controllers of premises, 266
    employees, 267
    employers, 263
    manufacturers, 266
  enforcement, 270
  Executive, 262
  factories, 274
  guidance notes, 271
  improvement notice, 272
  inquiries, 261
  inspectors, 270
  investigations, 262
  prohibition notice, 273
  prosecution, 273
  Robens Report, 261
  safety committees, 269
  safety representatives, 268
  so far as is reasonably practicable, 264
  written safety policy, 264
High Court, 22
Highways, 370
  advance payments code, 375
  builders' skips, 379
  building lines, 381
  closing of highways, 378
  dedication 371

Highways – *continued*
  excessive weight, 378
  extraordinary traffic, 378
  heavy traffic, 378
  highway authority, 373
  improvement lines, 381
  maintenance at public expense, 373
  obstruction of highways, 378
  ownership of the highway, 371
  private street works, 374
  scaffolding, 380
  standard of the highway, 376
  upkeep of the highways, 373
Hire purchase, 149
Holiday pay, 178
House of Commons, 37
House of Lords, 15, 18

Illegal contracts, 110
Implied terms, 92, 139
Improvement lines, 381
Improvement notices, 272
Incorporation, 62
Independent trade unions, 198
Indictable offences, 23
Industrial activities, 202
Injunctions, 2, 22, 50, 126, 214
Innocent misrepresentation, 106
Innominate terms, 94
Inquiries, 261
Insolvency practitioners, 71
Inspectors, 270
Insurance, 152
  assignment, 160
  average clauses, 159
  compulsory insurance, 156
  contribution, 159
  duty of disclosure, 152
  employers' liability insurance, 156
  exclusion of liability, 155
  indemnity, 159
  insurable interest, 153
  insurance contracts, 154
  material facts, 152
  motor insurance, 158
  Motor Insurers' Bureau, 158
  non-disclosure, 153
  property insurance, 154, 157
  public liability insurance, 157
  re-insurance, 159
  subrogation, 160
  utmost good faith, 152
  warranties, 155

Joint ownership, 293
Judgement, 20
Judicial immunity, 232
Judicial precedent, 13
Judicial review, 49
  abuse of power, 57
  bias, 54
  certiorari, 49
  damages, 50
  declaration, 50
  delay, 51
  directory, 53
  excess of power, 56
  fair hearing, 55
  fairness, 54
  improper purposes, 59
  injunction, 50
  leave to apply, 51
  mandamus, 50
  mandatory, 53
  natural justice, 54
  privative clauses, 51
  procedural impropriety, 53
  prohibition, 49
  public and private land, 50
  relevant/irrevelant matters, 58
  remedies, 49
  standing, 51
  unreasonableness, 57
Juries, 28
Justices of the Peace, 24

Land law, 281
  abstract of title, 289
  access to neighbouring land, 296
  adverse possession, 303
  airspace, 282
  animals, 285
  building schemes, 301
  chattel, 283
  classification of property, 285
  completion, 291, 293
  contract, 274
  contractual licence, 299
  conveyance, 291
  conveyancing, 297
  deeds, 286, 290, 293, 297, 302
  dominant land, 297
  easements, 286, 289, 292, 296

equitable interests, 286, 289, 292, 300
exchange of contracts, 298, 291
fee simple, 285
freehold estates, 285
fixtures, 283
insolvency, 283
interests in land, 286
joint ownership, 293
land, 282
land charges, 289
local land charges, 287, 291
leases, 285
legal interests, 286, 290, 292, 297
licences, 299
minor interests, 292
mortgages, 283, 286, 289, 292, 302
options, 289
overriding interests, 292
Prescription, 298
profits a prendre, 286, 289, 292, 299
registered land, 291
registration, 287, 289, 291, 300
restrictive covenants, 286, 289, 292, 300
rights and duties, 295
sale, 283
servient land, 297
support − rights of, 295
survey, 288, 291
tenancy in common, 294
term of years, 286
transfer, 293
treasure trove, 284
trusts, 286, 292, 293
unregistered land, 287
water, 285
Landlord and tenant, 305
assignments, 308
assured shorthold tenancies, 341
assured tenancies, 335
break clauses, 312, 319
business tenancies, 323
compensation, 332
continuation tenancies, 325
council tenants, 352
covenants in leases, 308
eviction, 322
exclusive possession, 305
express covenants, 310, 312
fitness for habitation, 315

fixed term leases, 305
fixtures, 284
forfeiture, 312, 314, 319
implied covenants, 310, 315
licence, 306
merger, 320
notice to quit, 322
offences, 322
options, 311
periodic tenancies, 305
premiums, 348
privity of contract, 308
privity of estate, 308
protected shorthold tenancies, 348
protected tenancies, 343
provisos, 311
public sector tenants, 352
quiet enjoyment, 310
rent control, 342, 348
rent review clause, 311
repairing covenants, 312
residential tenancies, 335, 451
restrictive covenants, 309
secure tenancies, 352
security of tenure, 306, 319
specific performance, 315
statutory tenancies, 344
subletting, 308
succession, 342, 348, 353
surrender, 321
tenants' right to purchase, 350, 351, 355
termination, 319
waste, 318
Language of the law, 2
Latent damage, 237
Law Commission, 6
Lay-off, 180
Legal Aid and Advice, 32
Legal Aid Board, 32
Legal personality, 63
Legislation, 5
Letters of intent, 163
Licence, 306
Lien, 17, 135, 146
Limitation, 146, 130, 164, 236
Limited liability, 62
Liquidation, 71
Listed buildings, 366
Literal rule, 12
Lobbying, 9
Local government, 43

Local government – *continued*
  bye-laws, 46
  Commissions for Local
    Administration, 45
  control of local authorities, 44, 46
  ultra vires, 44
Local plans, 357
Lord Chancellor, 18, 24

Magistrates, 24
Magistrates Courts, 18, 19, 23
Maintenance of highways, 373
Maladministration, 31
Mandamus, 23, 45, 50
Master of the Rolls, 18, 23
Material facts, 14, 152
Measure of damages, 127
Members of Parliament, 35
Membership of a trade union, 200
Memorandum of Association, 64
Merchantable quality, 139
Mischief rule, 13
Misrepresentation, 101
Mistake, 108
Monarch, 34
Mortgages, 283, 286, 289, 292, 302
Motor insurance, 158

Name of a company, 64
Natural justice, 23, 29, 45, 54, 364
Negligence, 223
  breach of duty, 237
  causation, 241
  contribution, 243
  contributory negligence, 243
  degree of risk, 238
  duty to take care, 224
  economic loss, 226
  intervening acts, 242
  judicial immunity, 232
  latent damage, 237
  limitation periods, 236
  negligent mis-statements, 229
  neighbour test, 224
  nervous shock, 232
  omissions to act, 237
  proximity test, 224
  public policy, 224
  reasonable care, 237
  remoteness of damage, 240
  res ipsa loquitur, 240
  seriousness of harm, 238
  statutory powers, 233
  volenti, 243
Negligent misrepresentation, 104
No-fault compensation, 210
Nominated sub-contractors, 171
Non-disclosure, 153
Non-natural use, 222
Non-occupiers' liability, 249
Noscitur a sociis, 11
Notice to quit, 322
Nuisance, 213
  abatement, 220
  act of a trespasser, 217
  consent, 218
  injunction, 214
  prescription, 218
  private nuisance, 214
  public nuisance, 213
  statutory authority, 217
  statutory nuisance, 219

Obiter dicta, 14
Objects clause, 64
Obstruction of highways, 378
Occupiers' liability, 244
  common duty of care, 245
  defences, 247
  exclusion of liability, 247
  lawful visitors, 245
  non-visitors, 245
  occupier, 244
  premises, 244
  trespassers, 245, 247
  warning notices, 246, 249
Ombudsman, 17, 31
Omissions to act 237
Options, 289
Orders in council, 10
Ordinary resolutions, 68
Outline planning permission, 362
Ownership of highways, 371

Parliament, 5
  business of the House, 37
  bye-laws, 10
  cabinet, 35
  Civil Service, 42
  constituencies, 35
  conventions, 36, 42
  Crown, 5
  delegated legislation, 10
  divisions, 36, 37

Index 393

elections, 35
financial procedures, 41
House of Commons, 34
House of Lords, 5, 35
lobbying, 9
Monarch, 34
MPs, 35
Prime Minister, 35
private members' bills, 6, 7, 39
procedure in the Commons, 37
question time, 37
Royal Assent, 36
select committees, 40
Speaker, 36
standing committees, 7, 40
statutory instruments, 10
whips, 8, 38, 40
Parliamentary Commissioner, 31, 38
Parliamentary counsel, 7
Partnership, 61
Part-time employees, 178
Patents, 182
Performance of contract, 114
Periodic tenancies, 305
Periods of interruption of employment, 185
Periods of notice, 188
Planning, 356
   advertisements, 368
   appeals, 364
   Article 4 directions, 361
   betterment, 368
   blight notices, 369
   building operations, 358
   change of use, 359
   compensation, 368
   conditional permission, 363
   conservation areas, 367
   demolition, 359
   development, 358
   development plans, 356
   enforcement notice, 365
   environmental assessment, 363
   fees, 361
   General Development Order, 360
   listed building consent, 366
   listed buildings, 366
   local planning authorities, 356
   local plans, 357
   outline planning permission, 362
   planning register, 362
   public inquiries, 364
   purchase notices, 369
   refusal of permission, 363
   simplified planning zones, 361
   stop notice, 366
   structure plans, 356
   tree preservation orders, 366
   Use Classes Order, 359
   waste land, 368
   written representations, 364
Polls, 69
Powers of attorney, 134
Practical completion, 169
Precedent, 13
Preferential creditors, 73
Premises, 244
Prescription, 218, 298
Pressure groups, 6, 9, 18
Pre-trial review, 20
Private street works, 374
Privity of contract, 88, 136, 308
Priority of estate, 308
Product liability, 253
Profits a Prendre, 286, 289, 292, 299
Prohibition, 45, 49
Prohibition notices, 273
Promissory estoppel, 86
Protected tenances, 343
Provisos, 311
Proximity test, 224
Public companies, 66
Public corporations, 47
Public liability insurance, 157
Public order, 207
Public policy, 86, 111, 224

Quantum meruit, 129, 164
Queen's Bench Division, 23
Queen's Speech, 36
Question time, 37

Ratio decidendi, 14
Reasonable care, 148, 237
Reasons for dismissal, 190
Recorders, 27
Redundancy, 195
Re-engagement, 193
References, 180
Registered title, 291
Re-instatement, 193
Re-insurance, 159
Remoteness of damage, 126, 240
Report stage, 8

Repudiation, 123
Requirement of reasonableness, 99
Res ipsa loquitur, 3, 240
Rescission, 106
Reservation of title, 145
Resolutions, 68
Restraint of trade, 111, 182
Restrictive covenants, 286, 289, 292, 300, 309
Robens Report, 261
Romalpa clauses, 145
Royal Assent, 8, 36
Royal Commissions, 6
Rule book, 199
Rylands v. Fletcher, 220
 escape, 221
 non-natural use, 222

Safety committees, 269
Safety policy, 264
Safety representatives, 268
Sale of goods, 139
 ascertained goods, 144
 dealing as a consumer, 143
 delivery, 142
 description, 141
 exclusion of liability, 143
 implied terms, 139
 merchantable quality, 139
 passing ownership, 144
 payment, 142
 reasonable fitness for purpose, 140
 remedies, 146
 reservation of title, 145
 specific goods, 144
 unascertained goods, 145
Sanctions, 1
Scaffolding, 380
Secondary action, 206
Secret profits, 181
Secure tenancies, 352
Security of tenure, 306, 319
Self-help, 17
Set-off, 17
Shares, 66
Skips, 379
Specific performance, 125, 315
Standard form contracts, 161, 164
Statutes, 6
Statutory authority, 217
Statutory instruments, 10

Statutory interpretation, 11
Statutory nuisance, 219
Statutory powers, 233
Statutory tenancies, 344
Stipendiary magistrates, 26
Structure plans, 356
Sub-contracts, 171
Subject to contract, 92
Subrogation, 160
Summary offences, 23
Supply of goods and services, 147
Suspension, 180

Tenders, 163
Term of years, 286
Termination of employment, 188
Terminology, 3
Tort, 208
 breach of statutory duty, 251
 conversion, 211
 defective products, 252
 economic torts, 203
 negligence, 223
 nuisance, 213
 occupiers' liability, 244
 product liability, 253
 Rylands v. Fletcher, 220
 trespass, 211
 vicarious liability, 255
Trade descriptions, 151
Trade unions, 197
 ballots, 206
 'beauty contests', 201
 Certification Officer, 198
 closed shop, 200
 collective bargaining, 197
 definition, 198
 economic torts, 203
 Employment Appeal Tribunal, 199
 expulsion, 202
 golden formula, 205
 immunity from liability, 205
 independent trade union, 198
 industrial action, 203
 industrial activities, 202
 joining a trade union, 201
 membership, 200
 picketing, 206
 public order, 207
 rule book, 199
 secondary action, 206
 trade dispute, 204

union membership agreements, 200
union only contracts, 202
Treasure trove , 284
Tree preservation orders, 366
Trespass to goods, 211
Trespass to land, 17, 212
Trespass to the person, 211
Trespassers, 245, 247
Tribunals, 29
Trusts, 286, 292, 293

Ultra vires, 44, 47, 64
Unascertained goods, 145
Undue influence, 110
Unfair dismissal, 188
Unincorporated associations, 62
Union membership agreements, 200
Union only contracts, 202
Upkeep of highway, 373

Use classes order, 359
Utmost good faith, 50, 104, 152

Valuation, 31
Variations, 167
Vicarious liability, 255
Volenti, 218, 243

Wages, 179
Warning notices, 246, 249
Warranties, 92, 155
Waste land, 368
Week's pay, 196
Whips, 8, 38, 40
Winding up, 71
Written safety policy, 264
Written standard terms of business, 98
Written statements, 177
Wrongful dismissal, 188